HOW ELECTRONIC THINGS WORK . . . AND WHAT TO DO WHEN THEY DON'T

HOW ELECTRONIC THINGS WORK . . . AND WHAT TO DO WHEN THEY DON'T

ROBERT L. GOODMAN

Second Edition

McGraw-Hill

New York   Chicago   San Francisco   Lisbon   London   Madrid
Mexico City   Milan   New Delhi   San Juan   Seoul
Singapore   Sydney   Toronto

The McGraw·Hill Companies

Cataloging-in-Publication Data is on file with the Library of Congress.

Copyright © 2003 by The McGraw-Hill Companies, Inc. All rights reserved.
Printed in the United States of America. Except as permitted under the United
States Copyright Act of 1976, no part of this publication may be reproduced or
distributed in any form or by any means, or stored in a data base or retrieval system,
without the prior written permission of the publisher.

1 2 3 4 5 6 7 8 9 0 DOC/DOC 0 8 7 6 5 4 3 2

ISBN 0-07-138745-5

*The sponsoring editor for this book was Scott Grillo. The editing supervisor was
Caroline Levine, and the production supervisor was Sherri Souffrance. It was set
in Times New Roman by Victoria Khavkina of McGraw-Hill Professional's
Hightstown, N.J., composition unit.*

Printed and bound by RR Donnelley

 This book is printed on recycled, acid-free paper containing a
minimum of 50% recycled, de-inked fiber.

McGraw-Hill books are available at special quantity discounts to use as premiums
and sales promotions, or for use in corporate training programs. For more
information, please write to the Director of Special Sales, Professional Publishing,
McGraw-Hill, Two Penn Plaza, New York, NY 10121-2298. Or contact your local
bookstore.

Dedication

This book is dedicated to my brother, Bill,
for giving me the idea to write this kind of book.

CONTENTS
AT A GLANCE

DETAILED CONTENTS

PREFACE

I think you will find this book unique in its simple explanations and its many easy-to-understand illustrated drawings and photos of how electronic equipment works in the home or office.

The brain storm for this type of book was started many years ago when my brother wanted to know how a picture was formed on a color TV. The planning, development, and portions of the drawings and writing for the first edition were in progress for eight years. The actual writing and production of the many photos and drawings took over two years.

The mission of the second edition remains to take the mystery out of how electronic consumer products work, for persons with little or no electronic background. Not only does this book give you simplified electronic equipment operations, but hints and tips about what to check when the device does not work properly or does not work at all. There's also information about how and what to clean, plus preventive maintenance that can be done to extend the life of these very expensive products. The book includes tips on how to protect products from voltage surges and lightning spike damage.

This is a basic "how electronics works" book for the consumer who buys and uses the many wondrous electronic product devices now found in most homes and offices. You now have in your hand a book with over 50 years of my electronic troubleshooting experience and information culled from over 60 of my published electronics books. Thus, this is a book that just about everyone needs to keep on their home or office bookshelf or desk.

The simplified technical electronics information and service tips you obtain from this book can help you in dealing with electronics technicians or service companies when you need professional service for the repair of your equipment. This might save you repair costs because service personnel will not be able to "pull the wool over your eyes," so to speak, since you will be better technically informed. Thus, service repair estimates and costs may swing in your favor. Also, the knowledge gained from this book might help to determine if you should repair a faulty device or purchase a new one.

Finally, this is a valuable book for the hobbyist, electronic experimenter, or any person interested in entering the wonderful world of electronics as a career.

Bob Goodman, CET
Hot Springs Village, AR

ACKNOWLEDGMENTS

Many thanks to the following electronics companies for furnishing some of the technical circuit information, drawings, and photos: Zenith Electronics Corp., Thomson Multimedia Corp., Sencore Electronics, Inc., and Bose Acoustic Wave Music Systems.

Many thanks to the electronics instructors and electronics service technicians that I have had the pleasure of meeting during the seminars that I have given for many years in all parts of the nation.

INTRODUCTION

This new edition is designed for anyone who wants simple explanations of how electronic equipment in the home and office works. Following is a chapter-by-chapter description of the wealth of information in this book that will take the mystery out of electronic consumer products.

Chapter 1 gives you a basic introduction to electronics—"Very Basic Electronics 101." The chapter contains photos and drawings of the components found in your electronic devices with explanations of what they do, how they are constructed, and how to test them. You'll be shown how to use a volt-ohm meter (or multimeter) to check the voltage and resistance found in electronic circuits. You'll learn how to build a simple circuit tester in order to check solid state devices such as transistors, diodes, and ICs.

Chapter 2 is an overview of how FM radio signals are developed and received on a stereo radio. You'll get tips on radio repair and a look at the Dolby audio system. You find out how loudspeakers work and how the advanced Bose Acoustics radio and speaker systems operate. The chapter concludes with an explanation of how cassette recorder/player machines work, audio cassette trouble symptoms, corrective action, and care and cleaning of these units.

Chapter 3 introduces you to the operation of audio and video laser disc players and compact discs (CDs) and how to clean them and perform minor repairs. You'll get hints on keeping your CD operating smoothly and a list of common CD problems and their solutions.

Chapter 4 contains an overview of color TV signal makeup, the components within the signal, and some of the various worldwide color TV standards. The stages that make up color TV set operation are explained via a block diagram that helps walk you through the circuit operations. You'll delve into horizontal and vertical sweep circuit operations, color picture tube operation, and how a color picture is developed on the screen. A preview of large-screen projection receiver operations follows. The chapter concludes with a list of typical color TV and PC computer monitor trouble symptoms and their solutions.

In Chapter 5 you'll learn about flat screen plasma TV/monitor devices, large screen projection sets, and the new digital HDTV system operations. You'll see how the plasma flat screen develops a TV picture and learn how to make adjustments. The chapter concludes with a series of HDTV questions and answers.

Chapter 6 has information on the new and exciting Digital TV DirecTV Satellite (DSS) transmission system and its operation, including an overview of the uplink earth station, the satellite that receives and retransmits the signals, and the dish/receiver that picks up the downlink signals. Detailed drawings will help you connect the DSS receiver to your TV receiver and VCR recorder.

In Chapter 7 you'll get a look at past and present video cameras and camcorders and review various features of this equipment, such as older models with vidicon pickup tubes and modern CCD solid-state image pickup chips and digital video cameras. You'll learn how camcorders work and how to perform minor repairs and clean recording heads.

Chapter 8 explains the telephone landline system and home phone operation and describes how the electronic phone works. You'll find out how to determine whether your phone or the phone company line to your residence is at fault. You'll learn how answering machines and cordless telephones work. All types of phone problems and their solutions are covered.

Chapter 9 covers the various remote control units used for operating TV receivers, CD players, DVD players, set-top boxes, cable control boxes, VCRs, and DSS satellite dish receivers.

Chapter 10 reviews basic printer, copier, and fax machine operation. You'll find out how the "Daisywheel," ink-jet, dot-matrix, laser, and color laser printers operate and how to troubleshoot them. The chapter concludes with information on the operation of copiers, scanners, and fax machines.

Chapter 11 gives you an inside look at DVD video player operation, DVD disc construction, and how the laser beam reads disc information.

Chapter 12 contains general electronic service and maintenance information that you will find useful for keeping your electronic devices in good working order.

1

INTRODUCTION
TO VERY BASIC
ELECTRONICS "101"

How Resistors Work

Resistors are made in various shapes, sizes, resistance values (in ohms) and wattage ratings. Resistors are the most common electronic circuits. In fact, ICs have many resistors inside them. Resistors are used as current-limiting devices and an electronic circuit will not work without them. You might think of a resistor as a control device that limits current flow to the circuit load. A circuit load provides the work; it can be a light bulb, motor, loud speaker, transistor, or IC. Resistor values are in ohms and are made of carbon or coils of resistance wire. Resistor values can be fixed or adjustable (as with a rheostat or like a variable volume control used on a radio or TV). The value in ohms of a resistor is what will determine the electron current. A low resistance will cause a large current to flow, and a high resistance will cause a small current flow.

RESISTOR TYPES

Many types, values, and sizes of resistors are used in electronic products. The photo in (Fig. 1-1) shows various wattage sizes of carbon and flameproof resistors from ¼-watt to 2-watt ratings. These fixed resistors are made to a specific resistance value and cannot be changed. The resistance value is indicated by color-coded bands or stamped numbers on the side of the resistors body. The symbol for a fixed resistor is shown in Fig. 1-2. The larger, 10- to 300-watt, power resistors are shown in Fig. 1-3.

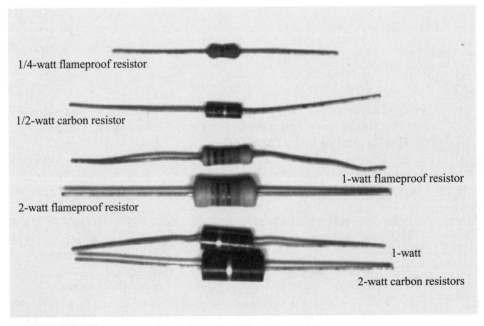

1/4-watt flameproof resistor

1/2-watt carbon resistor

1-watt flameproof resistor

2-watt flameproof resistor

1-watt

2-watt carbon resistors

FIGURE 1-1   **Various types and wattages of resistors.**

Carbon Resistor

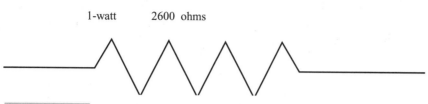

1-watt   2600 ohms

FIGURE 1-2   A schematic symbol of a fixed resistor.

.1 power resistor

1.6-ohm 300-watt resistor

50-ohm power resistor

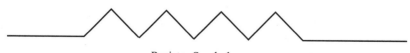

Resistor Symbol

FIGURE 1-3   Drawings of high-wattage resistors.

READING RESISTOR COLOR CODES

If you need to replace a resistor, you will need to be able to read the color code bands to determine its value because you might not have a schematic or the value might not be given on the circuit diagram. The standard resistor color code is:

Black   0
Brown   1
Red   2
Orange   3
Yellow   4
Green   5
Blue   6
Violet   7
Gray   8
White   9

Most fixed carbon resistors use the color band layout (as shown in Fig. 1-4) to indicate their value and tolerance. The first band color is for the first number of the resistor value. Band 2 indicates the second number. Band 3 is a multiplier to show how many zeros follow the first two color-band numbers. As an example, a 25,000 ohm (25 kΩ) resistor would have these band colors:

- Band #1: Red or 2
- Band #2: Green or 5
- Band #3: Orange or 3 for 3 zeros 000.

And the resistor would read as 25,000 ohms.

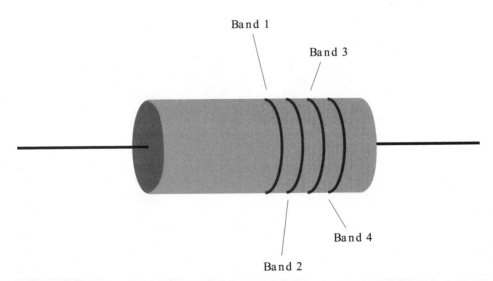

FIGURE 1-4   Resistor color-code position bands.

The fourth band is reserved for the tolerance band color. A silver color band shows that the resistance value indicated is within ±10 percent. If the band is gold, the tolerance is 5 percent of the stated value.

Variable resistors A variable resistor that you can rotate or slide is called a *potentiometer*. These are used for radio and TV volume controls or circuit adjustment controls, as shown in Fig. 1-5. The potentiometer is used in circuits where the voltage needs to be controlled from zero to the maximum, then back to zero. Figure 1-6 illustrates how a volume control would appear in a circuit schematic.

The small volt-ohm meter shown in Fig. 1-7 can be used to check fixed resistors for their correct value or opens, and also to check variable resistor controls for a bad spot when it is rotated. Sometimes these variable volume controls can be cleaned with a spray control cleaner and restored to proper operation.

Resistors/Pots

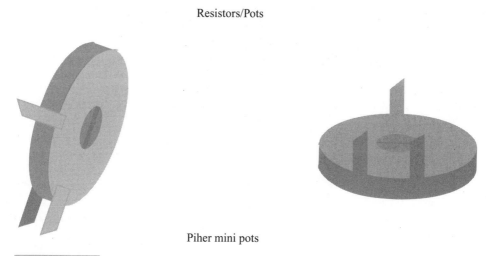

Piher mini pots

FIGURE 1-5   **Drawing of adjustable resistor mini controls.**

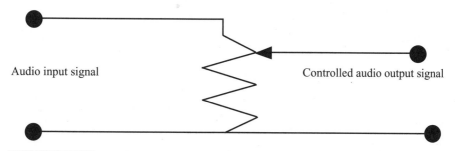

Audio input signal

Controlled audio output signal

FIGURE 1-6   **Circuit drawing of a volume-control resistor.**

FIGURE 1-7   A small volt-ohm multimeter that is
used to check resistance values in electronic equipment.

RESISTOR PROBLEMS

A resistor might be defective because of a manufacturing fault or even years of use in a
high-moisture area. But, the great majority of resistor failures are caused by circuit faults
or lightning/power surges. If you find a burnt resistor, it will probably be caused by one of
the following reasons:

■ A short on the circuit board or wiring, which might have had a liquid spilled onto it.
■ A shorted or very leaky capacitor.
■ A shorted transistor, diode, or IC.
■ A power-line surge or lightning spike damage.

   If a carbon resistor has been overheating slowly for a long time, the resistance will be
lower in value. If it has burned rapidly because of a short circuit, the resistor might go up
in a puff of smoke and the resistor will be open. In fact, you might only see a black, burnt
area with two wire leads sticking up. An intermittent resistor is a rarity and usually is of
the wirewound variety. The intermittent ones will usually look normal, but if you see one
that has a crack in it, replace it.
   Always replace resistors with the same type and value. You can replace a resistor with
one that has a larger wattage rating, if you have room to mount it on the circuit board.

Electronic Circuit-Protection Devices (Fuses)

When electronic equipment fails, depending on the fault, the power supply circuit will draw more current. To protect the other circuits from more damage, a fuse is installed to shut down the device. The fuse is placed in series with the current-drawing circuit, as indicated in Fig. 1-8. A blown fuse is, in effect, the same as turning off the power switch.

Many different types of fuses are used in consumer electronic devices. Figure 1-9 shows four types of fuses. Some of the different types of fuses used for electronic circuit protection are:

- Very small size microfuses.
- Fast-acting or quick-blow glass fuses.
- Slow-blow or lag-time fuses.
- Ceramic fuses.
- Slow-blowing glass fuses.

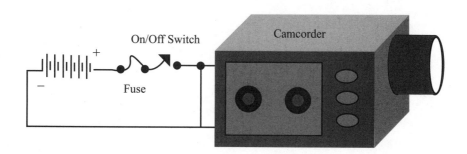

On/Off Switch

Fuse

Electronic Equipment

AC Power line (fused) operated equipment

On/Off Switch

Camcorder

Fuse

Fuse placement for battery operated equipment.

FIGURE 1-8   **Where protection fuses are installed for ac power line and battery operated electronic devices.**

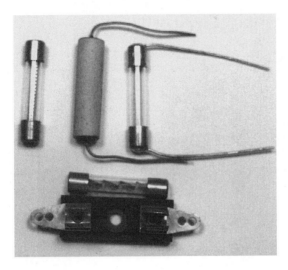

FIGURE 1-9   Various types of fuses used in consumer electronic devices. From the top left to right is a glass slow-blow fuse, a ceramic fuse, and a glass fuse with leads attached (pigtails) that can be soldered into a circuit board. At the bottom is a fast-blow fuse that can be snapped into a fuse holder.

- ■ Thermal-protection fuses.
- ■ Fusible resistor-type fuses.
- ■ Bimetal strip circuit breakers that can be reset.

TESTING THE FUSE

A fuse is either good or bad. With some glass fuses, you can see that the link has been blown or is missing. However, with other types of fuses, you cannot see inside or you cannot be sure if it is good or bad. To be sure, use an ohmmeter, as shown in Fig. 1-10, to

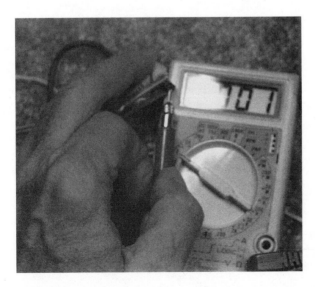

FIGURE 1-10   A digital ohmmeter being used to check a fuse.

check for continuity. The ohmmeter will quickly indicate if the fuse is good or bad. Usually, if the fuse has blown, the circuit is shorted or a high current is being drawn. Some fuses might fail from a defect, vibration link breakage or if loose in the fuse holder, might become very hot and will actually melt the solder alloy link and cause the fuse to be open. Always replace a fuse with one that has the same value.

How Capacitors Work

Capacitors are used in electronic circuits for isolation or blocking dc voltages, and used with coils to produce resonant or tuned circuits, transfer of ac signals, filtering out unwanted interference, and as smoothing filters in power supplies.

The construction of a basic capacitor is shown in Fig. 1-11. It consists of two plates (conductors) that are separated by an dielectric (insulator). The insulator material can be mica, paper, tantalum, plastic, fiber, or even air.

The capacity in (microfarad or picofarad) determines what size of electrical charge that the capacitor can store (hold). The capacity is determined by the size of the plates, the space between the plates, and the type of dielectric between the plates.

TYPE OF CAPACITORS

Figure 1-12 shows some disc ceramic capacitors on the right side and tubular electrolytic capacitors on the left side. Figure 1-13 shows a variety of teflon epoxy dipped capacitors. Figure 1-14 shows an assortment of mica trimmer capacitors and also several mini adjustable trimmer capacitors.

The names of various types of capacitors are as follows:

■ Can, tubular, and molded electrolytic capacitors are used for power-supply filtering.
■ High-voltage ceramic capacitors.

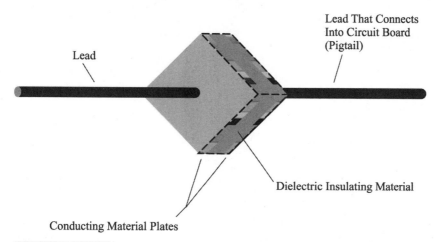

FIGURE 1-11   Illustration of how a capacitor is constructed.

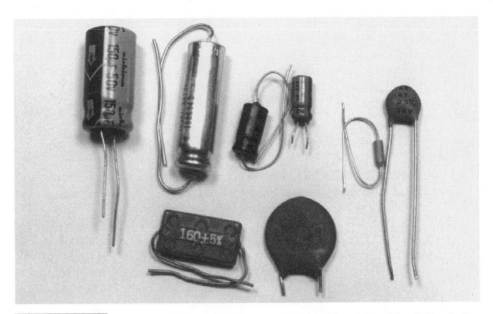

FIGURE 1-12   Disc ceramic capacitors are shown on the right side of the photo and tubular electrolytic capacitors are shown on the left side of photo.

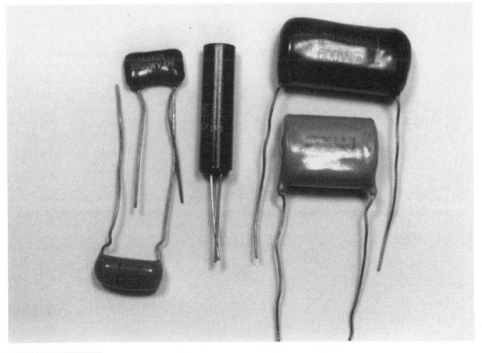

FIGURE 1-13   A variety of teflon epoxy-dipped capacitors.

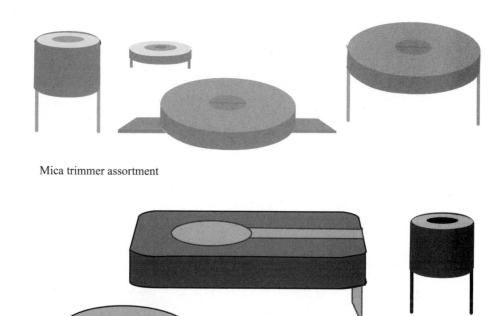

Mica trimmer assortment

Mini trimmers

FIGURE 1-14   **An assortment of mini mica trimmer capacitors.**

■ Tubular ceramic capacitors.
■ Feed-through capacitors.
■ Disc ceramic capacitors.
■ Variable air tuning capacitors.
■ Adjustable trimmer (mica and ceramic) capacitors.

CAPACITOR CIRCUIT DIAGRAM SYMBOLS

Figure 1-15 shows the circuit schematic symbols for a electrolytic filter capacitor, fixed capacitor and a variable capacitor. Figure 1-16 depicts electrolytic filter capacitors being used in a power-supply circuit. The tuned circuit shown in Fig. 1-17 shows a variable air capacitor with a small trimmer capacitor across it, as used in a radio to select the various station frequencies. A shaft is connected to the plates of the rotor, which rotate to change frequencies. The plate area is either decreased or increased, which, in turn, changes the capacitance value, and thus the frequency of the tuned circuit.

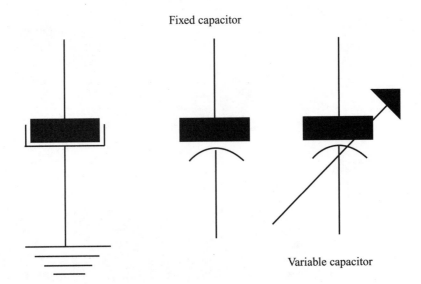

Fixed capacitor

Variable capacitor

Electrolytic Filter capacitor

FIGURE 1-15   Schematic circuit drawings of various types of fixed and variable capacitors.

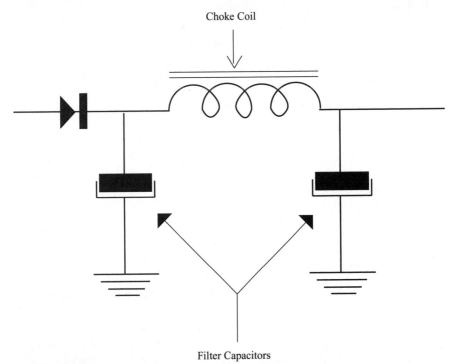

Choke Coil

Filter Capacitors

FIGURE 1-16   Filter capacitors are used in a power supply to smooth out the rectified dc pulse voltage.

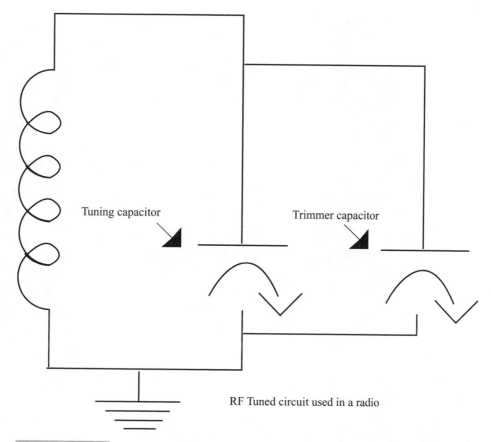

Tuning capacitor

Trimmer capacitor

RF Tuned circuit used in a radio

FIGURE 1-17   **A variable tuning capacitor circuit with a trimmer in parallel with it. This circuit is used to tune a radio to various frequencies (stations).**

TIPS FOR LOCATING FAULTY CAPACITORS

When a capacitor fails, it might become shorted, open, or leak. A capacitor checker can be used to find these problems, but this equipment is expensive. However, you can use a volt/ohm multimeter, like those shown in Fig. 1-18. You can use the ohmmeter range to see if the capacitor is shorted or open. If you measure a low-resistance reading, the capacitor is shorted or very leaky. If you obtain a high resistance reading with a "kick" of the meter needle, the capacitor is probably good. For an accurate, reliable test, the capacitor should be removed from the circuit.

In any circuits that contain solid-state devices (diodes, transistors, ICs, etc.), do not bridge a good capacitor across a suspected faulty one for a test. A spark will usually occur, which can damage the junctions within other solid-state devices mounted on the PC board.

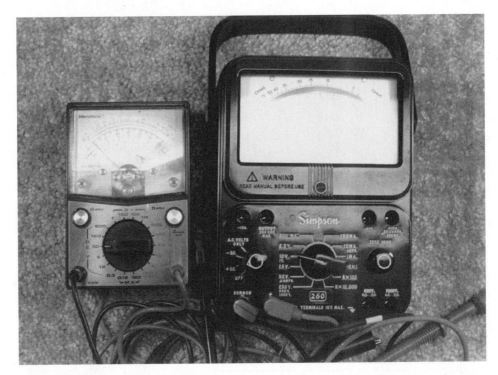

FIGURE 1-18   **Volt/ohmmeters that can be used to check capacitors and other electronic components.**

Transformer and Coil Operations

Transformers and coils are used in most electronic consumer devices. These can be power transformers in the power-supply section, radio-frequency (RF) transformers and coils in the RF and IF sections of TV and radio receivers, and chokes or coils used to eliminate various types of RF and electrical interference.

Figure 1-19 shows schematic symbols of transformers and choke coils that use iron, ferrite and air for the core forms. Transformers will usually have four or more leads and choke coils will only have two lead wires for connections. As the name implies, the transformer "transforms" pulsing dc or ac voltage up (to a higher voltage) or down (to a lower voltage) by induction from one winding to another adjacent near-by winding(s).

Transformer action can only occur when the voltage to the coil winding is changing, such as an alternating ac (alternating current) voltage. If a dc voltage is connected to a transformer winding primary, the secondary winding would only produce a voltage pulse for an instant when the input coil voltage is connected or disconnected. The magnetic field produced by the primary coil will (cut) go across the secondary winding and by this magnetic induction will induce an ac voltage into the secondary coil. Thus, a magnetic transfer occurs, which increase or decrease the ac output voltage, depending on the number of turns of the coil.

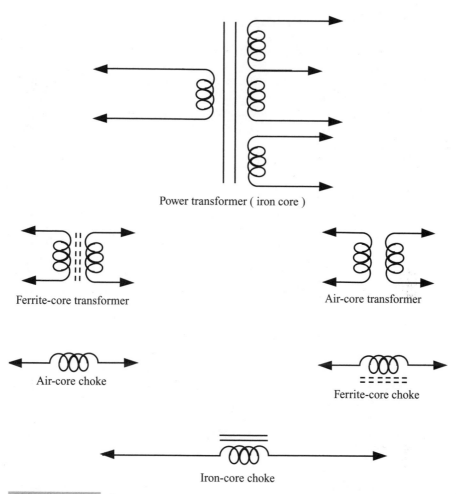

Power transformer (iron core)

Ferrite-core transformer   Air-core transformer

Air-core choke   Ferrite-core choke

Iron-core choke

FIGURE 1-19   **Schematic symbols for transformers and choke coils used in electronic equipment.**

Transformer types used in electronic equipment are:

- Power transformers.
- Audio transformers.
- Voltage regulator and smoothing transformers.
- Antenna matching transformers.
- Oscillator transformers.
- Adjustable core (slug) transformers.
- Radio-frequency (RF) transformers.
- High-voltage "flyback" transformers used in TVs.
- IF transformers.
- Interstage audio-matching transformers used in audio amplifiers.

TRANSFORMER TROUBLES AND CHECKS

Transformers can fail in many ways. Some of the various failures are:

■ The coil wire turns can open. This can occur where the copper wire coils are connected to the terminal lugs.
■ The windings can become shorted to adjacent turns of the coil wires.
■ The primary and secondary coil windings can become shorted to each other.
■ The primary and secondary coil windings can develop a high-resistance leakage.
■ The coil windings can become shorted due to insulation breakdown to the metal core, transformer case, or frame.
■ A power transformer might become hot, have a waxy material start leaking from the case and might actually smoke and burn up.

If you detect a burning odor from your equipment and see some melted wax coming from inside the transformer case located in the power-supply section, immediately turn it off or unplug the device. You might find the same symptoms if your equipment stops working and a fuse has blown. An overheated power transformer will usually not be damaged as the problem that caused this condition is some other component that has shorted out. These component faults could be a shorted diode rectifier, electrolytic filter capacitor, regulator transistor or a bypass capacitor. You can use an ohmmeter to check for any shorts or low resistance in the B+ supply lines.

Figure 1-20 illustrates how you can check for leakage between the primary and secondary of a transformer. With the device turned on and a dc voltage on the primary winding, measure for any voltage with your voltmeter at the points indicated on the two secondary windings. If you find even a very small voltage, the transformer has leakage and should be replaced. The ohmmeter is used to check across each winding for opens. Be sure that the device is turned off or unplugged for these ohmmeter checks.

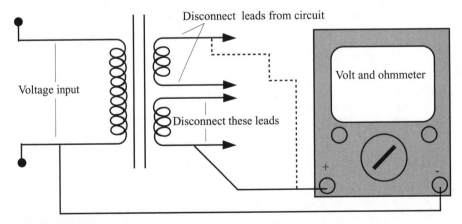

Power transformer

Disconnect leads from circuit

Voltage input

Volt and ohmmeter

Disconnect these leads

FIGURE 1-20   How to use an ohmmeter or voltmeter to check for a short or leakage between windings of a power transformer.

Because of the high-voltage involved with the high-voltage sweep or "flyback" transformers in color TV sets and monitors, they usually are the cause of the failure when they arc, smoke, or burn up.

Transistors, Integrated Circuits (ICs), and Diodes

This section shows how diodes, transistors, and ICs work, what they look like in your equipment, and some ways to check them out for failures. Figure 1-21 shows a 24-pin IC used in some camcorders and are sometimes called *microchips*, *chips*, or *ICs*. An IC consists of many solid-state transistors, diodes, resistors, coils, and capacitors. You can think of the transistor as the basic building block of which all IC chips are constructed. When ICs are used in computers, many transistor gates create binary data to either be: "off" or "on," which provide "0s" and "1s." Also, transistors make it possible to use a very small electrical current to control a much stronger second current. Transistors are called *semiconductor devices* (hence, solid-state) because they are actually made from materials, such as silicon and germanium, which are not perfect insulators nor good conductors. So, the current in these solid-state devices is controlled within a "solid-state" material.

DIODES

The diode is a solid-state device that will only permit current flow in one direction, or polarity, but not going in the opposite direction. It can be used as a protection device for dc-operated equipment. If equipment is connected accidentally to a battery with the wrong polarity, no damage would occur because no current would flow.

Diodes are very useful in power supplies to change ac voltage into pulsating dc voltage when used as a rectifier diode. The top drawing of Fig. 1-22 illustrates how the current will flow in one direction only through the diode. The bottom drawing of Fig. 1-22 is of a simple power supply, where the diode is used as a rectifier diode to change an ac voltage into a pulsating dc voltage, which is then smoothed out with filter capacitors for a dc voltage. Figure 1-23 shows the many various shapes and sizes of some common diodes in consumer electronic equipment.

With the correct polarity the voltage across a diode will let the current pass with no resistance or very easily. With the opposite polarity of voltage, the current will encounter a very high resistance and current will not flow. When the current cannot pass through the

FIGURE 1-21   **A 24-pin IC used in a camcorder.**

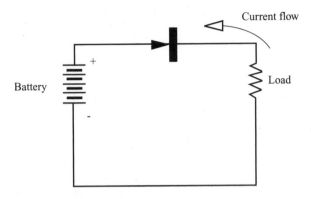

A diode will only let current flow in one direction.

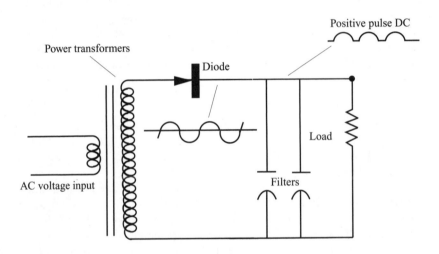

How a diode is used in a power supply circuit: The diode changes AC voltage to a DC voltage

FIGURE 1-22   **The top drawing shows how a diode will only let current flow in one direction. The bottom drawing illustrates how a diode is used as a rectifier in a power-supply circuit.**

diode, this is called *reverse bias*, but, when the current can easily flow through the diode, this is referred to as *forward bias*.

METAL-OXIDE VARISTOR (MOV) OPERATION

The MOV or varistor is used in many consumer electronic products. Figure 1-24 depicts the MOV or varistor in circuit diagrams. You can think of the varistor (voltage-variable resistor) as a device that has a high resistance at a low voltage, but with a certain higher voltage, the resistance drops to a much lower value.

MOVs usually consists of a zinc-oxide material. Because the varistor is not polarized, it is very useful in ac circuits. MOVs are used across the ac power line into electronic equipment and also in the telephone line input circuits (Fig. 1-25). The MOV is usually installed

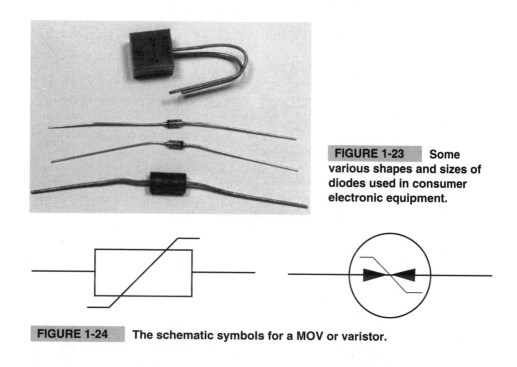

FIGURE 1-23   Some various shapes and sizes of diodes used in consumer electronic equipment.

FIGURE 1-24   The schematic symbols for a MOV or varistor.

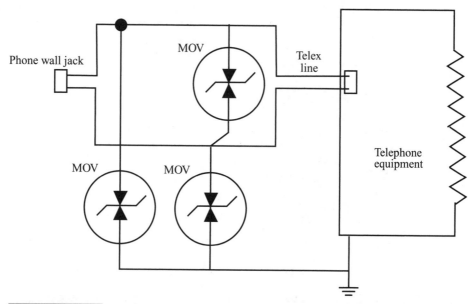

FIGURE 1-25   MOV spike-protection components installed on a home phone line.

for spike and voltage-surge suppression for circuit protection. The MOV is rated by its "breakdown" voltage, thus when installing or replacing an MOV, be sure that the rated voltage is a little higher than the voltage usually found at this circuit point.

How Transistors and ICs (Solid-State Devices) Work

You will find many transistors used in all electronic equipment. And, of course, integrated circuits (ICs) have lots of transistors inside their chips. The transistor package has three leads coming out (sometimes four leads) and is a solid-state electronics package that can perform amplification and switching of electronic signals. Figure 1-26 shows the various types and sizes of transistors used in consumer electronic products. There are two basic transistor designs. One type is the *bipolar* and the other is the *Metal-Oxide-Semiconductor Field Effect (MOSFET)*. Figure 1-27 illustrates a cross-section view of a NPN bipolar transistor structure. A bipolar transistor is usually made of a silicon material. Discrete transistor construction requires many complex steps that start with a blank wafer of silicon. Some of the steps include photographic masking, photo reduction of large-scale artwork, ultraviolet light to alter the chemical composition, and chemical solvent to remove unexposed photoresist.

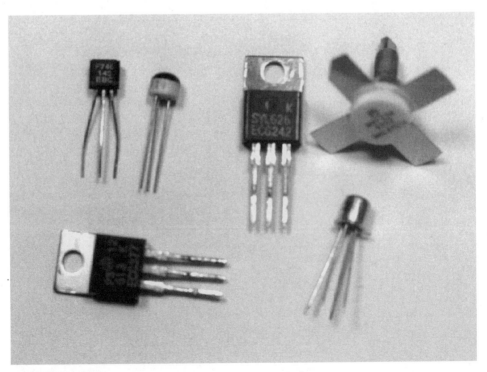

FIGURE 1-26   Various transistors used in consumer electronic products.

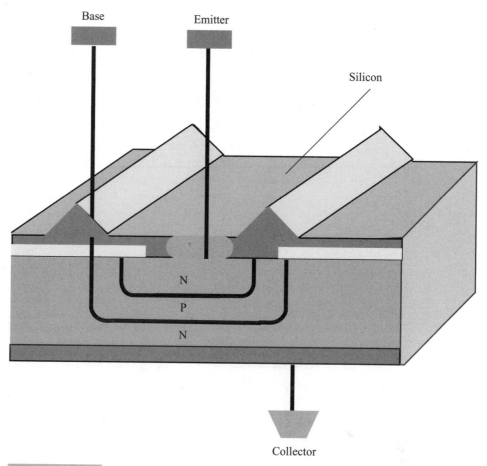

Base   Emitter   Silicon

N
P
N

Collector

FIGURE 1-27   A cutaway view of an NPN bipolar transistor.

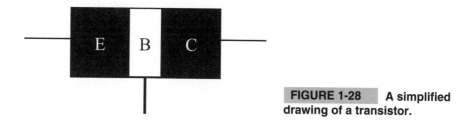

E   B   C

FIGURE 1-28   A simplified drawing of a transistor.

A transistor consists of thin layers of material with a collector on one side, a thin base layer in the middle, and the emitter on the other side. Notice the simplified drawing in Fig. 1-28. The material used for the emitter and collector sections are opposite of that used for the base.

For a PNP transistor, N-type material is used for the base, but the collector and emitter are made from P-type material. With an NPN transistor, the base is an P-type material and the collector and emitter are made of N-type material. A drawing of an NPN transistor with the circuit diagram below it is shown in Fig. 1-29. Figure 1-30 shows a drawing of a PNP transistor with its circuit diagram below it.

Review this one more time. With a NPN transistor, the base is a P-type material and the collector and emitter are both made from N-type material. Conversely, for a PNP transistor, the base is an N-type material, and the collector and emitter are of the P-type material.

In most cases, the NPN and PNP transistors work the same way in their circuits, except for the applied voltage polarities. A PNP transistor will have a negative collector voltage and a negative base bias voltage. An NPN transistor will use a positive voltage on the collector and a positive bias, thus it has a collector-to-emitter positive voltage.

THE INTEGRATED CIRCUIT (IC)

Inside an IC package is a small "chip" or microcircuit with many active and passive electronic parts interconnected on a small semiconductor substrate or wafer. The chip will perform many electronic circuit functions in a very small space. Figure 1-31 illustrates the many transistors, resistors, and capacitors found in a typical op-amp IC package.

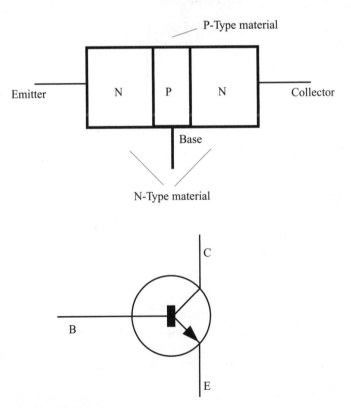

FIGURE 1-29   A circuit diagram of an NPN transistor.

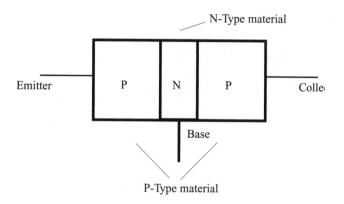

N-Type material

Emitter

P | N | P

Collec

Base

P-Type material

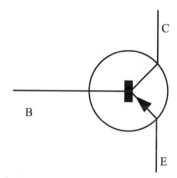

C

B

E

FIGURE 1-30   **A circuit diagram of a PNP transistor.**

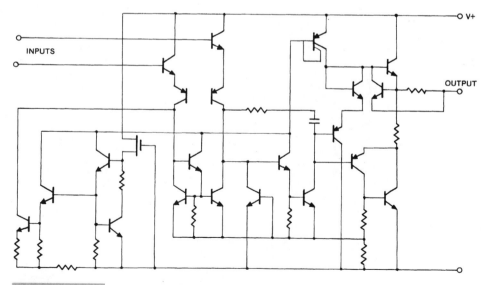

V+

INPUTS

OUTPUT

FIGURE 1-31   **The transistors, resistors, and capacitors inside of a typical IC op amp.**

Some of the advantages of ICs over conventional circuits are:

■ Because all circuit parts are on the same substrate, performance and temperature conditions will not vary.
■ Because of the microchip construction, more circuit operations can be mounted on smaller circuit boards.
■ The IC is the main reason that electronic products are more reliable because so many external electrical connections have been eliminated.
■ The IC has increased circuit performance and speed because of shorter lead interconnections. The invention of the IC caused the "great leap forward," which made possible the increased speed of computer computations and vast amounts of memory retention.
■ With lower power consumption and less heat loss, the IC has made electronic devices much more efficient.

The circuit diagrams for two types of ICs are shown in Fig. 1-32 and are the way you will find chips drawn on schematic diagrams. A photo of the round 8-pin IC is seen in Fig. 1-33 and its circuit drawing is shown on the right side of Fig. 1-32. A photo of some common 16- and 18-pin in-line ICs are displayed in Fig. 1-34.

Solid-State Scope Sweep Checker

You can build this simple checker that tests transistor, diode, Zener diode, SCRs, and even some ICs. This device connects to an oscilloscope to make a fast "go-no-go" test unit. This little sweep checker can even be used to check resistors, capacitors, and find shorts and open cir-

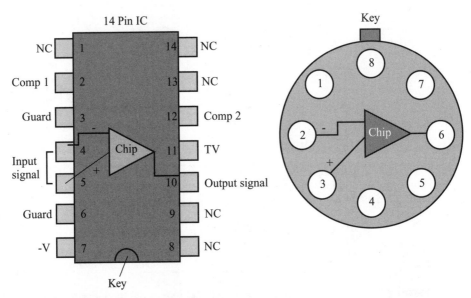

NC = No internal connection

FIGURE 1-32   **The circuit diagram and case layouts for two types of ICs.**

FIGURE 1-33   An 8-pin plug-in IC.

FIGURE 1-34   Some 16- and 18-pin in-line ICs.

cuits. I designed this "sweep junction" checker over 30 years ago; for some time, Texas Instruments (TI) used this device to sort out defective transistors and diodes on their production line.

This simple checker is easy to build-up and connect to your scope. You should find it to be a quick and easy, but reliable, test box for fast checking most solid-state devices. This tester, which connects to a scope, uses an ac sinewave to sweep the solid-state devices junction under test.

The simple circuit diagram for this checker is shown in Fig. 1-35 and also how to connect it to the scope's vertical and horizontal sweep inputs. Transformer T1 has a 120-Vac primary and either a 6.3-V or 12.6-Vac secondary. You can use red and black voltmeter leads with needle point tips for the test probes. The black lead is ground and the red lead is used for the positive test lead. The polarity of the leads will affect the scope waveforms by flipping the trace upside down when you reverse the leads to the component under test. The six scope pattern drawings shown in Fig. 1-36 are some typical traces you will find for the various components listed.

When checking a solid-state device out of the circuit, the main point of interest is the knee of the curve. A sharp bend usually indicates that the device is good. A straight horizontal line indicates an open junction and a straight vertical line on the scope pattern means a shorted component. If the supply voltage of the curve checker exceed the peak-inverse voltage (PIV) of the solid-state junction under test, Zener action might occur. This is indicated by a very short vertical line, see Fig. 1-37, at one end of the trace pattern and should be disregarded.

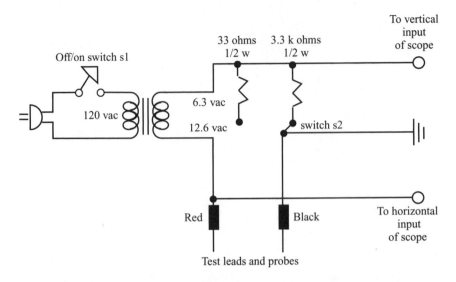

FIGURE 1-35   A circuit diagram for a solid-state sweep/junction checker.

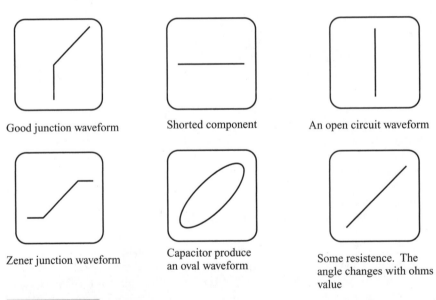

Good junction waveform

Shorted component

An open circuit waveform

Zener junction waveform

Capacitor produce
an oval waveform

Some resistence. The
angle changes with ohms
value

FIGURE 1-36   Some typical scope waveforms that you will find when using
the junction sweep checker.

Note:

Connect test probes across solid-state device to be checked with no power applied to
circuit under test.

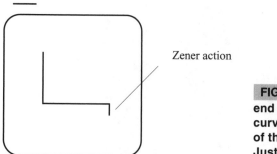

Zener action

FIGURE 1-37   The short line at end of the trace will occur when curve tracer voltage exceeds the PIV of the solid-state junction under test. Just disregard this line.

When solid-state devices are checked in circuit, the ideal out-of-circuit scope traces might not appear because other resistors, coils, and capacitors in the circuit might cause the trace patterns to vary. Thus, when checking in-circuit components, a comparative method must be used. Also, for a positive test, the component can be removed from the circuit.

When checking transistors, disconnect power from the device under test and connect the test probes. Always connect the test probes to the transistor terminals by the color code shown:

■ Base-emitter junction: Base red, emitter black.
■ Base-collector junction: Base red, collector black.
■ Collector-emitter junction: Collector red, emitter black.

The shape of the pattern shown in Fig. 1-38 is for a transistor with a high junction leakage. The pattern in Fig. 1-39 was obtained when the circuit under test had a very low resistance value.

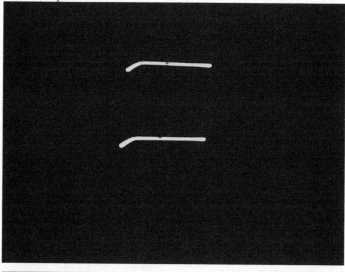

FIGURE 1-38   A scope pattern for a very leaky transistor.

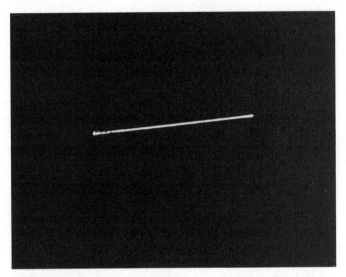

FIGURE 1-39   This scope pattern shows a circuit with a very low resistance.

Electronic Power Supplies

All electronic devices must have some type of power supply or source voltage to operate. Most draw power from an ac power line and use rectifiers and filters to produce a dc voltage. Some equipment operates from batteries and the power supply is used to recharge the batteries. Most electronic circuits require a dc or direct current in order to operate.

HALF-WAVE POWER SUPPLY

The circuit drawing in Fig. 1-40 is of a half-wave rectifier power supply. Notice at the top right, the negative going part of the sine-wave is missing and only the positive-going part is now available. The bottom waveform portion is removed by the diode rectifier because it only lets current pass in one direction. The pulsating dc is 60 times per second and is now smoothed out with a filter capacitor.

FULL-WAVE POWER SUPPLY

The full-wave power supply circuit shown in Fig. 1-41 lets both halves of the ac sine-wave be used, with an output ripple of 120 times per second, rather than 60 times, as with the half-wave power supply. The two diodes are connected so that one diode conducts on the other half of the cycle. Thus, the diodes are conducting on each half cycle. This 120-cycle ripple now must be smoothed out with a resistor or iron-core choke and two filter capacitors. The choke helps prevent sudden changes of current through it and a second electrolytic capacitor (C2) provides even more filtering.

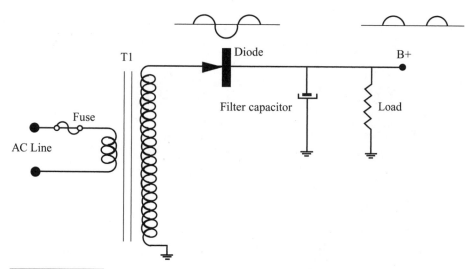

FIGURE 1-40   A half-wave rectifier power-supply circuit.

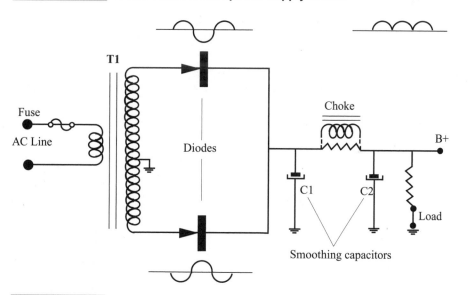

FIGURE 1-41   A full-wave rectifier power-supply circuit.

A BRIDGE-TYPE POWER SUPPLY

Bridge diode power supplies are used in many kinds of electronic equipment, such as TVs, video recorders, and stereo sound systems. The bridge circuit power supply is unique because it can produce a full-wave output without using a center-tapped transformer. The typical diamond-shaped diagram for this type power supply is shown in Fig. 1-42. You could think of the bridge-rectifier circuit as an electronic switching system. Think of the diode rectifiers as switching all of the positive ac pulses to the B+ line and all of the negative ac pulses to the B-line or to chassis ground.

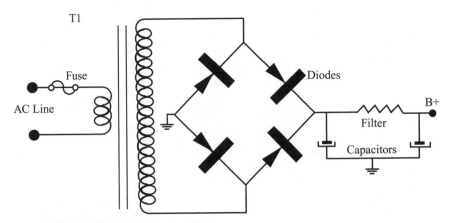

FIGURE 1-42   A typical bridge-rectifier power-supply circuit. Note the diamond shape of the diodes' layout.

THE VOLTAGE-DOUBLER POWER SUPPLY

A voltage-doubler power supply can have a transformer or it can be direct ac-line operated. The transformerless type is used in equipment that requires a higher dc voltage output and also to reduce the cost and weight of the device.

A basic transformerless doubler circuit is shown in Fig. 1-43. To see how it works, assume that the half-wave diode (X1) is connected to produce a positive voltage on the B+ line of

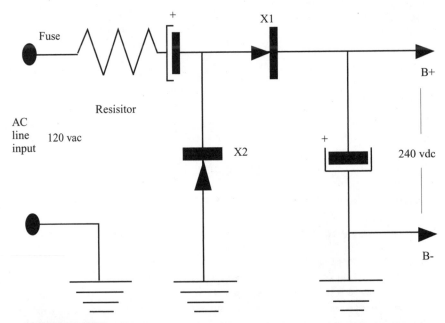

FIGURE 1-43   A power-supply voltage-doubler circuit that does not use a power transformer.

120 volts. Diode X2 is then added to the circuit, but is connected in the opposite polarity. This will make diode X2 –120 volts, with respect to ground. This will "add," then produce a voltage of approximately 240 volts between the B+ and B– points. The problem with this power supply is that the B– is connected to the chassis of the device. This makes it a "hot" chassis, which will create a shock hazard. When you take the case or cover off of this type of equipment, always be cautious and you should plug the device into an isolation transformer.

   Common power-supply problems are blown fuses, shorted diodes, burnt resistors, and open or shorted filter capacitors. Use your volt/ohm meter to check out the power-supply faults. A digital multimeter is being used in Fig. 1-44. You will find more power-supply information and circuits plus troubleshooting tips in other chapters of this book.

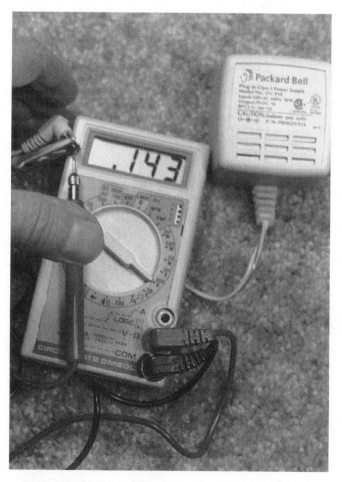

FIGURE 1-44   **A digital multimeter is being used to check a "block-type" plug-in power supply.**

Electronic Circuit Soldering Techniques

When removing or replacing parts on a PC board, you will need a soldering iron (20 to 25 watts), rosin "flux" or solder with a rosin core. A solder wick, which is a flat-braided copper strips is a useful aid for soaking-up solder when removing a part from the PC board. Figure 1-45 illustrates how the solder wick is used to remove solder from a part on the PC board.

Figure 1-46 shows two types of soldering irons. The top one is a 25 watt and should be used for all PC board soldering. The larger 45-watt iron is used for soldering chassis grounds and large-wire connector lugs. Figure 1-47 is of a soldering gun and it only heats when the trigger switch is pulled on. These guns will heat up in about five seconds and usually are high wattage. They should not be used for soldering on PC boards because you can damage one very quickly. Many of these guns are rated a 100 to 150 watts. Figure 1-48 shows how a small iron is used to solder in the pins of an IC.

ICs can be directly soldered onto the PC board or they might have a socket mounted onto the PC board and the chip will plug into the socket. In Fig. 1-49, an IC is being removed from its socket. Very carefully pry up each end of the IC, a little each time, so as not to bend or damage the pins. When installing the IC, be sure that the pins are straight and are lined up with the socket pin holders. Also, be sure the key or notch is correctly lined up with that marked on the PC board. A chip put in backwards can be very costly. Be cautious and recheck position of the chip key.

FIGURE 1-45   **Solder wick being used to "suck up" solder from a connection on a PC board.**

FIGURE 1-46   A 25-watt (bottom) and a 45-watt soldering iron.

FIGURE 1-47   A fast-heating soldering gun.

FIGURE 1-48   A 25-watt iron being used to solder the pins of an IC mounted on a PC board.

FIGURE 1-49   An IC chip being removed from a plug-in socket located on a PC board.

SURFACE-MOUNTED DEVICES AND THEIR SOLDERING TECHNIQUES

As surface-mounted devices (SMD) have evolved, the electronics industry have built SMD equivalents for most conventional electronic components. New electronic equipment contains SMD resistors, capacitors, diodes, transistors, and ICs. Even wire jumpers and 0-ohm resistors are used because they are more easily installed by automated assembly machines.

During assembly, the SMD unit is lightly glued to the circuit board with the metallic contacts lying on the copper path, where a circuit connection is to be made. Wave soldering then is used to join all SMDs electrically and mechanically to the board.

Some SMD basics   On most circuit diagram, an SMD device has an *M* following its part number. The *M* represents for (metal-electrode face bonding), which is the process used in producing chips.

Surface-mount components are available in various sizes and configurations, starting with large microprocessors, all the way down to single diode packages. Even single diodes and resistors are available in different sizes.

Surface-mounted resistors   A typical SMD resistor consists of a ceramic base with a film of resistive material on one surface. Refer to Fig. 1-50. Two electrodes are on the ends of the base, which is in contact with the resistance film. The contacts are used in making a solder connection to a PC board. The resistance of the device is determined by the amount of film material.

SMD resistors are typically in the 1/4- to 1/8-watt range. The regular color code is not used on SMD resistors. Three numbers are usually printed on the film and give the same information as the color code. The first two numbers represent the first two significant numbers of its value. The third number represents the number of zeros.

Surface-mounted capacitors   Chip capacitors are fabricated with layers of resistance film, separated by layers of a ceramic base material, which is the dielectric. Notice Fig. 1-51.

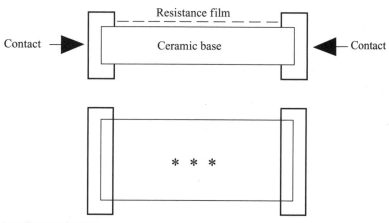

FIGURE 1-50   A surface-mounted device (SMD) resistor.

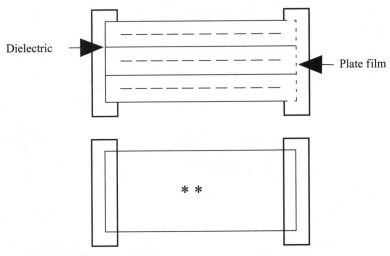

FIGURE 1-51 A surface-mounted device (SMD) capacitor.

The chip capacitor is very similar in appearance to the resistor. The body generally has a two-digit or two-letter code to show the capacitance of the device.

Surface-mounted diodes and transistors The SMD equivalent for solid-state devices are conventional silicon technology in new housings, again allowing for easier automated assembly. Refer to Fig. 1-52. The package for a diode is called an *SMC (single-melt component)*. The diode is marked on one end with a band to denote the cathode of the device.

The transistors are in packaging that corresponds to their purpose. The low-power device is in a SOT-23 (small-outline transistor) package. The transistors that function in heat-generating capacities are in a SOT-89 package that features a heatsink. The same packages are also used for FET and MOSFET devices.

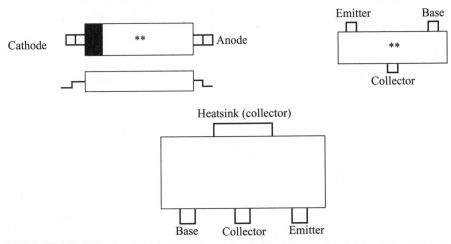

FIGURE 1-52 Drawings of SMD diode and transistor configurations.

Integrated circuits The SMD integrated circuit, like the diode and transistor, is conventional technology repackaged for automated insertion, as well as miniaturization of the circuit boards.

The SOIC (small-outline IC) is similar to the standard DIP packaging, except that the legs are designed for surface-mount soldering. Note layout of SMD IC in Fig. 1-53.

SMD-soldering techniques Soldering of and/or replacement of an SMD is different from a standard component in two ways. First, the reduced size of SMD components and circuit-board paths increase the need for care when repairing this type equipment. Secondly, the tools required for repair are more specialized. Excessive heat can easily damage not only the SMD, but also the PC board paths. A controlled-heat soldering iron in the 20- to 25-watt range is a must. Small-diameter rosin-core solder is also needed. Solder wick is needed in different sizes and can be cut in short pieces. A bottle of flux should be used as an aid in heat transfer. Small-tipped tweezers and dental picks are useful in handling the SMD parts. A magnifier with a light source is very useful for close-up inspections. And a grounded soldering iron and tip should be used along with an anti-static wrist band to prevent damage to static-sensitive SMD components.

Removing SMD resistors or capacitors In most cases, a SMD device is not reusable once it has been removed from the PC board. You should be sure that the device is defective using troubleshooting techniques before removing a SMD.

Now refer to Fig. 1-54. Add extra solder to the contact points to cause even solder flow. Grasp the component body with tweezers and gently rock back and forth while heating the solder on both ends. Remove the heat while continuing to rock the SMD contacts. Once leads are loose from the foil, quickly twist the SMD to break the epoxy or glue that was holding the SMD to the PC board.

SMD transistor removal Refer to Fig. 1-55. Add solder to all three terminals. Grasp the component body with tweezers or needlenose pliers. Heat terminal C and rock the body up

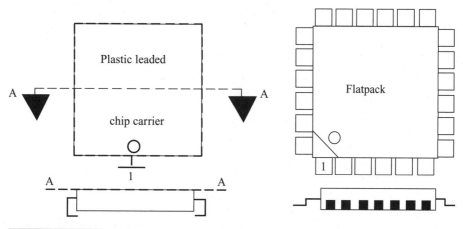

FIGURE 1-53   **Typical layouts of SMD ICs.**

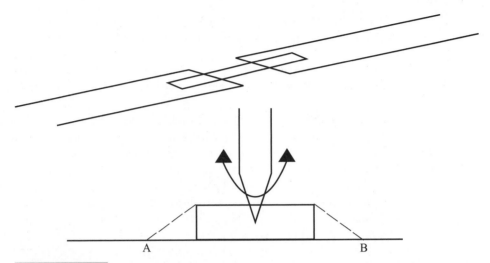

FIGURE 1-54   When installing an SMD, always add extra solder to all contact points for an even solder flow.

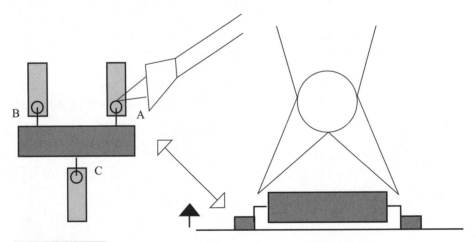

FIGURE 1-55   Add solder to all three terminals when starting to remove an SMD component.

until an open space exists between the terminal and pad. Now work on the other two terminals until loose.

Removing SMD integrated circuits   For IC removal, refer to Fig. 1-56. Apply solder liberally to all pins. Use a special soldering tip that will fit over the particular size of IC housing. This will allow all pins to heat up at the same time. Use a dental pick to lift the IC off as soon as the solder is molten.

SMD parts replacement   The replacement of any SMD follows a similar pattern. Be sure that the foil solder pads are free of any excess solder. Using short pieces of solder wick, clean

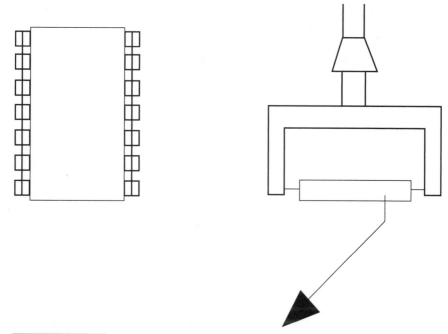

FIGURE 1-56   After all contacts of an SMD chip are heated equally, gently pry up the device for removal.

FIGURE 1-57   A camcorder PC board with several surface-mounted components.

the pads until they are smooth. Spray with board cleaner, if necessary, to remove any residue of rosin. Position the device on the pads and hold them, as necessary, with picks or tweezers. Melt a small amount of solder on the tip of the iron. Then apply it to the lead. This will hold the component in place. Then, using the proper size of solder, attach all remaining legs. The photo in Fig. 1-57 shows a camcorder PC board with several SMDs.

Electronic Test Meters (VOMs)

A volt/ohm test meter is a "must have" if you want to troubleshoot and repair any electronic equipment. These small, inexpensive meters can have an analog meter, which have a needle pointer that swings across the meter scale face plate, or a digital readout, which have the direct number readings on an LCD screen. Figure 1-58 shows some inexpensive digital read-out volt/ohm meters. You can find these meters at electronic parts supply stores, Radio Shack, Wal-Mart, and K-Mart stores. These test meters are called *multimeters* or *volt-ohm milliammeters (VOMs)*. These meters have pushbuttons or a switch to go from one function or rating to another.

You can use your voltmeter in the ac range to check voltage at wall sockets and where the ac line cord terminates in the equipment. You can check the ac power line voltage this way and see if the ac power is getting to the equipment power-supply section at the correct value. You can even locate open fuses and tripped circuit breakers with this ac voltage test. The dc range is used to check battery and charging voltage from any charger unit. Also check the dc voltage output from those small plug-in block power supplies. The dc range

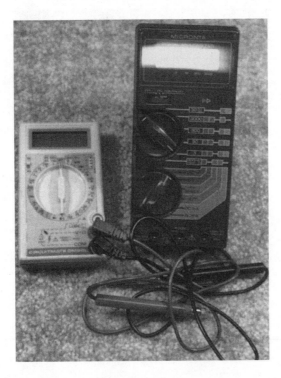

FIGURE 1-58   **Some inexpensive digital volt-ohm meters that are very useful for electronic circuit repairs.**

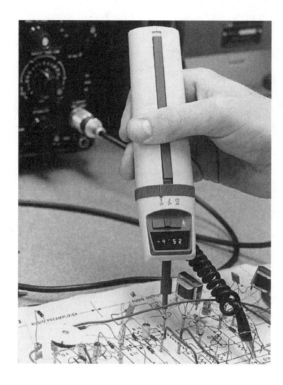

FIGURE 1-59   **A small, easy-to-use, battery-operated, portable volt-ohm meter. Some cost $45 to $85.**

is also used to check all of the other various dc voltage levels that are found in electronic equipment circuit boards. You can use the multimeter to look for low voltage, no voltage, or too high of a voltage.

A small, easy to use, portable, battery-operated volt-ohm meter is shown Fig. 1-59.

Tools for Electronic Circuit Repairs

Now for some information on some common small tools that are very useful for electrical and electronic circuit repairs.

Diagonal cutters, sometimes called *side cutters* or *dikes*, are used to cut wires and component leads. They are also useful for stripping insulation from wires that are to be connected or spliced together. You should have two sizes of diagonal cutters (4" and 6") and long-nose pliers. The long-nose (needle-nose) pliers are used to insert parts, position lead wires and shape wires for connections. The common "gas" and utility "slip joint" adjustable pliers are also very useful. Figure 1-60 shows some of these basic electronic tools needed for repair work.

The following is a list of basic tools you should find useful for electronic repairs:

- Long-nose pliers.
- Diagonal cutters.
- Needle-nose pliers.
- Long-nose pliers with side cutters.

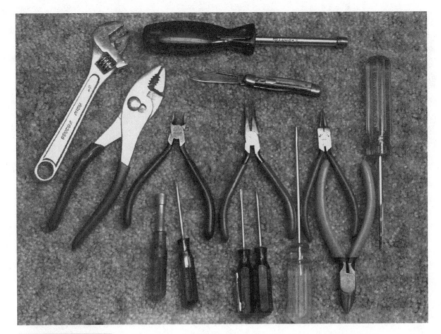

FIGURE 1-60   **Some basic tools you will need for electronic equipment repairs.**

- Utility pliers.
- Seizers, for holding and soldering small parts.
- Electrician's knife.
- Adjustable wrench (crescent).
- Various screwdriver sizes and tips.
- Nut drivers (spinners).
- A small set of jeweler screwdrivers.
- 20-watt and 45-watt soldering irons.

Some Service Repair Tips

When working on your electronic equipment, it is very helpful to have service information and diagrams to reference. Some new equipment has information packed in the box or you can write to the manufacturer for this information. Also, books and schematic folders are available for the various models of TVs, VCRs, camcorders, etc. You can usually find these books and folders at electronic parts stores, such as Radio Shack, Allied Radio, and MCM Electronics. You can also order from TAB/McGraw-Hill Electronics Book Club.

Before you start working on your equipment with a problem you might want to make some notes and review the problem(s):

- Notice when the problem occurs.
- Is the device cold or hot when the problem occurs?

- How does it perform or not perform?
- How often does it occur? Is it intermittent?
- Have you had the electronic equipment repaired for the same symptoms before?
- Does it have to operate a long or short time before the trouble appears?

Thus, as you see from this list, you need to note any type hint or clue to solve these mysterious electronic problems. It helps if you are a good detective.

You will find that with most electronic equipment, such as CD players, video recorders (VCRs), camcorders, cassette players, or telephone answering machines, the problem is generally mechanical and not electrical. All you need for these repairs is a set of common tools, a cleaner/degreaser solvent, lubricating oil or grease, some alcohol for cleaning, and then just use your common sense for repairs. Some of these simple-to-repair problems are:

- For a CD player, check and clean the lens because it might be dirty.
- Also, for a CD player, check the lubrication. Check for oily slide drawer belts and dirt on the sled tracks or gears. A defective or partially shorted spindle or sled motor.
- For a VCR, check for broken or loose belts or belts that need to be cleaned.
- Also, for a VCR, clean the heads, the tape travel tracks, and rubber idler wheels.
- For any kind of video or audio recorder, look for a defective cassette tape cartridge, broken or tangled tape, tape wrapped around the capstan, and jammed-up parts.
- For all VCRs, audio tape recorders, and TVs, check for blown fuses, loose plugs and connections, and power-supply problems.

INTERMITTENT TEMPERATURE PROBLEMS

Intermittent electronic problems are generally the toughest to pin down. Many of these faults show up after the equipment has warmed up. One trick you can try is using heat or a coolant spray (freon) to various small areas of the circuit board. A hair dryer is used in Fig. 1-61 to isolate a heat-sensitive component. This might take a little time, but you can solve the problem. The most common components to breakdown from heat or cold changes are ICs, transistors, diodes, and electrolytic capacitors. Also poor solder connections and PC board cracks can be located this way. And do not overlook small transformers, choke coil windings, and their connections.

NOISY ICS OR TRANSISTORS

Often, the noisy transistor or IC can be located in the input and output sound stages TVs, CD players, and cassette system audio circuits. The hissing or frying noise that occurs with low audio levels can indicate a noisy solid-state component failure. Lower the volume level and listen for the frying noise. If the noise is still present, you know that the defective component is between the volume control and speaker.

You can try isolating the noisy component by grounding the input terminal of the power-output IC or transistor with a 10-ohm resistor to ground. With other transistor stages you can ground the base with a 10-ohm resistor, as shown in Fig. 1-62. If the noise becomes lower or disappears, you know that the defective component is before this stage. If the noise is still present, replace the transistor or IC in this stage.

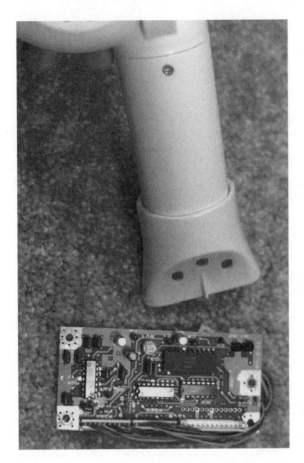

FIGURE 1-61   A hair dryer being used to heat one section of a PC board to locate an intermittent problem.

Sometimes spraying the suspected transistor or IC with a coolant spray will make the noise louder. Other times, the noise will disappear. At other times, again applying heat with a hair dryer on a suspected transistor or IC will make the noise reappear after applying another shot of coolant spray. Do not overlook the small ceramic bypass capacitors that can create noise when B+ voltage is on one side of this component. Replace the noisy component with a good part and then reheat or cool this same area again for a confirmation.

When the noise disappears with the volume control turned down, the noisy component will be ahead of the volume-control circuit. This transistor grounding technique can be used in other amplifier stages by jumping a 10-ohm resistor from the base to the emitter of the suspected transistor; if the noise stops, then the transistor is faulty. In a stereo audio amplifier system, start at the preamp input transistor and proceed through the circuit. If the noise is present after grounding out the first preamp signal, then the second preamp transistor must be noisy.

Usually, the noisy condition occurs in only one stereo channel. If both channels are noisy, suspect the stereo IC power output. The noise might disappear for several days,

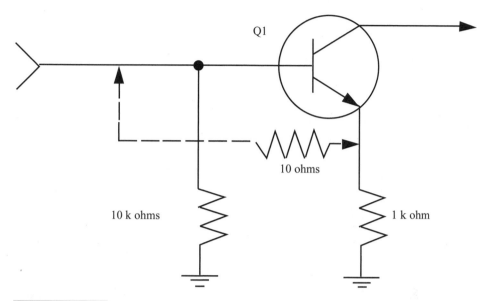

FIGURE 1-62   **A noisy transistor can be located by shorting a 10-ohm resistor between the base and emitter connections.**

then reappear again. Replace the power output IC if a loud frying or hissing noise is present at all times. A poor internal transistor or IC junction is generally the cause of this type of noise.

TESTING EQUIPMENT THAT INTERMITTENTLY BLOWS FUSES

Should you have an electronic device that blows fuses intermittently and eats fuses, then use the following tips:

To save money on blown fuses you can make up this tester from a blown fuse and a pilot lamp. Looking at the illustration shown in (Fig. 1-63), solder leads onto a pilot lamp that has a higher current rating than the fused circuit you are testing and also to a blown fuse that you have clipped into the fuse holder.

For checking the B+ power supply of a TV set, you would use a lamp rated at 250 mA or a no. 44 lamp. For a ½-amp (0.5 amp) current, use a no. 41 lamp, and for a 150-mA current drain use a no. 40 lamp. There are many more lamps with other current ratings that you can use as needed.

With the test lamp installed and the device turned ON, the lamp should glow at a medium brightness under normal conditions. Now keep an eye on the lamp as you twist the PC boards, move parts, tap the components, and heat or cool the various parts. If you find the defective part or circuit area the lamp will become very bright or may blow the lamp should a circuit short occur. In fact, over the years some electronic manufacturers have used pilot lights as fuses.

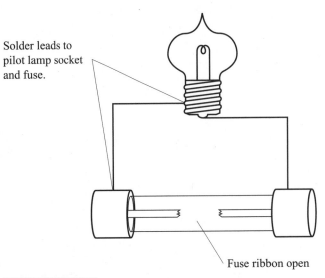

Solder leads to
pilot lamp socket
and fuse.

Fuse ribbon open

FIGURE 1-63   **Illustration of how to make a pilot light test device to check on circuit current being drawn without blowing a lot of fuses.**

Note:

If you replace one of these pilot lights that is used as a fuse, make sure it has the correct current rating. These "fuse lamps" worked very good in commercial two-way radio systems, as you could determine whether the circuit is functioning by looking to see if the fuse lamp was blown or if it was glowing.

POWER SUPPLY TROUBLE REPAIR TIPS

When you suspect problems in the power supply, to be safe, you should unplug the device and discharge to the chassis ground all large filter capacitors. The use of an ohm meter to check for resistance from B+ to ground is now called for. The resistance in ohms should be high at about 50k, or more. If you should find zero or a very few ohms, reading this would indicate a short circuit and call for component testing or removal. Should the ohm reading be around 15k or so, then the faulty component is nearby. The most common cause of low resistance readings in the power supply is one or more shorted or leaky capacitors.

You may encounter a TV set, stereo, or radio that will not operate even though the power supply checks out OK. In this case you need to use your volt meter to check out various dc voltage circuits. You should look for a dc voltage that is missing or too low, which indicates a problem in that stage you are measuring. You can now check each component to try to isolate the problem. To isolate the faulty stage, it's best to work from the input circuits on to the output stages. You will find in some tough cases that you may have to do this several times.

DIGITAL CIRCUIT POWER SUPPLIES

Many hard-to-locate and intermittent problems that occur in digital circuits are traceable to the power supply of these devices. Digital chips, because of their nature, seem to be very sensitive to any slight fault in the power supply and filtering system. The old TTL-type digital ICs do not give that much trouble, but the supply voltage to these devices must be between 4.75 and 5.25 volts. The now more popular CMOS devices require a wider range and more voltage tolerance but are affected more by noise, ripple, and power supply glitches. Power line voltage spikes and glitches can cause erratic equipment operation and may also damage the solid-state chips. This is a good reason to use an uninterruptible power supply (UPS) to plug in your more expensive electronic equipment.

Try to determine which part is faulty before replacing it, if at all possible. You don't want to start changing parts at random, also called *shotgunning*, like the fellow in Fig. 1-64, to solve a circuit problem.

You now know what components make up various electronic devices and how they work. You can now go onto the chapters of interest and solve the problems that occur in your equipment.

FIGURE 1-64   Do not start changing parts at random, which is also refered to as "shotgunning."

2

RADIO/AUDIO/STEREO/SPEAKERS/ MUSIC SYSTEMS AND CASSETTE PLAYER OPERATIONS

Broadcast Radio Transmitter Operation

For you to become familiar with AM/FM radio reception, start by reviewing how the FM radio signal is developed and transmitted. FM stereo signals must be compatible with monophonic FM radios, but they must also simultaneously carry other information, such as SCA background music, paging, and much more.

The two basic components needed for any stereo radio system are the right (R) and left (L) audio channel information. Refer to the basic stereo FM transmitter block diagram in Fig. 2-1. These left and right audio signals are matrixed, resulting in sum information (L + R) and difference information (L−R). Matrix is something within which something else originates or develops. To obtain sum information (L + R), +R was added to L; to obtain the difference information (L−R), a negative −R of the same magnitude as the +R (only 180 degrees out of phase) is added to L. Thus, L−R, the difference signal, was created. The composite L + R and L−R information is now used as FM modulating components in this system. Normally, the L + R information could immediately FM modulate the carrier. However, to be certain that the L + R information is in the same phase relationship to the L−R information, as they were when they came from the matrix when the FM modulated the carrier, it is necessary to insert a delay network in the L + R channel. The delay system is needed to shift the phase of the L + R modulating component in such a manner that it will be in phase with the L−R upper and lower 38-kHz sidebands when they also FM modulate the carrier.

In the FM stereo system of transmission, it is necessary that the L−R information AM modulate a subcarrier. To create this subcarrier, a very stable crystal oscillator produces a 19-kHz signal. The 19-kHz signal is doubled to obtain a 38-kHz subcarrier that is then AM

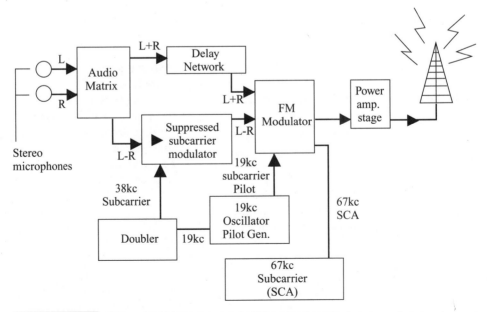

FIGURE 2-1   **A block diagram of an FM stereo transmitter.**

modulated by the L−R information. The 19-kHz signal is also used as a pilot signal or synchronization signal and it also FM modulates the carrier. Because all of the necessary signal information in the subcarrier system is contained in the upper and lower L−R 38-kHz sidebands of the AM-modulating envelope, the 38-kHz subcarrier need not FM modulate the carrier. Thus, the 38-kHz carrier is suppressed and only the remaining upper and lower L−R 38-kHz sidebands are used to FM modulate the radio carrier.

The FM broadcast system now has three carrier-modulating components: L + R audio information, two L−R upper and lower 38-kHz sidebands, and the 19-kHz pilot signal. As stated previously, it is necessary that these FM radio systems be compatible with facsimile or SCA. So, another modulating component, a 67-kHz subcarrier for SCA, needs to be added.

FM/AM Radio Receiver Operation

The Bose Wave Radio, shown in Fig. 2-2, delivers sound quality for its small size that can't be compared to conventional radios or to ordinary stereo systems.

Linking a special configuration of Bose's unique waveguide technology and the Acoustic Wave Music System to a top-quality radio receiver, the Wave Radio generates sound far more spectacular than its compact size or the sum of its component parts would indicate. Despite its small size, the Wave Radio provides full, rich sound to fill most size home listening rooms. This remarkable audio breakthrough in sound quality comes from the 34-inch single-ended waveguide inside the unit. More on the Bose waveguide speakers later in this chapter.

All functions on the Bose Wave Radio can be regulated by a credit card-sized remote-control unit included with the radio. The Wave Radio features AM and FM stereo radio and a dual alarm clock modes. It offers 12 radio presets, mute, scan and automatic sleep features, as well as battery back up, in case of a power failure. You can set the Wave Radio

FIGURE 2-2   **The Bose AM/FM Wave Radio.** Courtesy of Bose Corp.

so that you can fall asleep to one station and wake up to another. The volume will raise gradually to a volume level that you set.

RADIO CIRCUIT OPERATION

Now see how the various circuits of a radio receiver operate and the problems that can occur. Refer to the block diagram in Fig. 2-3 as various sections are covered.

The RF tuner section   The radio RF tuner selects the station you want to hear and also rejects any unwanted or undesirable radio signals or interference that is present at the antenna. The mixer stage is used to mix the RF station signal with the receiver's oscillator to produce the IF frequency. An AGC voltage is applied to the RF stage to reduce the amplification of this stage when a strong radio signal is being received. This AGC control voltage is developed at the detector and is usually used to control the amplification of stages in the IF and RF circuits of the receiver.

Automatic frequency control (AFC) circuitry   With any high-frequency oscillators, stability is very important feature and these circuits require some type of AFC control to compensate for oscillator frequency shift. This is accomplished by taking a sample voltage from the ratio detector and feeding it via a *varicap*, a voltage-controlled variable capacitor, to the oscillator stage. The varicap is connected across the oscillator tuned circuit and acts as a

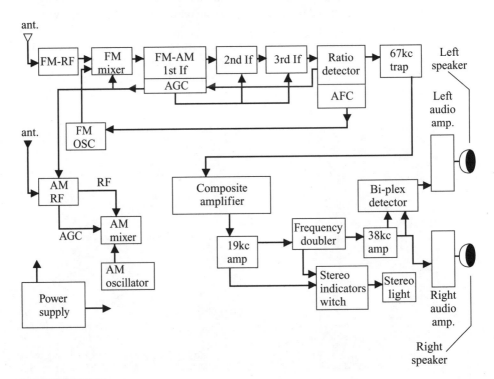

FIGURE 2-3   **The block diagram of an AM/FM stereo receiver.**

frequency-controlling device. If the oscillator should drift, a ratio detector unbalance occurs and a dc voltage is fed back to the varicap so that its changing capacitance will automatically adjust the oscillator frequency. Thus, it has an automatic oscillator frequency control that eliminates drift and simplifies station tuning. Analog tuners will usually have an on/off AFC switch. When tuning in a station, turn off the AFC switch to disable the AFC control to more accurately tune in the station. The newer receiver tuners are digitally logic-IC controlled and do not have an AFC switch.

Intermediate frequency (IF) amplifiers   The FM IF frequency is usually 10.7 MHz and the IF frequency for the AM section is 455 kHz. The IFs in a receiver are used to amplify the RF signal and, with the addition of traps, make the receiver much more selective. The gain of the IF amplifiers is controlled by an AGC control voltage. The better receivers will usually have four stages of IF amplification. The processed signal is then fed to the FM ratio detector.

Ratio detector AND composite amplifier   The 10.7-MHz amplified output signal from the last IF stage is fed to the ratio detector. The ratio detector is a standard FM circuit that consists of diodes or a special detector chip. Assuming that the FM station you are tuned to is transmitting in stereo and with an SCA program, the composite output signals from the ratio detector will be:

■ A 67-kHz SCA signal.
■ A 19-kHz pilot signal.
■ A L + R audio voltage signal.
■ Upper and lower 38-kHz sidebands.

The composite signal goes to the input of a 67-kHz trap. If the FM station you are listening to is also sending out a 67-kHz SCA signal, it cannot be allowed to enter the detector or the audio will be very distorted.

Composite amplifier function   With the 67-kHz SCA information trapped out, it is now necessary to amplify the remaining parts of the composite FM detected signal. The composite amplifier has a gain of nine or more times. The output of this composite amplifier is fed to two channels. The L + R audio voltage and the 38-kHz L–R upper and lower sidebands are fed directly into the biplex detector and are then recombined with the developed 38-kHz subcarrier, as well as simultaneous detection into L and R audio voltages. The 19-kHz signal is usually taken off of a transformer and fed to the 19-kHz pilot amplifier.

Other circuits in a stereo FM receiver consist of a 19-kHz pilot signal amplifier, 19-kHz doubler, 38-kHz amplifier, and a circuit to indicate when you are receiving a stereo radio broadcast. This is called the *stereo indicator switch circuit.*

Biplex detector operation   Some receivers use a bilateral transistor in the biplex detector circuit to accomplish stereo signal separation.

For biplex detector operation, the (L + R) audio signal appears at the "L" and "R" output circuits in equal amplitude of the same polarity. With only a few turns in the 38-kHz transformer secondary winding, there is only a low-resistance path for the (L + R) signal.

The (L–R) 38-kHz sidebands are demodulated by the action of a transistor into two equal amplitudes, but with opposite polarity (L–R) regular audio signals in the same L and R output circuits. The biplex solid-state circuit thus acts to reinsert the 38-kHz contiguous wave (CW), which is a subcarrier into the (L–R) 38-kHz sidebands. At the same time, it demodulates this signal into the (L–R) audio signal and also provides the matrixing of the two sets of audio signals.

The demodulation efficiency of the multiplex "average-type" detectors is about 30 percent. The demodulation efficiency of the biplex detector circuit is near 60 percent. Furthermore, the L and R channel separation is improved to better than 6 dB at the higher audio frequencies between 8 kHz and 15 kHz. The biplex circuit is designed to provide about 25 dB of separation between the L and R channel signals at 1000 Hz.

One of the most desirable features of the biplex detector is that when tuning across the dial, both stereo and non-stereo (monophonic) stations are received at approximately the same volume level. During monophonic FM program transmissions, the 19-kHz pilot signal is not transmitted. If the 38-kHz switching signal is not applied to a switching transistor, it will remain turned off. In this case, the L + R audio signal will be divided between the two channels and fed to both the left and right audio amplifier channels.

The two stereo audio amplifier stages boost the signal level high enough to drive loudspeakers. They can be two or more speakers for each channel. The stereo amplifier stages will also have tone, loudness (volume), and balance adjustment circuits and controls for you to adjust to various room arrangements and to your listening preference.

The Dolby recording technique   First, see how an ordinary standard audio recording is produced.

Making a standard audio recording   Figure 2-4 illustrates how music consists of different loudnesses, separated by intervals of silence.

Loud and soft sounds are shown here as long and short lines. The music represented by this drawing starts loud and gradually becomes very soft and quiet.

Figure 2-5 represents noise. Any recording tape, even of the highest quality, makes a constant hissing noise when played. At very slow speeds and narrow track widths (used in cassette players), tape noise is much more noticeable than with a professional tape recording and CDs (although some noise is on these recordings, also).

Figure 2-6 depicts both noise and music on a tape recording. When a tape recording is played, the noise of the tape conceals the quietest musical sounds and fills the silence when

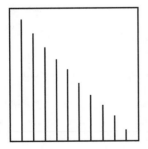

FIGURE 2-4   **The music, represented by this drawing, starts loud and gradually becomes very quiet.**

FIGURE 2-5   A blank tape will make a hissing noise when played back.

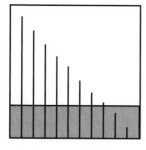

FIGURE 2-6   Tape background hiss can even be heard on some quiet music selections.

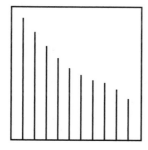

FIGURE 2-7   The Dolby system "listens" to the music first and adjusts the music level accordingly.

no sound should be heard. Only when the music is loud will the noise be masked and usually not heard.

However, tape noise is so much different from musical sounds that it sometimes can be heard even at these times.

How a Dolby recording is produced   Let's now see how the Dolby recording is made and what happens during tape playback.

The Dolby system "first" listens   Before the tape recording is made, as shown in Fig. 2-7, the Dolby system "listens" to the music to find the places where a listener might later be able to hear the noise of the tape surface. This happens mainly where the quietest parts of the music are recorded. When it finds such a place, the Dolby system automatically increases the volume being recorded so that the music is recorded louder than it would be normally.

Figure 2-8 gives you an indication of what the Dolby system is doing during recordings. In a Dolby system, recording the parts of the music that have been made louder, stand out clearly from the noise.

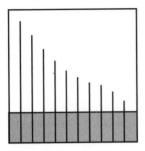

FIGURE 2-8   When Dolby is used for recording, it makes the louder music stand out with brilliant sound.

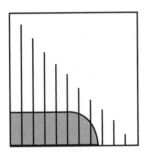

FIGURE 2-9   When a Dolby recorded tape is played back on a Dolby machine, the loudness is automatically reduced in all places that it was increased before.

As a result, the Dolby system recordings sound brilliant and usually clearer—even when played back without the special Dolby system circuit.

What the Dolby system does during playback is illustrated in Fig. 2-9.

When the tapes are played on a high-fidelity (hi-fi) tape recorder equipped with the Dolby system circuitry, the loudness is automatically reduced in all of the places at which it was increased before recording. This restores the music to its original loudness once again.

At the same time, the noise that has been mixed with the music is reduced in loudness by the same amount, which is usually enough to make it inaudible.

Tips for Making Your Audio Sound Better

Of course, the placement of your stereo speakers is a very personal matter, depending mainly on the arrangement and layout of your listening room, speaker positions, and the way you listen to music. Where you place your speakers does make a difference in how your system will sound. Before settling on a final arrangement, try several arrangements.

Bass response is very dependent on speaker location. For maximum bass, place the speakers in the corners of your room. Placing the speakers directly on the floor will produce an even stronger bass response. If the bass sounds boomy and exaggerated, move the speakers away from the corners slightly, pull them out from the wall, or slightly raise them up off the floor.

POSITIONING YOUR STEREO SPEAKERS

Stereo speakers should be placed from 6 to 8 feet apart. Putting them too close together reduces the stereo effect, but placing them too far apart reduces bass response and creates a "hole effect" in the middle of your room. Generally, most speakers have a tweeter dispersion angle of close to 60 degrees. For this reason, your listening position should be in the overlap zone, so you want to angle the speakers toward you for better stereo sound.

FM Radio Antennas

Usually, the built-in antennas in most receivers are adequate for good reception. However, if you are having reception problems, try the following hints.

For better FM reception, you can build the folded dipole shown in Fig. 2-10. Just splice together 300-ohm TV twin-lead, as shown. Apply a small amount of solder and heat to the

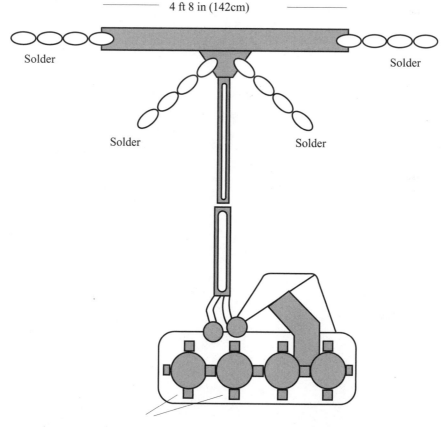

Antenna terminals FM 300 ohms

FIGURE 2-10   An FM dipole antenna that you can build up.

twisted ends until the solder flows over each wire strand. Attach the lead-in to the 300-ohm terminals on back of the receiver. The antenna can be stapled or tied to the back of the receiver or placed on a wall. Turn or move the antenna around for the best reception.

You can use a set of TV rabbit ears, or buy an FM antenna. Some deluxe antennas feature electronic "tuning" for a more directional station reception.

An outside VHF/UHF/TV antenna will also work well for your receiver. A "splitter" will let you connect a TV and FM receiver to the same antenna. If you live in the country side, a specially designed FM antenna can receive an FM station from greater than 100 miles away.

Some Receiver Trouble Checks and Tips

Now go over some radio receiver problems.

RECEIVER WILL NOT OPERATE AT ALL

If your receiver is completely dead, check the power supply with a dc voltmeter for a presence of B+ voltage. If there is no B+ voltage, check for an open fuse (Fig. 2-11). If your radio has a built-in cassette tape deck and/or CD player and they are working ok, then the

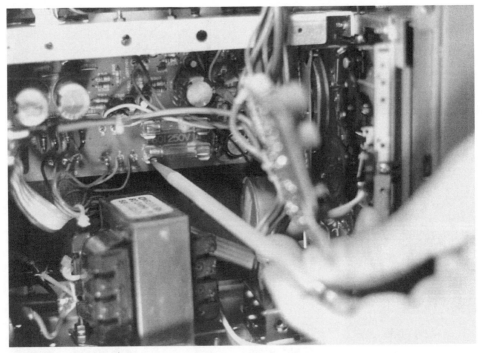

FIGURE 2-11   Check for a blown fuse if the receiver and audio amplifiers are dead.

power supply should be working and the problem is in the radio RF tuner, IF stages, or detector/multiplex circuit stages. To repair these stages, you need a voltmeter (VTVM), transistor/diode checker, signal tracer, and oscilloscope. These repairs require a professional electronics shop or technician.

If the receiver or cassette/CD player has no audio, then the problem would be in the audio power amplifiers or speakers if the power supply checks out OK. Amplifier problems could be caused by poor solder connections, cable plug-in sockets, defective ICs or transistors, open coupling capacitors, or burned (open) resistors and coils. An audio signal tracer can be used to isolate a loss of audio in these amplifiers. When you are signal tracing in either the RF, IF, or audio stages, the dead stage will become apparent when a signal is found at the input of a stage, but not at the output of that stage.

Another quick check is to place the tracer probe at the speaker-output coupling capacitors. The same signal should appear at both ends of the capacitor if it is good (or shorted), but an open capacitor will have an input signal, but no signal at its output connector. Open electrolytic coupling capacitors between the output stages and the speaker are fairly common. So, if the left or right audio channels are dead, you should check this out first.

If no signal is found at either end of the speaker coupling capacitor of the dead channel, move the probe to the driver stage output and then to the input. A signal at the output but not at the input proves that the stage is defective.

INTERMITTENT RECEIVER PROBLEMS

Signal tracing is effective if the intermittent condition can be induced. To speed up the break down, you can use a heat gun (hair dryer) or some cooling spray to make the intermittent condition start or stop. After you find that a thermal condition triggers the fault, the heating and cooling should be applied to a small circuit area until the trouble can be pinpointed to one component.

Capacitors are a common cause of intermittents. They can become intermittently leaky, shorted, or open. A leaky capacitor can change the bias on a transistor or IC and cause it (and other components) to fail. Some intermittent problems will change the B+ voltage, so closely check this to determine if voltage change might be the cause or the effect.

Other receiver intermittents are caused by various controls and switches that need cleaning. Check the front-panel controls and switches. Notice the push-button switches in Fig. 2-12 and, while operating them, listen for any intermittents. These controls and switches can be cleaned with a special spray contact cleaner. If spraying with a cleaner does not correct the intermittent problem, the control or switches will have to be replaced.

If the station tuning dial will not move the pointer, the cord is probably broken. If it is broken, you can replace the cord by restringing it. The tuning cord is shown in Fig. 2-13. If the cord is slipping, you can apply some anti-slip liquid or stick rosin compound.

SOME RECEIVER SERVICE DON'TS

When working on solid-state (transistors and ICs) receivers, key voltage and resistance checks can usually be used to find the fault. However, before you start probing around,

FIGURE 2-12   Clean the selector switches for intermittent or noisy operation.

FIGURE 2-13   If the station selector will not turn, check for a broken or slipping string or belt. Replace broken string and apply antislip stick or liquid to the string.

taking measurements, and replacing components, you should look over the following rules.

- Don't probe around in a receiver plugged into the ac socket. A short from base to collector will usually destroy a transistor or IC. Many stages are direct coupled, thus lots of components can be damaged. Always turn off power to the receiver before connecting or disconnecting the test leads.
- Don't change components with the power applied to the receiver. Always turn off power to the receiver, except when taking voltage readings and then be very careful.
- Don't use test instruments that are not well isolated from the ac line when making measurements on equipment connected to the same power source (even if the equipment is turned off). This prevents cross grounds. Check all test instruments and use an isolation transformer on the receiver under test.
- Don't solder or unsolder transistor or IC leads without using some type of heatsink clip. This prevents damage to heat-sensitive solid-state components. Long-nose or needle-nose pliers make a good heatsink for soldering.
- Don't arc B+ voltage to ground because transient voltage spikes can ruin ICs and solid-state devices fast.
- Don't short capacitors across another capacitor or circuit component for a test. This can also cause ICs and transistors to be damaged.
- Don't forget that many solid-state stages are directly coupled. A fault in one stage can cause failure in another stage.
- Don't use just any ohmmeter for resistance checks. The voltage at the test probes can exceed the current or voltage limits of the solid-state device under test. The lower resistance scales on 20,000 ohms-per-volt meters are usually safe for short- or open-circuit checks.
- Don't forget to reverse the leads when making in-circuit resistance checks. The readings should be the same either way. A different reading usually means that a solid-state junction is affecting the reading. You are actually making a check across a junction in a transistor or IC.
- Don't forget to use extra caution when checking, unsoldering, or inserting and resoldering MOSFET transistors or MOS ICs.

Loudspeaker Concepts and Precautions

One of the most important components of a good audio system is the speaker and its enclosure. This section covers speaker operation, how they are connected, speaker enclosures, and tips on hooking up your speaker system.

HOW SPEAKERS ARE CONNECTED

Figure 2-14 shows connections for a woofer, mid-range horn, high-frequency tweeter, and a crossover network for a typical speaker system. In this set up, the crossover coil (1-mH choke) is installed in parallel with the 1-kHz exponential horn after the series crossover capacitor. In other systems, you might find the horn connected in series with the tweeter. The crossover point is usually at the 1-kHz frequency point.

The purpose of the crossover coil is to shunt the heavy bass frequencies around the 1-kHz horn and tweeter, thus affording added protection for the voice coils of the horn and tweeter. The crossover coil also serves to smooth the horn's acoustic crossover point and improves the speaker's sound reproduction.

A brief analysis of the speaker system (shown in Fig. 2-14) is as follows: Assume that the complete audio spectrum appears at the input of the 8-ohm speaker system. All frequencies will appear across the 12-inch woofer. The 12-inch woofer will reproduce audio frequencies from 30 Hz to 1 kHz. The crossover capacitor will block virtually all frequencies below 1 kHz and pass the audio frequencies above the 1 kHz (crossover) point. The 1-mH choke will act as a very low impedance to any frequencies below 1 kHz that might still be present after the blocking capacitor while acting as a high impedance to the frequencies above 1 kHz. Audio frequencies above 1 kHz will now be present across the 1-kHz horn. The acoustic audio output of the 1-kHz signal is essential flat, out to approximately 8 kHz. The capacitor blocks the frequencies below 8 kHz and passes the higher frequencies across to the 3-inch tweeter for a smooth acoustic output to approximately a frequency of 16 kHz.

If you are connecting new speaker to your audio system, be sure that they match for impedance (such as 8 or 16 ohms) and have proper power-handling capability for your power amplifier. It is possible to damage a speaker system—even if its power-handling capacity is the same as, or higher than, the power output rating of the amplifier to which it is connected. Damage can occur to the speaker system because almost all power amplifiers deliver more than their rated power output. This is especially true if the amplifier is operated at maximum, or very high volume settings while the tone controls are set at, or near, maximum boost. A safety margin should be allowed between the amplifier's rated output and

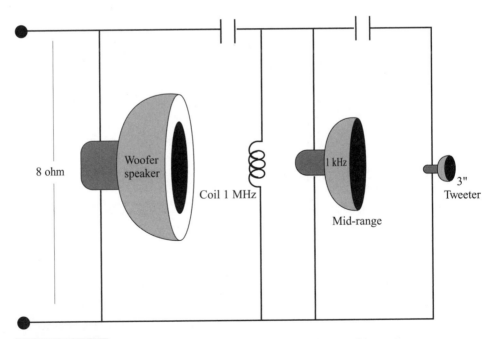

FIGURE 2-14   **Crossover network circuit for a speaker and horn.**

the speaker system's rated handling capability if the amplifier is to be operated at very high volume levels. If distortion is noticed at high volume levels, it is recommended that the volume level be reduced because you might be reaching the safe operating limits of the amplifier or the speaker system.

HOW TUNED-PORT SPEAKER SYSTEMS WORK

The speaker system shown in the (Fig. 2-15) drawing is of the tuned port type. The cross section view of this speaker enclosure is shown in (Fig. 2-16) and will be used for the following explanation. This enclosure has four openings in the front panel (one for each of the three speakers and one for the port), and the remaining panels are of solid construction.

A tuned-port speaker can be described as a tuned enclosure in which the air in the port will resonate with the air in the main area of the cabinet, at a given frequency. This frequency determines the effective low-frequency cutoff of the system (cabinet and enclosure combined). Below the selected frequency (30 Hz), the response drops very rapidly (approximately 24 db/octave).

This system could also be described as an acoustic phase inverter. That is, at some frequency, within its normal operating range, the air in the port is moving in an outward direction (to the front) while the speaker cone is also moving in an outward direction (to the front). These two movements would occur at the same time, and in phase.

Basic advantage of a tuned-port enclosure, as used in this speaker system, over a typical acoustic suspension (closed box) enclosure are:

■ Reduced low frequency distortion.
■ Increased efficiency. This requires less amplifier driving power for equal loudness level.

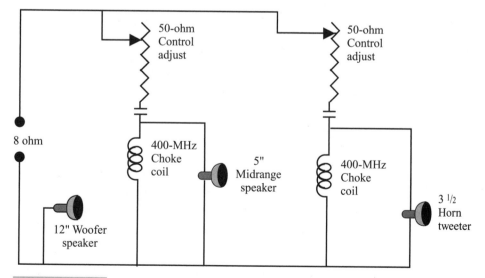

FIGURE 2-15   **A speaker circuit system with an adjustable crossover network.**

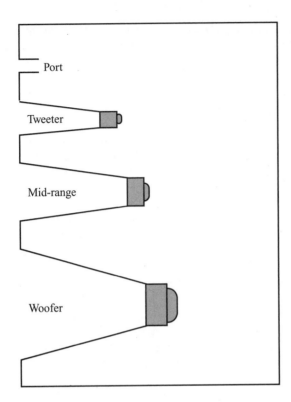

Port

Tweeter

Mid-range

Woofer

FIGURE 2-16   A cross-sectional view of a typical multispeaker enclosure.

Sound level of signals radiated through the port (in the 30- to 70-Hz range) is comparable to the sound level radiated by the woofer (in the range of 70 Hz to 1 kHz). For the 15-inch woofer, a 3½-inch diameter port is required and it must "pulsate" air at a much higher velocity than a woofer with a diameter of 12 inches. Several factors must be considered to maintain the required port velocity.

- The woofer uses a highly efficient magnetic structure, making it comparable to a powerful electric motor. This forces air in the port to move at high velocities—even though the air in the box is attempting to stop the motion of the speaker
- The internal air pressure in a tuned-port enclosure is much higher than that in a conventional closed-box enclosure. The mechanical construction of a tuned port enclosure must be built better than for an air-suspension type.
- Speakers (and other components) must be securely fastened to prevent air leaks. Leaks or loose components can result in losses, which cause a deterioration of performance. To increase effective cabinet volume and also to dampen internal resonances of the enclosure, acoustic padding is placed on the inside surfaces. These pads must not obstruct the port.

When servicing or connecting the amplifier systems, always be sure that the speakers are properly phased. If both speakers are phased in the same way, poor channel separation will result. Also, some loss in midfrequency response will probably be noticed. If the two input speaker signals are not 180 degrees out of phase, the bass frequencies will cancel, resulting in poor bass performance.

The size (gauge) of the speaker wires is also important for proper speaker sound reproduction. This will be very noticeable on high power amplifiers at high volume levels. A very small, 28-gauge speaker wire could distort and, in some cases, damage the amplifier output stage or even the speakers.

The Bose Acoustic Wave speaker system   Two of the chief differences between the technology used in the Bose Acoustic Wave stereo introduced in 1985 and that used in the Wave Radio today is the positioning of the loudspeaker and the length of the waveguide. The speaker in the Acoustic Wave is located so that one-third of the tube is behind the speaker and two-thirds are in front of it. In the Wave Radio, the speaker is nestled at one end of the waveguide, a less-efficient position, but the only one possible in a device this size. The Acoustic Wave houses 80 inches of waveguide; the Wave Radio cabinet enfolds 34 inches of ductwork. The longer waveguide in the Acoustic Wave system makes it possible to reach deeper into bass, to perhaps 40 Hz.

Another example of the Bose advanced systems is the all-in-one system format called "Lifestyle 20," which represents a quantum leap in both performance and convenience over other systems. A photo of the Bose "Lifestyle" system, including the jewel cube speakers is shown in Fig. 2-17.

Bose Series III Music System   Another great Bose audio center, shown in Fig. 2-18, is the Acoustic Wave Music System III. The system, measuring about 10 inches high by 18 inches wide and 6 inches deep, includes a full-featured CD player, AM/FM stereo tuner with 10 presets, and all the speakers, amplification, and equalization technology to fill a room with concert hall sound.

The newest version of this system provides even smoother audio performance and a slim remote that can operate the unit from 20 feet away. Other user-friendly features are color-coded, one-touch button operation, volume protection, and a continuous music option.

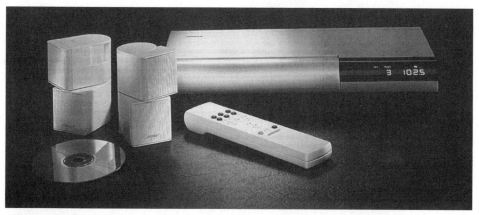

FIGURE 2-17   **Bose Lifestyle 20 music system and jewel cube speakers.**
Courtesy of Bose Corp.

FIGURE 2-18   **Bose Acoustic Wave music system.** Courtesy of Bose Corp.

FIGURE 2-19   **A cutaway view of the Bose waveguide chamber design.** Courtesy of Bose Corp.

Waveguide technology is based on controlled interaction of acoustical waves with a moving surface. This interaction occurs inside the precision waveguide—a mathematically formulated tube inside of which a loudspeaker is placed. The waveguide inside the Acoustic Wave Music System is nearly seven feet long and folded numerous times to fit inside the enclosure.

The cut-away view drawing in Fig. 2-19 shows the sound-channel configurations and the 36-inch-long waveguide inside a 14-inch case enclosure. This Bose acoustic wave system, which is about the size of a briefcase, has a tube length of 80 inches. The waveguide precisely matches the specifications of the speaker and skillfully controls the flow of air. This is how Bose is able to produce rich, full sound from unassuming small equipment.

Bose Lifestyle 901 System   Combining Bose's best loudspeaker with its most advanced systems technology, the Lifestyle 901 music system, shown in Fig. 2-20, is intended to come as close as of today to the sound of the original live performance.

The Lifestyle 901 system resulted from 12 years of physical acoustics and psychoacoustic research at the Massachusetts Institute of Technology. Many of the design improvements represented technological challenges for Bose engineers. The desire for high power handling and better efficiency produced two major achievements: the helical voice coil (HVC) driver and the Acoustic Matrix enclosure.

The HVC driver uses aluminum edgewound on an aluminum bobbin. The design allows significantly more windings on the bobbin without the air gaps caused by round-wire windings. The result is an efficient driver with high power handling.

Furthering efficiency and enhancing bass performance became the challenge of the Acoustic Matrix enclosure. By porting each of the nine drivers, air from the back of the cone could be used to increase efficiency and provide even deeper bass. Original designs produced the desired effect, but with undesired port noise. The final design is an injection-molded plastic enclosure that ports each of the drivers into a separate chamber, which, in turn, is ported through one of three reactive air columns. This sophisticated approach again required Bose engineers to design a manufacturing process from scratch.

The 901 speaker performance is optimized through integrated electronics, including amplification, signal processing, and active electronic equalization. Knowing the performance parameters of the 901 speaker allowed Bose engineers to match the ideal amplifier to achieve the renown room-filling sound of the speaker. Highly sophisticated system-protection circuitry ensures that the speaker is never over-driven and prevents interruption of the radio programs or music.

The system is controlled by the music center and integrated signal-processing provides deep, well-defined bass at all listening levels—even background levels—by compensating for your ear's decreased sensitivity at low volumes.

FIGURE 2-20   Bose Lifestyle 901 music system. Courtesy of Bose Corp.

FIGURE 2-21   Bose Acoustimass 10 Home Theater system.
Courtesy of Bose Corp.

Bose Home Theater system   The Bose 10 home theater speaker system is shown in Fig. 2-21. The Acoustimass 10 home theater speaker system includes five Bose signature double cube speaker arrays, a single Acoustimass module that can be hidden anywhere in the room, and unique, easy-to-use connectors. The system is compatible with all digital and analog surround sound electronic formats.

Bose technology allows a single unobtrusive Acoustimass module to provide pure low-frequency sound to the front and rear channels in the system. Virtually invisible cube speaker arrays produce consistent spectral and spatial perspective for front, right, center and rear channel sound.

If the Acoustimass 10-cube arrays are practically invisible, the system's low-frequency module can be also. It is small enough to be hidden anywhere in the room, but its deep bass performance is sure to be noticed. The module launches sound waves from three high-performance $5\frac{1}{4}$-inch drivers into a room in the form of a moving air mass, unlike conventional systems that rely on the vibration of a speaker cone. The result: pure sound, wider dynamic range, and virtually no audible distortion. A built-in protection circuitry guard's system components against excessive volume input levels.

Cassette Players—Operation and Maintenance

The audio cassette players use a cartridge with two reels mounted inside a plastic holder. The tape is slightly more than $\frac{1}{8}$-inch wide and is used for monaural and stereo audio recordings. The cassette tape is a thin plastic film coated with a layer of brown metallic dust (oxide). During recording, the oxide is given a detailed magnetic code. During playback, the tape passes over the record/play head with a head gap between two small magnets. The magnetic code on the tape changes the magnetic field at the head gap and the recording is decoded. When the head gap is clean, the tape undamaged and its speed is correct, you will hear the "live" sound intended by the musicians and recorded by the sound engineer.

A mono recording consists of two tracks, each 0.59 inches wide, separated by a guard band of 0.011 inches. In addition, a 0.032-inch guard band separates each pair of tracks, thus ensuring playback compatibility of stereo tape recordings. Total track width of each stereo pair, plus their guard band, is equal to one mono track width. Stereo prerecorded tapes will be reproduced in mono on a mono tape recorder unit. Left and right track signals will be combined by the playback head and be reproduced as a mono program. Recordings that are made on a mono unit will be reproduced only as mono sound—even on a stereo playback unit.

GENERAL CASSETTE CARE

Cassette operation is very much the same as for the original reel-to-reel tape recorders, except that the cassette reels are smaller and enclosed within a molded plastic cassette housing. When being operated, the tape will be unwound from one reel supply, move past the tape heads and pressure pads, between the capstan and pinch roller, and finally on to the take-up reel. Movement of the tape by the drive system will stop when the tape is fully wound onto either reel in the cassette cartridge. A simplified drawing of the cassette recorder and tape path is shown in Fig. 2-22. The direction of the tape motion is determined by the function buttons. The cassette also has a window, through which you can estimate how much tape and playing time remains.

To prolong tape life, store the cassettes in a clean, cool, dry area in a closed container that will protect them from dust and moisture. Each cassette should be stored in its original container because this will help prevent dust and other materials from entering and causing possible tape damage. Avoid storing tapes after running at fast forward or rewind because this tends to create an unevenly wound tape.

Layers of tape will be compressed or loose, and wavy tape in addition to creating extra segments of tape by stretching it slightly its structure. All of this creates extra wow and flutter. For the same reason, it is good to play the cassette at least a few times a year. Do not store cassettes next to a heat source or stray magnetic fields. In warm summer climates, do not store in an auto or in direct sunlight.

CASSETTE TAPE CIRCUIT OPERATION

A block diagram of a typical cassette tape unit is shown in Fig. 2-23. Notice the audio signal input and output jacks on the left side of the drawing.

The Play mode   With the cassette unit in the Play mode (Dolby noise circuit off), the audio signal moves from the Record/Play head via the Record/Play switch and into the

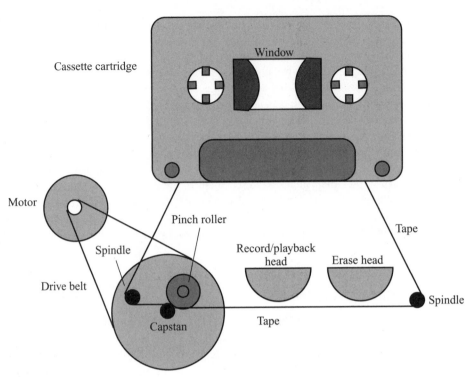

FIGURE 2-22   The tape path in a typical cassette recorder.

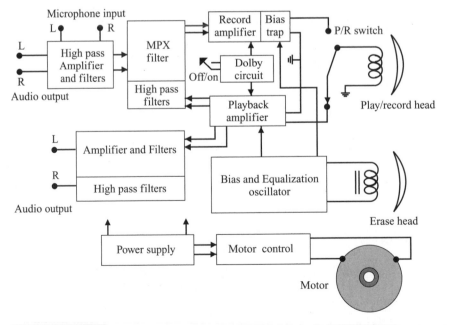

FIGURE 2-23   A block diagram of a typical cassette deck.

playback amplifier block. A desired signal feedback then goes to the equalization block during playback operation. The audio processed signal then goes into more amplifier and filter block circuits and then is fed via jacks and cables to the audio power amplifier to the speakers of your audio system.

Switching the "Dolby noise-reduction circuits" ON, while in the Play position, will result in added operating circuits which control the dynamic processing characteristics and reduces the tape noise level. The audio signal is routed through the noise-reduction circuits and some high-pass filter stages. The high-pass filter attenuates the low and midrange frequencies.

The Record mode   During the Record mode, microphone audio is processed by an IC amplifier and other external audio by another set of input jacks and IC pre-amplifiers.

In the Record mode, when the Dolby noise-reduction circuits are switched on the audio goes to the high-pass filter stage. The high-pass filter functions in a similar way as in the Play mode, but is part of a positive feedback loop, instead of a negative feedback loop used in the Play mode. This will result in record circuit characteristics that are complementary to that of tape playback.

From the record amplifier, the audio signal goes through a bias trap. More on the bias oscillator later. This trap prevents the bias oscillator signal from getting back into the record amplifier or other circuits where the bias frequency could cause some undesired effects.

Equalization circuits   Equalization circuits are needed because of the different types of recording tape available. Some tape decks have equalization provisions for recording and playback for several types of tapes, such as ferric oxide, ferri chrome, and chromium dioxide. Your more-expensive tape decks will have switches or buttons on the front panel to adjust the unit for proper bias and equalization to match these various types of tapes. This tape-type equalization should not be confused with the normal record and playback equalization provided on all machines for proper reproduction.

Tape player electronics   Most modern cassette players now have all of the electronic components mounted on one PC board. These components will consist of capacitors, resistors, diodes, and ICs. The power supply might be found on this board or be located in another section of the audio system. If the tape unit is dead, then check out the power supply for correct voltages. Also, check any fuses in the power-supply section that might be blown.

On some tape recorders, you will find a automatic level control (ALC) circuitry located on the main board and it will have an adjustment marked (ALC Adj). An ALC circuit not working or not adjusted properly should be suspected when audio playback has distortion or changes in recording levels are being noticed. A frequency-compensating network is incorporated within the amplifier PC board circuitry, providing equalization required for proper record/playback response of the tape composition.

Bias oscillator operation   The bias oscillator circuitry serves two functions in a recorder. One is to supply erase current to the erase head while the second function is to supply record bias current to the play/record head. A pre-recorded tape must be cleanly erased to make another good tape recording on it. Bias current varies with the recording-level bias adjustments. The bias current is combined with the audio output signal from an IC amplifier, after which it is fed to the respective left and right windings in the play/record head. This bias current signal is required to make an magnetic audio tape recording. The bias oscillator current is usually generated from an IC on the PC board. Some recorders will have two bias control adjustments to establish the correct recording bias level and playback level.

Cassette belt and rubber pulley drive systems   You will usually find several belts and rubber drive wheels within any cassette tape mechanism. Most will have a motor drive belt to the capstan and flywheel assembly (Fig. 2-24). The drive belt is very small in some tape players. Some motor drive belts are only two or three inches long. The belts can be flat, round, or square in shape. Besides the motor drive belt, another belt runs from the flywheel to the take-up reel. You might find a fast-forward belt drive on some cassette players. Some of these belts are slim and not very thick, so they can stretch and cause erratic speed and wow. Clean each belt and drive wheel when you encounter speed problems with alcohol on a cloth. When the belts have been cleaned and you still have an erratic speed problem, then you need to replace the motor drive belt. A photo of a belt drive and flywheel is shown in Fig. 2-25. A typical cassette belt and drive arrangement is shown in Fig. 2-26.

Slow tape speed can be caused by a slick or dirty motor drive pulley. A dry capstan/flywheel bearing can cause the tape to run slow. A worn, stretched, or greasy belt can be the cause

FIGURE 2-24   Cassette drive belts and gears.

FIGURE 2-25   The flywheel and belt drive assembly.

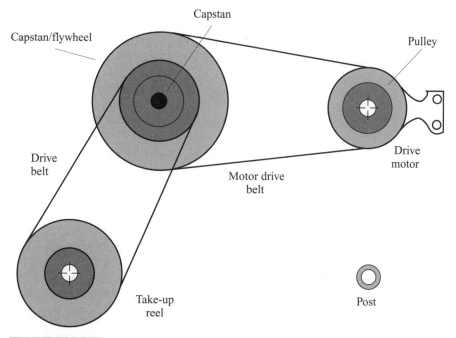

FIGURE 2-26   A belt from the motor drives the capstan/flywheel and the take-up reel assembly is belt driven from a small pulley on the capstan hub.

FIGURE 2-27   The tape can wrap around the capstan. Remove the tape and clean the spindle with alcohol.

of slow speed, also. Clean any of the packed tape oxide from capstan, belts, drive wheels, pinch roller, or the play/record heads. Slower speeds can result if excessive tape is wrapped around the pinch roller and in between the rubber roller and bearing mount. Also, if the tape has broken, you might find it wound around the capstan spindle Fig. 2-27. Dig the tape out and clean the capstan with alcohol. Suspect a faulty drive motor if you have erratic or slow speed after all drive parts are clean. Do not overlook a defective cassette tape. To be sure, replace it with a new tape.

Fast forward not working   Most cassette players will push the idler wheel over to rotate the take-up reel hub. The idler wheel is rotated by friction against a wheel that is connected to the capstan/flywheel shaft. If the tape unit operates normally in Play and slow in Fast Forward mode, suspect slippage within the idler wheel area (Fig. 2-28). Be sure that all drive wheels are clean. With a belt drive fast forward, clean the belt and drive pulley. If the cassette plays slow as well as fast forwarding, clean the motor drive belt and the flywheel area.

On some recorders, the fast forward and rewind are driven by plastic gears. The plastic gears will mesh together when placed into fast forward. The capstan gear drives a larger idler wheel gear that drives another shifting idler gear wheel. The idler gear wheel is

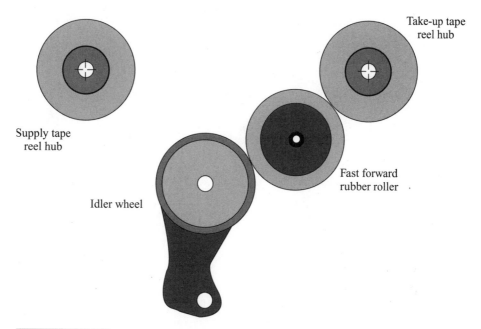

FIGURE 2-28   **The idler wheel is pulled toward the fast forward rubber wheel, which drives the take-up reel.**

moved toward the take-up spindle, which engages two small gear wheels. The plastic gear wheels then rotate when in Fast Forward and Play modes.

These gear-type players will not slip, but might jam if a gear tooth is broken. A misplaced washer or clip might cause gear misalignment and cause the loss of Fast Forward or the Play mode.

Tape will not rewind properly   The fast forward and rewind speed is very slow. In the Rewind mode, the idler wheel is shifted when the Rewind button is pressed (Fig. 2-29). Check for worn or slick surfaces on the idler or drive surface area. Clean them with alcohol. Keep in mind that the pinch roller does not rotate in either the Rewind or Fast-Forward mode, only during Record and Play.

The need to demagnetize tape heads   Tapes that have been used many times for prolonged periods of time are induced with residual magnetism found in heads, guides, and capstans. A magnetized component (especially heads) anywhere in the tape path will create some hiss and permanent loss of high frequencies on the recorded tape, whether you are recording or just playing the tape. To demagnetize the cassette deck, use a commercially available head demagnetizer. Keep all tapes away from the immediate vicinity of any demagnetizer to avoid accidentally erasing the recorded tapes.

While holding the head demagnetizer away from the tape unit, connect the demagnetizer to an ac outlet and turn it on. Slowly bring the demagnetizer close to each of the surfaces that normally contact the tape. With the demagnetizer still on, slowly withdraw it from the unit (two feet or so), and turn it off.

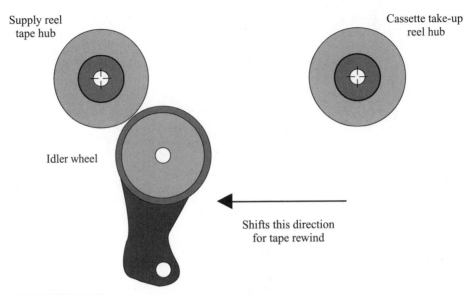

FIGURE 2-29 **The rubber idler wheel is pulled toward the supply reel hub in the tape rewind mode.**

 When servicing the tape deck, do not use any magnetized screwdrivers or other tools near the head or other metal parts that the tape travels around or near. This could magnetize those parts and erase your tape.

TAPE HEAD CLEANING AND MAINTENANCE

During normal cassette operation, oxide particles are loosened from the tape and build up on the tape head, erase head, capstan shaft, and rubber pinch roller. The erase head is pointed out in Fig. 2-30. The tape player should be cleaned at regular intervals because oxide accumulation can cause distortion and possibly affect tape playback and recording.

Clean the head as follows:

■ Press the Stop/Eject button to open the cassette compartment.
■ Remove the cassette.
■ In some older machines, you might want to press the play lever. The various points that need to be cleaned on a cassette machine are shown in Fig. 2-31.
■ While holding the tape door or lid open, use a long cotton swab to clean heads, capstan shaft, and pinch roller with tape head cleaner or pure isopropyl alcohol (Fig. 2-32).

Both mechanical and electronic deck parts affect sound quality. Today's electronic parts are largely unaffected by dirt and have a very long life. However, the mechanical parts that guide the tape and control its speed for accurate decoding will accumulate dirt and dust. Routine maintenance of these parts will extend the useful life of your recorded tape.

FIGURE 2-30   Location of the erase head.

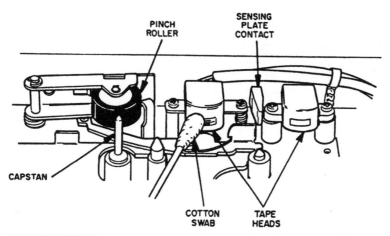

FIGURE 2-31   The various points that need to be cleaned on a cassette player.

The play/record head (Fig. 2-33) is both mechanical (guides the tape) and electronic (decodes at the head gap). Microscopic dirt caught in the head gap will immediately change the magnetic field and affect the sound quality.

The capstan and pinchroller control the tape speed. As dirt collects, tape slippage and tracking errors occur. The speed becomes erratic and the music sounds slow and warbly.

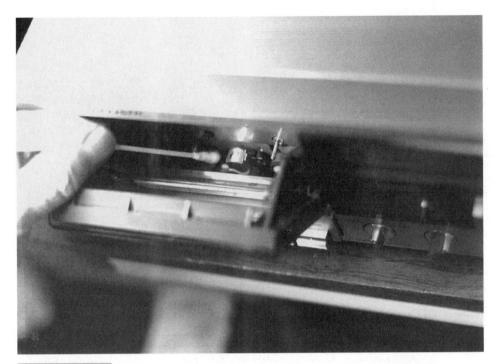

FIGURE 2-32   Use a cotton swab with alcohol to clean the tape record/ play heads.

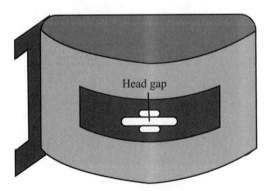

Head gap

FIGURE 2-33   The record/play head, showing the head gap detail.

In severe cases, the tape can stick and unwind into the deck mechanical parts. Dirt rarely collects immediately under the moving tape. As the tape rubs across the mechanical parts the dirt shifts above and below the tape's path, or into grooves and gaps, collecting to cause future problems. A primary cause of tape failure is dirt carried on the tape from the deck and wound up under tension in the cassette. Sandwiched between layers of tape, this dirt scratches the metallic oxide, damaging the recorded sound. The "live" sound reproduction is reduced with each play. Regular cleaning prevents dirt from collecting in the deck. Tapes will stay cleaner and last longer.

A habit of routine maintenance and cleaning prevents these problems. Irreplaceable tapes last longer, and you enjoy all of the "live" sound quality that your costly sound system can provide.

Operation of the Trackmate cleaning cassette   The Trackmate system has engineered quality cleaning into a single, easy-to-use cassette (Fig. 2-34). Other cleaning cassettes are technically dependent on fabric tape or felt for cleaning. Tapes are ineffective and do not reach where dirt collects, beyond the tape path. Felts touch only a narrow portion of the record/play head, capstan, and pinchroller, missing the erase head, tape guides, and stud posts, leaving them dirty. The Trackmate brushes form fit all these mechanical parts. The 32,000 absorbent, flexible, fibers seek and remove dirt from all of the surfaces and gaps where it collects (Fig. 2-35). Static-control fibers inhibit the attraction of further dust. These high-tech cotton buds have more than 100 times the active cleaning surface area of some earlier products. They automatically clean deck parts from top to bottom, leaving a dirt-free path for the recording tape to safely track around on. Figure 2-36 shows the special cleaning fluid being applied the Trackmate fiber brushes.

AUDIOCASSETTE PROBLEMS, SOLUTIONS, AND CORRECTIONS

The following information includes cassette problems that you may have and tips on what to do to solve these problems:

Portable cassette—no tape movement or sound   Always check the batteries first for any portable cassette problems. Replace if dead or weak. Some portable cassettes will

FIGURE 2-34   The Trackmate cleaning cassette device.

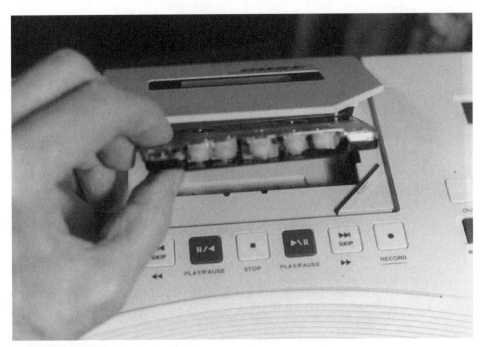

FIGURE 2-35   The Trackmate cleaning cassette has 32,000 absorbent, flexible cleaning fibers.

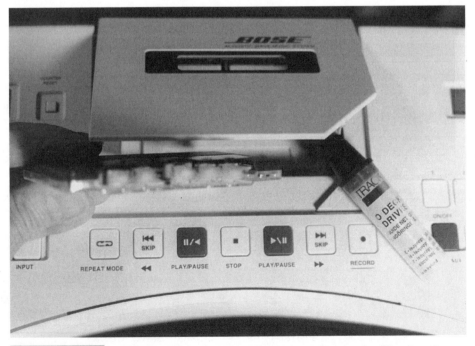

FIGURE 2-36   Cleaning fluid is being applied to the Trackmate fiber brushes.

have a small leaf spring that makes contact when record, play, fast forward, and rewind are switched into action. Check and clean these copper spring-type contacts with a cleaning fluid, or pull a very fine piece of sandpaper between the closed contacts. If the batteries and contacts are good, then the motor is probably defective.

Sluggish tape rewind   You will find that rewind and fast forward run faster than the play/record modes. In some older-model players in the rewind mode, the shifting idler wheel is shifted when the rewind button is pushed against the turntable reel assembly as shown in Fig. 2-37. Check for worn or slick, shinny surfaces on the idler or drive wheel area. Clean well with alcohol. Note that the pinch roller does not rotate in either direction for rewind or fast forward.

With a gear drive system, the idler is shifted against the gear of the supply spindle. Usually, you will find that the rewind speed is slower than the fast forward speed. In rewind, the capstan gear rotates the large drive gear, which in turn rotates the shifter idler gear, and the idler drives the gear on the bottom of the supply spindle.

No fast forward action   With most surface drive tape systems, the idler wheel is flipped over to rotate the take-up reel. The idler wheel is rotated by friction driving against a wheel that is attached to the capstan/flywheel shaft. If the cassette player works normally in play mode and slow in fast forward, then suspect slippage on the idler drive area. Note Fig. 2-38. Clean all drive surfaces. When the fast forward is belt driven, clean the belts and drive pulley. Should both play and fast forward operate slowly, clean the motor belt and flywheel surfaces.

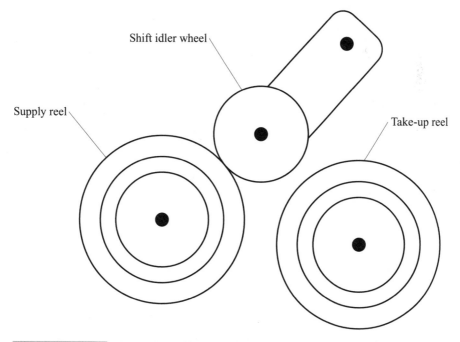

Shift idler wheel

Supply reel

Take-up reel

FIGURE 2-37   **The idler wheel will shift toward the supply reel in the rewind mode.**

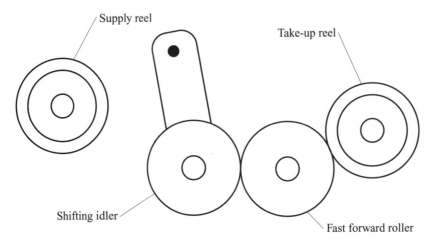

FIGURE 2-38   **The idler wheel is shifted toward the fast forward roller. It will then drive the take-up reel at a faster speed.**

In some model cassettes, the fast forward and rewind are driven from small plastic gears. These small plastic teeth mesh when switched to fast forward. The capstan gear rotates a larger idler wheel and drives another shifting idler gear wheel. Refer to Fig. 2-39. The idler gear is shifted toward the take-up spindle, which engages two small gear wheels. At the bottom of the take-up reel is a plastic gear wheel that rotates in fast forward and play.

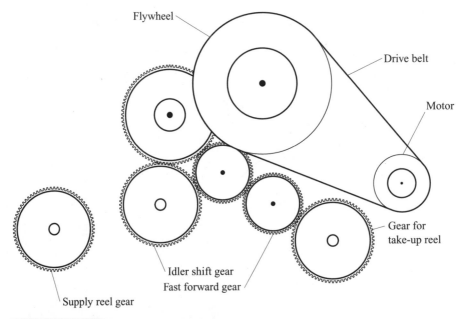

FIGURE 2-39   **The small plastic gears are shifted into different positions for various cassette functions. In this drawing the idler gear is positioned to spin the take-up spindle at a faster forward speed.**

These gear-type assemblies generally will not operate slowly or slip while rotating. Check for broken gear teeth or jammed gears when fast forward does not rotate. A missing C washer can let the small gears fall out of line, and this will disable the fast forward and play modes.

Auto shutoff not working   When the cassette is out of tape, increased tension on the tape triggers a small ejection lever that mechanically releases the play/record assembly and turns the drive or motor to off. In more expensive units, mechanical and electronic automatic shutoff systems are utilized. In many units the ejection lever is referred to as a detection or contact piece, as shown in Fig. 2-40.

The automatic stop-eject or detection piece has a plastic cover over a metal angle lever that can be adjusted at the end where it triggers the play/record assembly and the automatic stop. The ejection piece is mounted alongside the tape head. When the end of the tape has been reached, the tape exerts pressure against the ejection piece and mechanically triggers the play/record mechanism.

Check the adjustment of the auto stop mechanism when the tape will not shut the machine off automatically. Check and see if the lever is bent out of line. The eject or detection piece should ride against the tape at its end. Straighten up the lever or replace it to correct the auto shutoff problem. You can carefully place a drop of oil at the bearing if the ejection piece is binding or difficult to move.

Checking the belt drives   There are various belt drive systems found to operate cassette machines. A majority have a belt drive to the capstan assembly. The drive belt is very small in the mini/microcassette players. The motor drive belt in some models is very short. These drive belts are usually flat or square in shape.

In addition to the motor drive belt, another belt runs from the flywheel to the take-up reel. A few models of the mini-cassette players have a fast forward drive belt. Because these belts are very small and thin, they have a problem of stretching and will then cause slow speeds. Clean each of these belts when you have a speed or "wow" problem with alcohol and a clean cloth. After these belts are cleaned and you still have a speed problem, replace the motor drive belt.

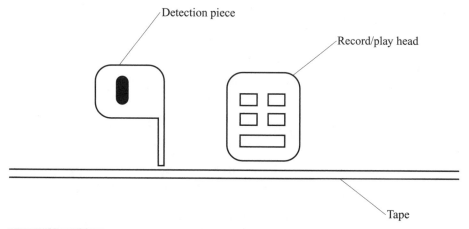

FIGURE 2-40   The detection piece is located close to the tape path to shut down the unit when end of the tape is reached.

Cassette switch problem notes   You will usually find many small switches used in these personal cassette player/recorders. The sound-level equalizer (SLE) switch improves recording in locations away from the source, such as auditoriums and conference rooms. The pause and VAS are slide switches. Usually, the radio-tape switch is a slide switch. When these switches do not work, or are erratic or intermittent, squirt a switch cleaning spray down into the switch contact. Try not to get any of this cleaning fluid on the belts or idler wheels.

The on/off switch that controls power to the motor and amplifier circuits could be a leaf switch that is pressed together for record, play, rewind, or fast forward. These small switch contacts could become dirty; if so, clean them with a switch cleaning fluid. Suspect a defective or dirty leaf switch when you have intermittent operation. These switches are usually squeezed together with a metal lever that may need to be adjusted.

Unit will not load cassette cartridge   Check inside the tape holder for any dirt or foreign material. Check to make sure the record safety lever will release. Check for any cracked or broken holder. Also, the cassette may be cracked or broken. Try a new cassette tape.

Inspect for proper door closing. Check to make sure the unit is in the play mode. The mechanism may be misaligned and will not let the cassette load. Look for any small items that may be inside the cassette holder that can prevent proper cassette loading.

Cassette blows fuses   In the larger-model cassettes, suspect a blown fuse if the unit is dead and nothing will light up. A good place to start is to look for a shorted silicon diode rectifier in the power supply. Also, a shorted filter capacitor, IC, or output transistor could be the culprit. Remove these components one at a time, and if the fuse does not blow, then you have probably found the defective component.

The deluxe stereo cassette players with higher-power audio output may have four large transistors. Usually, two are located for each audio output channel. You can check out which channel is blowing the fuse by taking a low-ohm resistance measurement between the collector of these transistors and ground. Now test each one while in circuit for leakage. Next, test them out of circuit for leakage. You may find one transistor shorted and the other one open.

While the transistors are out of the circuit, check for burned or open bias resistors. Usually, when a power-output transistor is shorted, the bias resistor will open up. Also, when the two transistors are out of the circuit, check the driver transistor. In some cases the driver transistor becomes leaky and this can damage the directly coupled power-output transistors. Most all power-output transistors can be replaced with universal types. Leaky power-output ICs may also blow out the fuse.

Deck shuts down after a few seconds   If you have a case where the tape deck keeps shutting off after only a few seconds, suspect that the automatic shutoff circuits are not working. In these units with automatic shutoff, a magnet is fastened to the end of a pulley on the counter assembly. Some models have a magnetic switch behind the magnet or IC. The magnet must keep rotating to keep the cassette player operating. When the magnet or tape stops, the magnetic switch or IC will shut down the operation of the cassette unit.

Should the drive belt to the counter be broken, the cassette will start up and shut down immediately. Check for a broken belt from the counter pulley. Note if the tape counter is rotating. If the belt is operating and the counter pulley is also, but the unit shuts down, suspect a defective switch or IC. An IC is used in some units while other models have a magnetic switch. The magnetic switch and ICs are special components.

A smoking cassette unit   Quickly pull the ac plug on any cassette or other electronic equipment if it's smoking. With an ohm meter check the primary winding of the power transformer for an open condition. If it's OK, then check the B+ supply voltage or make a resistance measurement across the large filter capacitor. An ohm reading below 100 indicates a short circuit.

If the transformer is overheating, check each silicon diode for a short. Also, a shorted or high leakage of the output IC or transistor will cause the power transformer to overheat.

If the transformer has been overheating and the above checks are OK, then remove all other secondary transformer connections. Now plug the AC cord of the cassette into the wall unit and if the transformer still runs hot or makes a noise, then it must be replaced. Refer to Fig. 2-41 for this power supply circuit.

Noise problems   If you hear loud mechanical noises, shut the cassette down and check it out. Should the noise be a crackling, fuzzy, or frying noise, then suspect an IC or transistor fault. Now check for noise in each speaker. If you hear frying noise in only one channel with the volume turned down, then the noise is being developed in that audio output channel.

To isolate the faulty component, spray each transistor or IC with coolant and note if the noise stops or becomes louder. You should spray each component several times before moving onto another one. At times when the coolant hits the faulty component the noise will quiet immediately.

If the coolant test does not indicate a problem, you can try shorting the base of transistors or input of ICs to ground with an electrolytic capacitor. Start at the volume control and

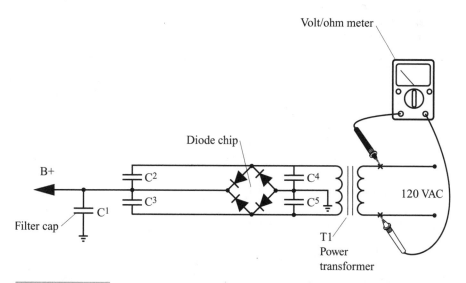

FIGURE 2-41   Use a volt-ohm meter to check for proper voltage and any shorted components. Replace transformer T1 if it runs hot after all connections and loads are removed.

work to the output stages and speaker. When the noise goes low in volume or quiets, then you have located the faulty stage. Testing may not reveal the defective component so it's best to replace it. You may also find a defective bypass capacitor in the audio circuits that can be causing a frying noise.

Rewind and fast forward problems   Usually the more expensive cassette decks have two motors. One motor is used for regular playback and the other for faster forward and rewind speeds. Suspect a defective high-speed motor or circuit when the deck does not rewind or go into the fast forward mode.

Check for proper voltage into the motor terminals. To do this connect your volt meter across the motor leads and then push the fast forward button. If it still does not operate, push the rewind button. If there still is no tape movement, then suspect a faulty motor if voltage is OK across its terminals. Also, check the possibility that a diode or resistor may be open in series with the motor leads if voltage is found at the terminals. Also, look for broken belts if the rewind or fast forward is not performing properly.

Erratic tape speed   In some cases erratic speed could be caused by a loose motor drive belt, an oily belt, or a dry capstan bearing. For this speed problem clean the motor drive belt, motor pulley, and capstan/flywheel.

Uneven tape speed can also be caused by a pinch or pressure roller that is not perfectly round or is worn. Check the rubber pinch roller area for broken tape. In many cases when the tape spills out and breaks, excess tape is wound around the pinch roller. The pressure roller can be removed for cleaning and removal of tape that is wound around it. Before replacing the pressure roller, put a drop of oil on its bearing. Do not get any oil on the rubber parts. Make sure the roller will move freely.

Check to make sure the pressure roller spring has enough tension. The pressure roller helps pull the tape, along with the capstan, across the tape heads and feeds it onto the take-up reel table. Make sure the pinch roller runs smoothly and evenly. Replace the pinch roller if it's lopsided or has worn edges.

Poor recordings   You can be sure that the erase head is not working if you hear several recordings during playback. The erase head erases any previous recording before the tape passes the record/play head(s). To check, place the unit in the record mode but do not have any audio input. Operate the unit in the record mode for a few minutes. Now, rewind the tape and put it into the play mode. All of the previous recordings should be removed.

Should the tape still have recordings on it, the erase circuits are not working. Clean the erase head and any other heads with alcohol. Also, clean the record/play switch with a spray switch cleaner. Now recheck again.

If the recordings are still on the tape after cleaning, check the erase head for an open winding or broken lead wires to a head. Also check for a good ground connection to the erase head. Check for a dc voltage to the erase head. Use a scope to check for the bias oscillator waveform, as that is required for the erase head to function.

Fast forward problems (single-motor unit)   The fast forward function in a single-motor deck is done with mechanical idlers or gears to increase the speed. In some of these decks, another winding runs at a faster rate of speed. When the player is placed in fast forward, the normal running motor wire is out of the circuit and the fast forward winding is switched into action.

With normal-speed operation, the B+ voltage is fed to the main winding and through a resistor and capacitor. In fast forward, the switch places the fast forward winding to the B+ and removes the resistor and capacitor from the circuit. In these cassette systems, the motor speed is changed by switching the windings at the motor instead of using a mechanical scheme.

Following is a list of symptoms, causes, and solutions:

Symptom: Cassette tape will not move.
What to do: Clean and/or adjust the control switches.
Probable cause: Motor not running.
What to do: Replace motor.
Probable cause: Drive belt worn or broken.
What to do: Replace with new belt.

Symptom: Tape movement erratic or slow.
Probable cause: Motor bearing dry or drive belt worn.
What to do: Replace motor or drive belt.
Probable cause: Oil or grease on capstan.
What to do: Clean capstan with alcohol.
Probable cause: Pinch roller dirty or cassette defective.
What to do: Clean pinch roller and try a new cassette tape.

Symptom: Tape tears or jams.
Probable cause: Take-up reel torque is too high.
What to do: Adjust or clean turntable clutch assembly.
Probable cause: Bent tape guide or misaligned head.
What to do: Replace head or readjust.

Symptom: Tape will not wind properly.
Probable cause: Tape torque is too low.
What to do: Adjust clutch subassembly.
Probable cause: Clutch arm assembly worn.
What to do: Replace clutch arm assembly.
Probable cause: Pinch roller out of alignment with capstan.
What to do: Adjust pinch roller or replace it.
Probable cause: Belt loose or off clutch assembly.
What to do: Clean belt and/or replace it.
Probable cause: Take-up idler wheel is worn.
What to do: Replace idler wheel.

Symptom: Tape speed is too slow.
Probable cause: Voltage to motor is low.
What to do: Check power supply.
Probable cause: Drive belt is slipping.
What to do: Clean or replace drive belt.
Probable cause: Motor stalls.
What to do: Replace motor.
Probable cause: Pinch roller is dirty.

What to do: Clean pinch roller with alcohol.
Probable cause: Oil or grease on capstan.
What to do: Clean capstan with alcohol.

Symptom: Wow and flutter during playback.
Probable cause: Cassette pad pressure is too high.
What to do: Replace with a new cassette.
Probable cause: Pinch roller is dirty or worn.
What to do: Clean or replace pinch roller.
Probable cause: Oil on capstan or other moving parts.
What to do: Clean all of these parts.
Probable cause: Capstan shaft is eccentric.
What to do: Replace flywheel.
Probable cause: Tape not following (tracking) in the proper path.
What to do: Check all components and realign the tape path.

Symptom: Fast forward is inoperative.
Probable cause: Fast forward torque is low. Clean or replace fast-forward clutch assembly. Replace spring in fast-forward clutch if pressure is low.
Probable cause: Defective motor.
What to do: Replace motor.

Symptom: Tape will not rewind.
Probable cause: Idler arm damaged.
What to do: Replace idler arm.
Probable cause: Rewind torque is weak.
What to do: Clean fast-forward clutch, idler assembly, and drive reel surfaces from oil, grease, or other impurities. Replace any rubber surfaces that are worn or uneven.
Probable cause: Brake assembly is still in contact with drive reels.
What to do: Adjust, repair, or replace the brake assembly.

Symptom: Rewind speed is slow.
Probable cause: Supply voltage is low or motor is defective.
What to do: Check power supply or install new motor.
Probable cause: Idler is slipping.
What to do: Replace or clean idler wheel.

Symptom: Tape climbs up capstan.
Probable cause: Shaft of pinch roller assembly is bent or loose.
What to do: Replace pinch roller assembly.

Symptom: No audio when playing a tape back.
Probable cause: Defective play/record head.
What to do: Replace or clean play/record head.
Probable cause: Defective power supply or playback amplifier circuits.
What to do: Check the power supply and recorder electronic playback circuits.
Probable cause: Defective cables or cable connections to the power amplifiers or speakers.
What to do: Check all connections and cables. Check power amplifiers for proper operation.

Symptom: No sound, only noise comes from the speakers.
Probable cause: Record/play head open.
What to do: Replace the play/record head.
Probable cause: Open or short circuit in cable to head or faulty plug connection.
What to do: Clean connections and cable or replace cable assembly.
Probable cause: Shielded wire between record/play head and circuitry is pinched, cut, or shorted.
What to do: Replace this shielded wire or repair.

Symptom: Weak playback audio sound.
Probable cause: Dirty play/record head. Check and clean.
Probable cause: Defective amplifier components.
What to do: Check voltages and repair amplifier stages.
Probable cause: Cassette is defective.
What to do: Try a known-good cassette.

Symptom: Poor high-frequency audio response on playback.
Probable cause: Record/play head is dirty.
What to do: Clean the head.
Probable cause: Azimuth adjustment is wrong.
What to do: Check and correct azimuth adjustment if it is wrong.
Probable cause: Record/play head is magnetized.
What to do: Demagnetize the head.

Symptom: The volume varies on tape playback.
Probable cause: Improper pressure of record/play head against the tape.
What to do: Adjust for proper head penetration.
Probable cause: The tape is not following the proper path.
What to do: Check and adjust mechanical components in the tape path.

Symptom: Audio is distorted during tape playback.
Probable cause: Defective components such, as transistors and ICs, in the playback amplifiers.
What to do: Repair audio amplifiers.
Probable cause: Defective speakers or connections.
What to do: Check for poor speaker lead connections or rubbing voice coils and warped speaker cones. Replace speakers if defective.
Probable cause: Record/play head is dirty.
What to do: Clean the head and adjust if necessary.

Symptom: Tape not being recorded in Record mode.
Probable cause: Defective play/record head.
What to do: Replace defective head.
Probable cause: Defective bias oscillator circuit.
What to do: Repair bias oscillator circuit. A typical bias oscillator circuit is shown in Fig. 2-42.

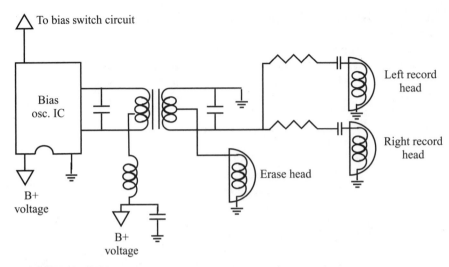

FIGURE 2-42 A simplified bias oscillator circuit is shown. The bias oscillator signal is fed to the left and right recording heads and to the erase head.

Symptom: No recording be made with microphones.
Probable cause: Defective microphone or microphone plug in jacks.
What to do: Replace microphone or repair the plug-in microphone jacks.

Symptom: Tape cannot be erased.
Probable cause: Defective erase head, dirty erase head, or bias oscillator not working.
What to do: Replace or clean erase head. Repair bias oscillator circuit.

3

AUDIO/VIDEO AND
CD PLAYER
OPERATIONS

How CD and Laserdisc Players Work

To explain CD player operation, this chapter uses a Zenith LDP510 multi laserdisc player. This laserdisc player produces very good quality video and audio.

Figure 3-1 shows the cartridge for a Pioneer audio CD player that uses a six-disc plug-in CD holder. In Fig 3-2, you can see how these CDs swing out to load and also play them. Fully loaded, these cartridges will give you six hours of uninterrupted music in your home or auto.

The Zenith laserdisc and CD player can play back five different kinds of discs:

- *12-inch laserdiscs (LD)*   These can contain up to 120 minutes (60 minutes per side) of high-quality video and digital or analog audio.
- *8-inch laser discs (LD single)*   Plays back up to 40 minutes (20 minutes per side) of high-quality video and digital or analog audio.
- *5-inch compact disc video (CDV)*   This plays up to 5 minutes of high-quality video and up to 20 minutes of digital audio.
- *3-inch compact disc (CD single)*   This plays up to 20 minutes of digital audio.

FIGURE 3-1   A six-CD cartridge that slides into a Pioneer home or auto CD player device.

FIGURE 3-2   Another view of the Pioneer six-CD cartridge, showing how the discs swing out for replacement. This cartridge, when fully loaded, can provide approximately 6 hours of music.

This Zenith multidisc player is a remarkably versatile player unit for both video and audio playback modes. Unlike video and audio tape players, these players can quickly find a specific point on the recorded CD.

SKIP, SEARCH, AND SCAN OPERATION

Now look at how the disc players perform these search and scan operations.

Chapter/program skip   Most laserdiscs are divided into segments or chapters, and compact discs are divided into programs. Each program on an audio CD is usually an individual song. With either a laser or compact disc loaded in the CD player machine, pressing the Skip button, forward or reverse, will cause it to skip almost instantly to the beginning of the next or previous chapter or program.

Chapter/program search mode   Laserdiscs and CDs with individual chapter or program numbers and descriptions written on the jacket label make it even easier to access specific segments directly. Simply press the Chapter/program key on the remote control and then the desired number of the chapter or program to be played back.

Disc scan mode   Both laserdiscs and CDs can be scanned by the Zenith laserdisc player to access a section within a chapter or program. Laserdiscs recorded in CAV (standard play) will show high-speed playback of the video on the TV screen. Compact discs, too, will deliver high-speed audio playback when scanning.

The laserdisc players all use the same operating principles as CDs and CVDs. This Zenith player is capable of playing these discs in addition to laserdiscs.

The recording of the master disc is accomplished (as shown in the description blocks of Fig. 3-3). The video is FM modulated on a 8.5-MHz carrier frequency. Audio signals are

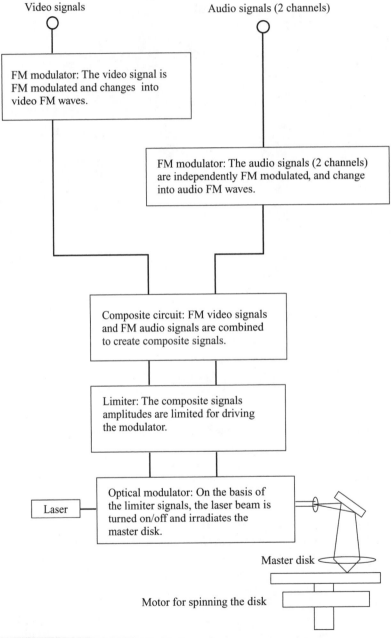

FIGURE 3-3   **A block diagram of how a laser master disc is produced.** (Zenith Corp.)

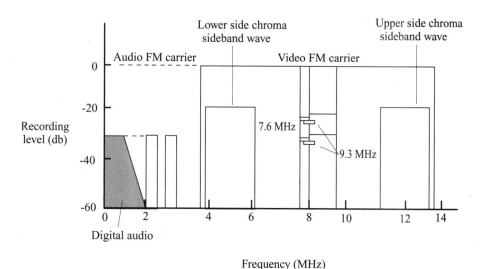

FIGURE 3-4   **The frequency spectrum chart of the laser CD beam.** (Zenith Corp.)

also FM (frequency modulated) on two carrier frequencies at 2.3 MHz and 2.8 MHz. The video and audio signals are combined to create a composite FM signal. This is passed through a limiter before going onto the modulator. In the limiter, the FM signal wave is formed into square waves that are coupled to the modulator to turn the laser beam on and off to create pits in the disc surface.

The master disc is composed of a photoresist, that is exposed by the laser beam, to create pits, in accordance with the video and audio information from the FM carrier wave. The master disc is then used to produce a completed laser vision disc.

The frequency spectrum of a recorded signal is shown in Fig. 3-4. The video frequency spectrum is from 4 to 14 MHz, thus giving it a very wide bandwidth. The chrome signal is comprised of a low frequency and is then frequency converted to produce sidebands as indicated. In this spectrum is a vacant space from 0 to 2 MHz, allowing for digital audio signal recording.

HOW THE LASERDISC IS MADE

Precision recording and playback of the CD in the very small micron scale is possible because laser beams are used for cutting. The rows of audio and video signals (called *pits*, recorded as bumps on the disc surface) are actually tracks impressed on one side of the CD disc. Each of these tracks contains one video frame.

One picture after another is picked up at the rate of 30 frames per second to produce a full TV picture during playback.

Disc construction information Of the six types of CD discs, the 5-inch size is the most popular today. Refer to Fig. 3-10 for more details on these disc types. The 12-inch disc consists of two one-sided discs (300 mm in size) that are bonded together and are 1.2 mm thick. Program recordings begin at the 110-mm diameter point and end wherever the program ends,

(1)Shape and dimension

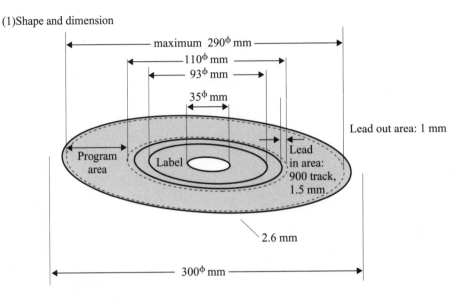

(2)Cross-sectional dimension

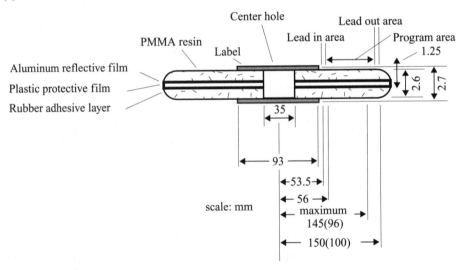

FIGURE 3-5   **The shape and dimensions of a cross-sectional drawing of a laser disc.** (Zenith Corp.)

up to a maximum diameter point of 290 mm. The shape, dimensions and cross-section view of the CD is illustrated in Fig. 3-5. Disc playback systems are classed as either standard (CAV) discs, which play for 30 minutes on one side, or long-playing (CLV) discs, which play for 60 minutes on one side. In either case, playback begins on the inner circumference and ends at the outer circumference.

A lead in about 1.5 mm wide is placed at the inner circumference, which is before the program starting point. The lead in serves as the intro for the program and contains information, such as the trademark of the company that made the disc.

A lead out at least 1 mm wide is located at the outer circumference, which is after the program ending point. It is used to display information, such as the end mark. The installed CD discs rotate clockwise when viewed from the top of the machine.

Standard discs rotate at a constant 1800 RPM (rotations per minute), but because long-playing discs are recorded at a constant linear velocity of 11 m/s, the number of rotations vary with pickup location from the inner circumference (1800 rotations) to the outer circumference (600 rotations). The center hole of the disc is 35 mm in diameter to provide stable support and suppress warping—even when the disc rotates at such high speeds.

How the disc is constructed   A resin protective-film coating is applied above an aluminum reflective film deposited on the pit transfer surface of the 1.2-mm thick PMMA base (transparent acrylic resin disc).

Because the protective films are bonded together, the pits of recorded information on finished discs are fully protected by embedding them at least 1.2 mm from the disc surface. Refer to Fig. 3-6 for the construction of the disc's internal structure.

This structure protects against scratches and dust on the signal pickup surface, and prevents scratches and fingerprints from deteriorating the audio or video quality during playback. This is also because the laser beam is precisely focused on the signal (pit) and the disk surface is outside the focal point. Discs have a very long operating life because you can wipe them clean with a soft cloth when they get very dirty. Handling and storage of the CD discs are very easy and convenient.

Signal (pit) detection scheme   Now see how the disc signal (pit) and signal pickup is accomplished. The pickup of a laserdisc player extracts signals impressed on the disc for amplification and playback, and generates an output that can be used as audio and video signals for TV picture reproduction.

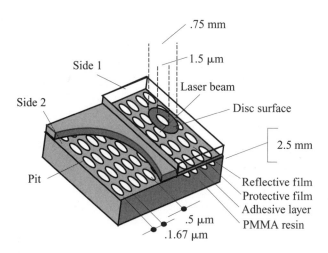

FIGURE 3-6   A drawing of the internal structure of a laser disc. (Zenith Corp.)

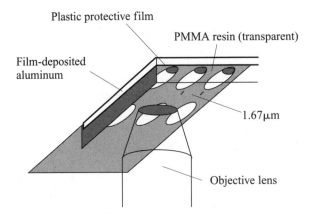

Plastic protective film

PMMA resin (transparent)

Film-deposited aluminum

1.67 μm

Objective lens

FIGURE 3-7   A drawing of how the laser signal operates and picks up data from the CD disc. (Zenith Corp.)

Each laserdisc signal is recorded as bump-like pits in varying sizes. The signal recording method is exactly like that for CDs (compact discs), so the method for extracting signals is also the same.

A semiconductor laser emits a pin point of light on a laserdisc. Only light that strikes a pit and is reflected is picked up and converted to an electrical signal. This operation is depicted in Fig. 3-7. The slight distance maintained between the pickup and the signal surface prevents damage to the pickup and disc. These are the advantages of the noncontact-type laserdiscs.

Optical pickup and detection via the pit signal   Almost all light emitted from the semiconductor laser is reflected at locations without the pits. A reflected light differential (power) is generated at the pit locations because only a portion of the light is reflected.

Because the length of each pit differs according to the impressed information, the reflected light differential (power) based on the varying length is converted to an electrical signal by the photodiode. Eventually, it becomes the audio and video signals, as shown in Fig. 3-8.

Light shone on the disc and reflected back just as it would return to its place of origin during the process of signal extraction, so an extraction method that can identify the light is required. For this job, a half mirror is used because it reflects 50% and passes the other 50%.

Besides the route just described, the light also passes through the grating, collimator lens, and objective lens. Each of these items are designed to control the direction in which the light advances and then to assign the correct signal to the pit.

The laserdisc pits   The size of a disc pit that represents a recorded signal is extremely small. An enlarged view of these pits are shown in Fig. 3-9.

A standard CD contains approximately 12 to 15 billion pits on one side. To get some perspective on this number, you can think of it as roughly equivalent to the number of brain cells in an adult person. The large 12-inch disc would contain much more information.

These tiny pits lined up on a single circuit of the disc is called a *track*. One track contains information for a single picture or screen full on a TV. Two fields, like that on a TV, are formed from 30-frame screens every second.

Movies are 24 frames and one laserdisc track equals one frame. Consequently, one side of a laserdisc records 54,000 tracks. Because 30 tracks form a one-second image, one side of a disc records 30 minutes of video.

Various types of CDs The types of discs that can be played back on laser players are shown in Fig. 3-10. Nearly all discs are LD, CD, or CDV compatible. Let's now look at the specifications of these various laserdiscs.

Standard (CAV) discs These standard discs have a constant angular velocity. A disc spins with a constant rotational speed of 1800 RPM. Playback time on one side of a 12 inch disc is 30 minutes, recording a maximum of 54,000 frames of picture information. As the

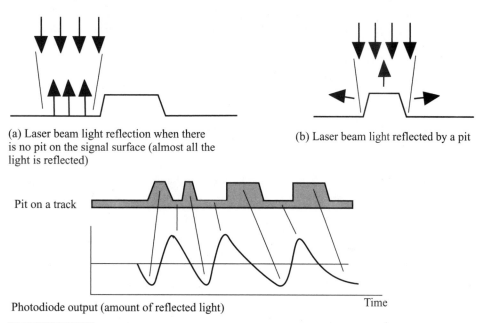

(a) Laser beam light reflection when there is no pit on the signal surface (almost all the light is reflected)

(b) Laser beam light reflected by a pit

Pit on a track

Photodiode output (amount of reflected light)

Time

FIGURE 3-8   **Signal extraction from the disc by the optical pickup device.** (Zenith Corp.)

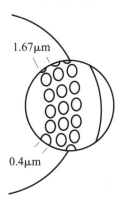

1.67μm

0.4μm

FIGURE 3-9   **A magnified view of the pit found on a CD disc.** (Zenith Corp.)

disc spins once, it turns each picture frame, which are each provided with frame numbers from 1 to 54,000.

Long play (CLV) These discs have a constant linear velocity. The rotational speed changes accordingly, when the signal in the inner circumference is being read, it spins at 1800 rpm. When the signal in the outer circumference is being read, it spins at 600 rpm. The playback time for one side of a 12-inch disc is 60 minutes. The playback elapsed time is recorded on the disc from the beginning.

Compact disc with video (CDV) Pictures associated with digital sound are recorded on the outer tracks (video part) five minutes long. Digital audio on the inner tracks (audio) 20 minutes: A normal playback begins at the video part through to the audio part.

   The main component of this CD equipment is the mechanical operating portion. This part of the CD player is broadly divided into the pickup carriage portion and the mechanical subchassis section that handles the loading and motor elevation operations. This CD player section is shown in Fig. 3-11.

How the pickup carriage functions

■ The pickup carriage, which features all tilt mechanisms, including a tilt sensor and tilt motor, is driven in the feed direction along the feed rack. A drawing of this pickup carriage unit is shown in Fig. 3-12.

■ This pickup carriage is equipped with a mechanism to adjust the tilt sensor mounting angle (screw) so that the sensor is constantly vertical in the radial direction with respect

Types of discs

The following types of discs can be played back on laser players because nearly all discs are LD, CD, or CDV compatible

Disc	CD		CDV	LD		
Disc size	CD single 3-inch	5-inch	5-inch B A	LD single 8-inch	8-inch	12-inch
Maximum recording time	20 minutes (one side only)	74 minutes (one side only)	25 minutes (one side only)	CAV: 14 mins. CLV: 20 mins. (one side only)	CAV: 28 mins. CLV: 40 mins.	CAV: 60 mins. CLV: 120 mins.
Recorded contents	Audio only: digital audio		A: Video&Digital audio (5 min.) B: Digital audio (20 min.)	Video & digital / analog audio		
Indication symbol	CD SINGLE	COMPACT DISC DIGITAL AUDIO	CD VIDEO	CD VIDEO	LD LASER DISC	

FIGURE 3-10   **The various types and sizes of laser discs.**

MECHANICAL PRINCIPLE

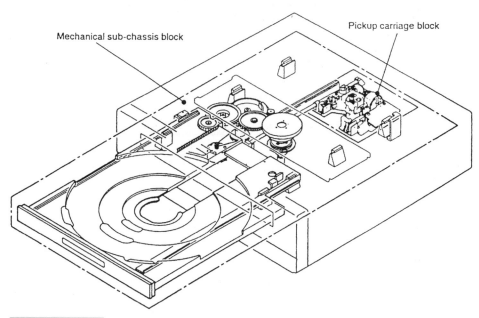

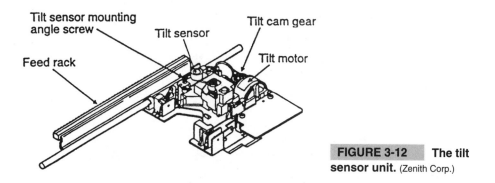

to the disc position. The tilt cam gear, which controls tilt, is equipped with an overtilt mechanism to lower the position of the pickup when tray loading, and a limiter mechanism to prevent the tilt cam gear from over rotating.

■ The pickup base to which the pickup is attached is adjusted in the normal direction by a screw to maintain positional accuracy of the pickup in the normal adjustment direction.

How the mechanical subchassis works

■ The loading mechanism, spindle motor elevation mechanism that clamps the disc, and the feed mechanism for the pickup carriage are located on the same chassis. All of these operations are handled by a single loading motor.

■ The loading mechanism uses an auto loading system so that a light push on the tray or a press of the Open/close button triggers the loading motor to pull in the drawer. On some PCs, the CD drawer is pulled in (loaded) as the program is being run.

■ The clamp mechanism raises the spindle motor to clamp a disc between the turntable and the clamper mounted on the subchassis.

Protection, setting mode, playback, and disc ejection of each disc is performed by each mechanical sequence on the subchassis that has just been described.

Mechanical tray operations   Refer to Fig. 3-13. The tray has a guide groove on the right side to engage the resin guide on the subchassis, as well as a drive transmission rack for horizontal movement that engages the slide on the right bottom surfaces and drive gears on the left bottom surfaces.

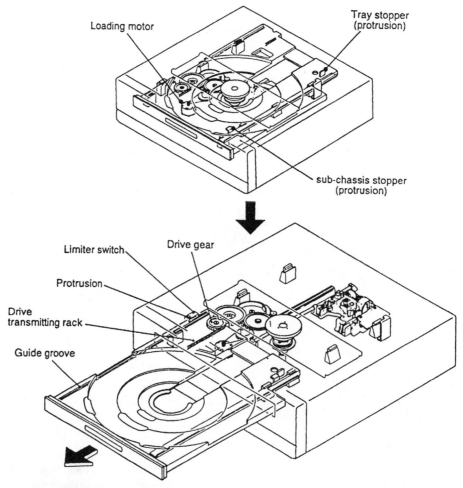

FIGURE 3-13   **Component layout and parts location inside the laser player drawer assembly.** (Zenith Corp.)

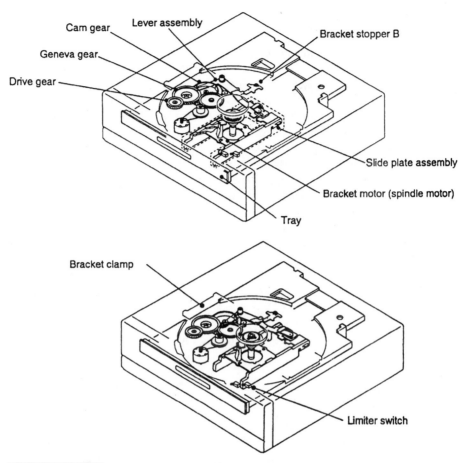

Cam gear
Lever assembly
Geneva gear
Drive gear
Bracket stopper B
Slide plate **assembly**
Bracket motor (spindle **motor)**
Tray

Bracket clamp
Limiter switch

FIGURE 3-14   **Location of the bracket motor and Geneva gear assembly located in the disk drawer.** (Zenith Corp.)

A protrusion at the rear of the tray serves as a stopper. When the tray is finished unloading, the stopper contacts the protrusion on the subchassis to prevent the tray from moving too far forward.

The end of slide tray unloading operations is detected when the protrusion at the left rear of the tray presses the limiter switch mounted on the subchassis side. In other words, the loading motor is off when the tray is unloading, but a slight press on the tray breaks contact between the protrusion on the tray and limiter switch. This causes the loading motor to turn on, and the tray is automatically pulled into the player. Normal tray operation can be performed by the player operating buttons.

The disc loading operations   Refer to Fig. 3-14 for how the disc loading is accomplished. The loading mechanism is made up of the following items:

■ Tray.
■ Slide plate assembly.

- Lever assembly.
- Bracket stopper "B."
- Bracket motor (spindle motor).

When the tray is housed in the player, the Geneva gear that engages the cam and drive gears locks to prevent the tray from moving back and forth. Grooves at the rear of the cam gear engage the lever assembly pin, causing the lever assembly to rotate. The other end of the lever pin pulls the slide plate assembly to slide the assembly horizontally.

Three angled grooves on the inner side of each side plate of the slide plate pair on the assembly engage protrusions from either side of the bracket motor (spindle motor). Two shafts set up on the rear surface of the subchassis restrict the bracket motor to the vertical direction and align it in the horizontal direction to ensure that the bracket motor operates vertically along the angled grooves. When the bracket motor is firmly pressed against the rear surface of the subchassis, the slide plates cut off the limiter switch, and the movement stops. This engages the lever assembly and bracket stopper B, which locks the lever assembly into position.

The spindle motor rises, causing the turntable to raise the clamper and the disc in the tray. The clamper clamps the disc with magnetic force between the bracket clamps and plate spring force.

Pickup carriage operation The pickup carriage operates as follows in each mode of the loading operation. Refer to Fig. 3-15 for the carriage operation.

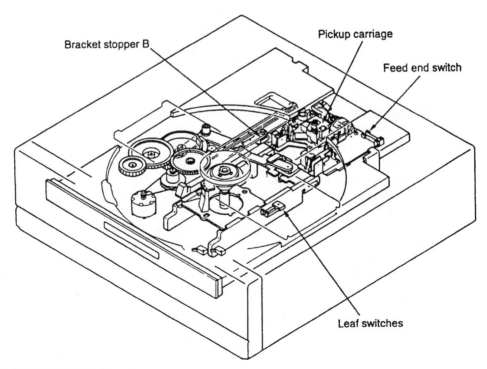

FIGURE 3-15 **Location of the leaf switches, feed end switches, and pickup carriage.** (Zenith Corp.)

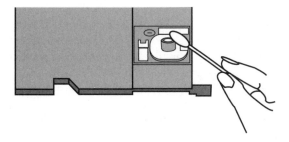

FIGURE 3-16   **Illustration of how to use a cotton swab to lightly wipe the lens opposite the pick-up carriage.** (Zenith Corp.)

Tray out/tray in During this time, the pickup carriage is positioned farther out than the outer periphery of the disc. The feed end switch is pressed and the carriage is locked in the feed direction by bracket stopper B. The pickup reaches overtilt by moving to the maximum tilt operating range, lowering the pickup so that it does not interfere with the tray.

Spindle down/spindle up During this time, tilt mechanism operations positions the pickup nearly horizontally. Overtilt and horizontal tilt are determined by detecting the rotation position of the cam gear. This position is detected by the reflector plate mounted on the rear side of the cam gears that operate tilt, and by two photo reflectors, which are mounted on the subchassis directly across from the reflector plate.

Spindle-up complete-feed operations When the spindle is completely up, the pickup carriage is released by the movement of the bracket stopper B. The feed rack drive gears, which were locked, rotate and the pickup carriage moves toward the inner periphery. The feed rack then engages the gears closest to the loading motor for feed operations with minimum backlash in the range applicable for actual disc playback.

The inner periphery of each disc is detected when the pickup carriage turns the tandem leaf switches mounted on the subchassis on and off. This way, the positions necessary for playback of each disc, such as the lead in for CDs and DCVs, the video part of CDVs, and the lead in for LDs, is detected. During playback, the tilt sensor signal keeps the tilt mechanism operating as an ordinary tilt servo, and keeps the pickup horizontal to the disc surface. It also ensures stable playback—even with extreme warping by setting the tilt fulcrum point closer to the inner side of the pickup.

Pickup lens cleaning of the laserdisc player Depending on the environment where the equipment is used, the pickup lens will become dirty (dirt/dust) after an extended period of time. Dirt will deteriorate picture and sound quality during playback, as well as destabilize the playback. When this happens, clean the lens:

1 Remove the laser player cabinet.
2 Use a new cotton swab to lightly wipe the lens located opposite the pickup carriage section. Wipe two or three times in a spiral moving from the center to the outer periphery (Fig. 3-16).

■ If the pickup carriage is not at the feed end (far peripheral standby) position, turn on the power, position the pickup carriage at the feed end, and turn off the power before cleaning.
■ Do not leave the laser player out of the cabinet for any extended period of time.

- If you accidentally get dirt on the lens, such as fingerprints during cleaning, wipe with a cotton swab. If the dirt does not come off by wiping, place a small amount of isopropyl alcohol on a new cotton swab and wipe two or three times in a spiral motion, moving from the center to the outer periphery (edges).
- Do not use any type of alcohol besides isopropyl alcohol. Other types of alcohol might damage the lens.
- Do not wipe the lens forcefully because this might cause scratches on the lens.

DVD Discs

These newer dual-layer high-density CDs, called *DVDs*, produce good-looking video and great sound quality. However, the entertainment industry does not know exactly what DVD stands for. It could be "Digital Video Disk" or "Digital Versatile Disk," but no one seems to know for sure at this writing.

Some of the first users of these DVD disks will be the group of video connoisseurs that have for many years been buying the 12-inch laser discs that can match the DVD's video and audio quality. These DVD disks hold much more information, cost less, and are easier to use and store than the 12-inch laser discs.

DVD TECHNOLOGY

With compression technology you can squeeze as much as 17 gigabytes of data onto a disk the size of a 5-inch CD, which can only hold about 650 megabytes of data. The pits that actually make up the present standard CD, means they can be jammed closer together and are read by a more-accurate laser beam. And a more important feature is that the material that is recorded is compressed and requires a lot less space.

On the CDs now being used, you have to turn a disc over to play the other side or have one laser on top and another laser on the bottom of the discs. On these new laser discs, so as not to turn them over for a long movie, two separate program layers are recorded on one side of the DVD. When playing this type disc, the laser will first focus on one layer and then onto the other layer without any interruption. Some discs use double layers on both sides of the disc for an even longer playing time. This dual-layer technique is illustrated in Fig. 3-17.

LASER LIGHT AND LASER DIODE INFORMATION

Laser is an acronym for *light amplification by simulated emission of radiation*. The laser device produces coherent radiation in the visible light range. The radiation of the laser beam is very narrow and does not spread out over a great distance, unlike light from a standard light bulb. Thus, the beam can be controlled to pin-point accuracy and can also be modulated. There are many different types and power ranges of laser devices.

The laser diode is a type of semiconductor device that emits a coherent light beam. The diode has an internal reflection and reinforcement capabilities. The laser diode is the size of a crystal of table salt.

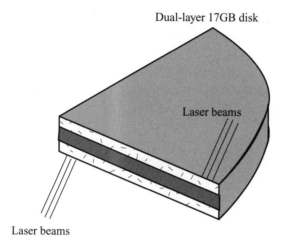

Dual-layer 17GB disk

Laser beams

Laser beams

FIGURE 3-17   **A cutaway view of a dual-layer disc. Compression squeezes as much as 17 gigabytes of data into a disc the size of a 5-inch CD.**

Laser diodes are also used as emitters in short- and long-range fiberoptics cable systems and sensors in your CD players. The typical laser diode uses 100 to 200 mW to power it and develops about 3 to 5 mW of output power to produce one very thin light beam.

CD PLAYER SECTIONS

Now review and summarize the various sections of a typical CD player.

The power supply   The CD player does not require a large-current power supply and the portable units use batteries. Most late-model CD players use the switching type of power supply and it is a sealed unit and is not recommended for repairs. Most develop three levels of voltage from the power supply. For the logic power, the V_{cc} is +5 Vdc. Analog and motor disc drive players use ± 15 V).

The optical deck section   All of the components located on the optical deck are used to load and spin the CD and some type of motor-driven loading device. On portable units, the drawer usually has to be pulled or pushed manually.

Some probable troubles   Troubles in the optical deck could be an loose or oil slick belt that causes the drawer not to open or close. It might also stick and not go through its cycle. Also, look for worn or damaged gears and dirty control switches.

THE ELECTRONICS PC BOARD

Problems in this board call for a professional electronics technician to find and repair any problem. This PC board also has many adjustment controls that should not be changed unless you are a qualified CD service technician.

THE DISC MOTOR

The disc motor is called the *spindle motor* and it spins the CD disc.

Some probable troubles   A defective motor with an open or shorted motor winding. Worn, dirty, or dry motor bearings. If your unit has a drive belt, it could be dirty, worn, or loose.

THE SPINDLE PLATFORM TABLE

The CD disc sets on this spindle turntable.

Some probable troubles   A dry or worn spindle table bearing can cause the CD to wobble and cause intermittent loss of the playback video. The spindle could also be bent and the table could be dirty. Lubricate the bearing and carefully clean the turntable and all other components in this area.

THE SLED MECHANISM

The sled is the device where the optical pickup is mounted. The optical device is mounted on the sled so that it can be moved across the disc to play and/or locate specific portions of data. The sled is on guide rails and is moved by a worm gear and a linear motor. It works very much like the hard dive in a PC.

Some probable troubles   Check for dirt and grease on the slide rails. They might be gummed up, have damaged gears, or need lubrication.

THE PICKUP MOTOR

The optical pickup is mounted on the sled which zips across the disc to access the various sections for playback. It uses a dc motor that can be belt or gear driven.

Some probable troubles   The sled motor could be defective and not run. The gears might be worn or jammed or dry. The belt might be broken or very loose.

THE DISC CLAMPER

In most machines, the clamper is a magnet on the opposite of the disc. The clamper magnet keeps the disc from slipping on the spindle platform.

Some probable troubles   The clamper will not completely engage; when this occurs, the disc will slip on the spindle. This problem is usually caused by a mechanical trouble in the drawer-closing components.

THE OPTICAL PICKUP UNIT

You can compare this device to a record player, which uses a needle (stylus) to contact the grooves in the record to play audio. In this case, the optical pickup reads the information encoded on the CD. This unit includes the laser diode, optical device, focus and tracking components, and the photodiode assembly. All of the optical pickup components are mounted on the sled, which is connected to a servo and reading processing electronics with flat, flexible cables.

CD Player Problems and Solutions

The following information is an accumulation of CD player problems that have been found and the solutions that have fixed the problems.

SOME CD PLAYER COMMANDS WILL NOT FUNCTION

The CD display will come on when power is applied but all or some commands will not function. Check out the following service tips:

■ *Control panel faults*   If one or more buttons will not work, check for poor connections on the flat cable between the front control panel and the main PC board. Clean or replace the switch buttons that do not work. If the player is equipped with remote control, check the remote for proper operation and batteries.

■ *May not be resetting*   If the CD player will not reset and will not accept user input, check power supply voltages and reseat all internal flat cable connections.

DEAD CD PLAYER

For a dead CD player, always check out the power supply and proper ac power plug voltage and fuse connections first. Check for proper operation of the off/on switch. Check for a fuse blown on the PC board that is usually mounted close to the power supply diodes and filter capacitors.

DRAWER WILL NOT OPEN OR CLOSE

When the drawer will not open with a push of the front panel button, listen for a motor hum that sounds like it is trying to push open the drawer. If you can hear the motor humming but see no drawer movement, check for a slick or loose belt, jammed slide, or other mechanical problem. You can clean the belt and see if that solves the problem, but this will only be temporary. The belt should be replaced for a lasting, troublefree solution. If no motor sound is detected, the problem could be a bad motor or front panel push button or faulty plug connections or motor control chip. And if the CD unit is remote controlled, use the remote and see if the drawer will function. If it will now operate then the front panel push button is the prime suspect.

UNPREDICTABLE DRAWER OPERATION

Some CD drawer problems may be erratic; the drawer may slide open when not asked to, open and quickly replay the CD, or quickly reverses course part way through its normal operation. And you may hear the motor still running even after the drawer has fully closed or opened.

All of the above erratic problems are generally caused by dirty or misadjusted contacts on the drawer position sense switches. In most of these units you will find three sets of switch contacts that control drawer functions. When these switches get dirty, worn, or misadjusted the drawer will operate erratically.

These switches are as follows:

- Drawer-pushed sense switch
- Drawer-open sense switch
- Drawer-closed sense switch

Check out, clean, and adjust if needed all of these switches and erratic drawer operation should be solved.

DRAWER WILL NOT CLOSE PROPERLY

This is a problem that you may not notice when the drawer closes. If the belt is worn or loose, it may not pull the drawer to its completely closed cycle. This will also result in an erratic CD operation. The cause could be the disc not being clamped, a tracking problem, speed variation, or the disc not being recognized. Clean the belt or better yet replace it with a new one. Also, clean the drawer slides, which may be gummed up, and lubricate any moving parts.

If the above checks do not solve the problem, then check out any of the gears that control the drawer action. Look for a gear that has jumped a tooth or has broken gear teeth. If a gear problem is found, this could result in a mistimed condition that will cause incomplete cycle operation. If the motor continues to run after the drawer cycle stops, then look for some type of mechanical damage.

VARIOUS INTERMITTENT OPERATION MODES

Some intermittent CD player operations could be as follows:

- Plays good for some time but will only play part of the disc.
- May play part of a disc and then completely shut down.

For these CD player symptoms check the following:

- A defective disc
- Various mechanical troubles
- Bad interlock connection switch
- A dirty lens
- Flex cables that may be cracked
- Loose or dirty cable connections
- Power supply problems

A CD player that will play some discs but not others may have a dirty lens glass that only needs to be cleaned.

PROBLEMS DEVELOP AFTER UNIT HEATS UP

A CD unit that develops problems after operating a while may have a component that fails with heat. This could be a unit that does not recognize a disc or becomes noisy after several minutes of operation. To locate these faulty heat-sensitive components you can use a hair dryer or a coolant spray on each component, and if the problem disappears then you have found the faulty component.

PLAYER AUDIO PROBLEMS

Some of the audio symptoms are as follows:

- Distorted sound, noise, or hum
- No audio or low volume level
- Audio only on one channel
- Intermittent audio

Let's now check out the more simple problems with audio faults. Check for dirty or loose contacts of the RCA jacks or poor solder connection where they are mounted on the PC board. You can check this by moving the cable connectors and probing the circuit board around the RCA jack area. The cables could also be bad from flexing and can be checked by replacing with a known good cable.

The audio problem may not be in the CD player but in other related audio equipment. Check for any bad cables and connections to audio amplifiers and speakers. Repair, clean, or replace any of the above items that seem to be faulty.

A REVIEW OF COMMON CD PLAYER PROBLEMS

- *The CD player will operate OK, then stop in same location of the disc* Your unit might have a transport lock screw. Check to be sure that it is in the Operate position.
- *Drawer loading problem because of belts* Belts loose, very worn and slipping, oil on them and cracked due to age.
- *Poor video or audio* The optics are dirty. Clean the lens, prism and turning mirrors.
- *Mechanical section not working properly* Parts dirty and need to be cleaned and lubricated, grit or sand in the moving parts.
- *Broken parts* This includes brackets, mountings and gears.
- *Intermittent interlock or limit switches* They are dirty and need to be cleaned or need adjustment.
- *Intermittent operation because of poor electrical connections* Check poor cracks or poor solder joints on the PC board, poorly contacting connectors, or broken flex cable trace leads.
- *Motors* The winding could become open or shorted. The motor bearing can become worn, loose, or dry and need lubrication.
- *Electronic servo problem* The servo requires focus, tracking, or PLL (phase-lock loop) adjustments.
- *Laser defect* The laser diode might be dead, weak, or not be receiving correct dc power.

The laser diode has a very low failure rate.

■ *Photodiode array problem*   The diode might be weak, defective, or have shorted segments. Also look for faulty heat-sensitive components.

Checking and Cleaning the Laser Player

If the CD laser player is operating erratically, you need to check the drawer components and sled drive unit to see if it needs to be cleaned and lubricated. Also check and clean the objective lens.

Carefully clean the lens because it is very delicate. Start by blowing out any dust or dirt around the lens. Then clean the lens with special lens cleaners. It is made of plastic, so do not use any strong solvents. Pure isopropyl alcohol is also effective for cleaning the lens.

A CD lens-cleaning disc is not as effective because it does not remove grime and grease, and can sometimes cause the performance to be worse.

Check the spindle bearing because this can cause noise in the audio. There should not be any play in the CD platform.

Check the drawer mechanism for smooth operation. Clean and lubricate if it needs it. Check the belts to see if they are worn or loose. Also check the motor and gears for proper lubrication.

Check out the various components in the sled drive that moves the pickup device. Look for worn belts, worm gears, and slide bearings. They might need to be cleaned and/or lubricated.

CD PLAYER WILL NOT OPERATE (START-UP)

A start-up problem for a CD player is when the player is not reading the disc directory properly for various reasons. And for a changer-type CD player the discs will keep loading and unloading but will not play the CD. In some cases the unit will not even load the disc from the carousel at all.

Some possible start-up causes are as follows:

■ A dirty disc
■ A defective disc
■ Dirty lens
■ Faulty focus or tracking actuator or driver
■ Gummed up track
■ Defective laser or photodiode array
■ Lubrication dried up on track and other moving parts
■ Damaged parts
■ Optical alignment faulty

You can take the case covers off the CD unit and with a visual inspection may be able to determine one of the above problems. One service item that can eliminate many problems is to carefully clean the photo diode or laser diode LED and optical pickup lens. However, for most of the other listed problems you should seek out a professional service center that specializes in CD or DVD repairs.

THE CD SEQUENCE START-UP ROUTINE

Let's now look at the start-up sequence of a typical CD player to help you determine what action you need to pursue.

Of course the sequence of these start-up events and problems that occur at failure, will vary depending on the model and player design. For most units, the display callout "---:---" or "disc" or "error" tells you the machine will not play properly. Thus, when you see these readouts, a blank screen, or a flashing display, you know the unit will not play a CD.

We will present an overview of these CD player actions that should give you a better clue of its operation.

For normal start-up, the CD is inserted (for a carousel the play button is pushed) and the following operations should occur.

1 The drawer closes (carousel will rotate to correct CD) and the CD is clamped to the turning spindle.
2 In some models the interlock engages. If no interlock is used, the optical sensor or optical pickup may act as the disc sensor.
3 The pickup resets to the starting (index) location, which is toward the disc center. This is usually located by using a limit switch or the optical sensor.

If these start-up sequences do function in this order, then perform the above cleaning and checks.

If you have checked out the mechanical operation and verified that the drawer is performing OK, then you need to make sure the optical lens is clean. In most units the lens should appear shiny with a blue tinge. Any dirt or dust on the lens will degrade the unit's performance or even stop its operation. In some models you may have to remove part of the clamping mechanism to clean the lens. Use caution in cleaning the lens and make sure it is perfectly shiny when you are finished.

A dirty lens, even one that looks clean, can result in a number of start-up problems. Thus, cleaning the lens should be performed first before looking into other, not-so-obvious, problems.

If the CD player still does not perform or operate properly then you should consider taking the machine to a professional CD service repair center.

NOTES ON CD READOUT FAILURES

The readout of the digital audio or data of a CD depends on proper focus and proper functioning of the tracking servos and system controller. When your CD unit is playing, searching, or seeking, the laser beam's focus must be sharp even for warped discs, some bumps, or vibrations. Total failures or slight faults of any of these systems can result in skipping, sticking, audio noise, or total failure of the seek and search operations. Note again that you should check for a warped or smudged disc or dirty laser lens.

CD SKIPPING PROBLEMS

If the CD player is skipping, the following symptoms will usually occur:

- The CD becomes stuck, jumps back, and then repeats the same information a few seconds later.
- The CD disc becomes stuck and repeats a fraction of every second (one CD rotation).
- The CD starts skipping continuously, or maybe every few seconds (either forward or backwards).
- The CD player starts having repetitive noise at the disc rotation rate. This is about a 200 to 500 rpm rate which comes out to about 3- to an 8-Hz low-frequency sound.

If your CDs are good, then you need to make the following checks. If the lens is clean, then look for dirt on the optical pickup worm gear or if the gear is dry then it should be lubricated. If the spindle bearing is worn or electronics adjustments are needed then it's best to bring the CD player into a service center.

NOISE PROBLEM

A CD with a repetitive noise problem can have several possible causes. Some of the most common noise problems are a dirty lens, loose spindle, dirt on the disc table, disc not firmly clamped, bent spindle, misadjusted focus, worn spindle bearing, and a weak laser.

OPTICAL PICKUP SLED COMMENTS

The sled moves the optical pickup device across the disc in order to retrieve the information data. Generally, any sled problems are a mechanical fault. When the sled is binding, you will probably have long-distance skipping, repeating, jumping, or failure to search past a certain location on the disc. Defective or erratic limit switches can cause jamming or overrunning at the start or end of the disc, or may not reset during the start-up mode.

Check for free movement of the optical pickup sled on its tracks or bearings. You can do this by manually rotating the sled motor or gear assembly. Clean, lubricate, or adjust the sled as required.

4

HOW COLOR TVs, DIGITAL HDTV RECEIVERS, AND PC MONITORS WORK

CONTENTS AT A GLANCE

The Color TV Signal

Before looking at the block diagram operation of a typical color TV, see what the color TV signal (which comes to your TV via antenna, cable or DBS dish) energy components contains.

The video signal that comes from the TV transmitter is an electrical form of energy that enters the free space in some form of electromagnetic waves. No one at this time has an explanation of what makes up this electromagnetic form of energy. It travels at the speed of light. For this transmitted wave to carry intelligence, it must be varied (modulated) in some way.

Your color TV is thus receiving from the TV transmitter visual images in full color, as well as audio sound. The U.S. color TV system must be compatible with black-and-white television standards. Compatibility is when the system produces programs in color on color TVs and black and white on monochrome TVs. Conversely, color TVs will receive B&W pictures when they are being transmitted.

The color TV signal contains two main components, luminance (black & white or brightness) information, and chrominance (color) information, which is added to the luminance signal within the color receiver circuits to produce full-color pictures.

The transmitted color TV signal contains all of the information required to accurately reproduce a full-color picture. The U.S. standard-frequency TV channel width for a color picture is 6 MHz.

The color signal contains not only picture detail information, but equalizing pulses, horizontal sync pulses, vertical sync pulses, 3.58-MHz color-burst pulses, blanking pulses, and VITS and VIRS test pulses. The horizontal blanking pulse is used to turn off the electron scanning beam from the gun in the picture tube, at the end of each scan sweep line. Vertical blanking pulses are used to blank out the beam at the top and bottom of the picture, so you will not see any of the transmitted pulses used for picture control and testing.

The color (chrome) is phase-and-amplitude encoded relative to the 3.58-MHz color burst and is superimposed on the black-and-white signal level. This video information is used to control the three electron beams (red, green, and blue) that scans from left to right with 525 horizontal lines across the face of the picture tube. How the color picture tube works is explained later in this chapter.

COLOR TV SIGNAL STANDARDS

The signals from the color TV transmitter are reproduced on the screen of the TV to closely match those of the original scene. The black-and-white and color signals have an FM (frequency modulated) sound carrier and an AM (amplitude modulated) video carrier with a channel bandwidth of 6 MHz.

The portion of the black-and-white signal that carries the video picture information is called the *carrier amplitude*, which is the brightness or darkness of the original picture information modulation. Now, a portion of the color video signal that carries picture information consists of a composite of color information and amplitude variations. A unique feature is the 3.58-MHz color syncburst, which is located right back of the horizontal sync pulse. This color pulse is often referred to as "setting on the back porch of the horizontal sync pulse." Every horizontal scan line of video color information contains the picture information, horizontal blanking, sync pulses, and a color burst of at least eight cycles.

The picture information in a color TV signal is then obtained from the red, green, and blue video signals that the color TV camera generates as it scans the scene to be transmitted. The luminance (Y) and chrominance (C) signals derived from the basic red, green, and blue video color signals have all of the picture information required for transmission. They are then combined into a single signal by algebraic addition. The final color product contains a chrominance signal, which provides the color variations for the picture, and the luminance signal, which provides the variations in intensity or the brightness of the colors.

The three different color TV broadcast standards in the world are not compatible with each other. In the United States, the NTSC standard was devised by the National Television System Committee and in 1953, was approved by the FCC (Federal Communications Committee). The NTSC system is now used in the USA, Canada, Japan, and other countries. Most of Europe uses the PAL system, and the SECAM standard is used in Russia. Before the year 2006, all TV video signal transmissions will be changed over gradually to compressed digital video and sound format. This system was approved by the FCC in 1997.

In Chap. 5 you will find more information on how flat screen TV/Monitors, high-definition TV (HDTV), and home theater "large screen" receivers operate.

At the conclusion of this chapter you will find an overview of the operation and some changeover notes for the new digital HDTV television system that is now being introduced nationwide.

Color TV Receiver Operation

Now, follow a block diagram of a typical modern color TV receiver and see how the various sections work together to produce a color picture.

THE "HEAD END" OR TUNER SECTION

The color TV signal enters the TV receiver at the electronic tuner. This is where you change stations on your TV, via an antenna, cable system, or a DBS satellite dish. Referring to the overall TV block diagram in (Fig. 4-1), you will note the electronic tuner has RF amplifier, oscillator, and mixer stages that convert the RF carrier signal to a 45.75-MHz signal IF that is fed to the video IF stages or strip. Most modern TV tuners are remotely controlled and have AFC (Automatic frequency control) to lock in the TV. Figure 4-2 shows the electronic tuner mounted in a late-model color TV.

THE IF STAGES AND VIDEO AMPLIFIER/DETECTORS

After amplification and signal processing in the IF stages, the video and audio signals are found at the output of their detectors. The audio has now been detected in the 4.5-MHz audio IF section and stereo audio is fed to the right and left audio power amplifiers, then onto the speakers. Some TVs also have MTS/SAP and DBX decoders.

VIDEO DETECTOR

An AM (amplitude modulation) detector, also called a *diode detector*, converts the picture frequency within the video IF stages down to a video frequency. Right after this detector is the 4.5-MHz sound trap to remove a frequency that would result in a heterodyning of the

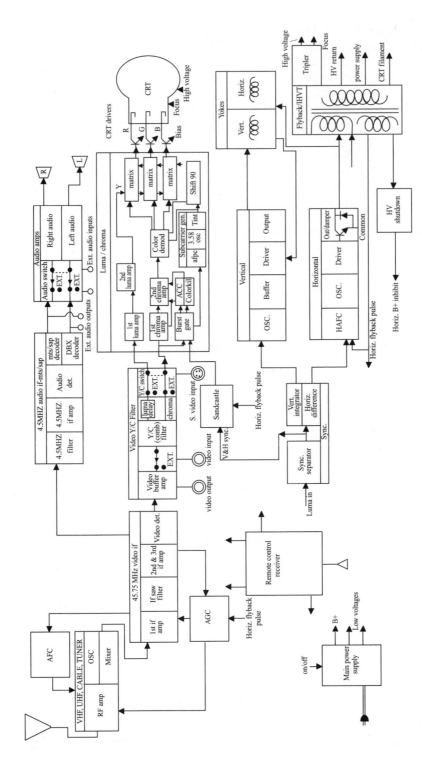

FIGURE 4-1 A block diagram of color TV/monitor system.

FIGURE 4-2   The electronic tuner "head end" used in a TV.

picture and audio frequencies of the video IF circuits. Without this trap, dark bars would appear across the picture and cause hum in the audio.

VIDEO AMPLIFIERS

The video amplifiers amplify the video from the detector to levels great enough to drive other circuits.

LUMA DELAY LINE

This is a device that electronically delays the Y signal (monochrome video), or luminance signal about 0.8 microseconds to match the delay in the chrome circuits. The reason for this is that the color signal is delayed more than the luminance because more circuits are used to process the color signal and they are tuned to a narrower-frequency bandwidth.

THE CHROMA PROCESSING CIRCUITS

From the video detector, the signal goes to some video-processing stages. You will find a comb filter, used to produce sharper pictures and the delay line. Chroma and video signals along with vertical and horizontal sync signals are also obtained from the video stages.

THE CHROMA AND LUMINANCE STAGES

The composite video signal from the detector is fed to the first chroma amplifier and also to the luminance (B&W) amplifier in the luma/chroma block. The 3.58-MHz oscillator and subcarrier generator is used to extract the chrominance signals. This is accomplished in the color demodulator stage. These signals go to the matrix circuits and then onto the cathodes of the color (CRT) picture tube guns. This block also contains blanking circuits, burst gate, and color killer stages.

COLOR-KILLER CIRCUIT OPERATION

The color-killer circuit has a couple of basic functions. Its performance during black-and-white picture transmission is to keep high-frequency signals or noise from being amplified via the chroma amplifiers. This keeps the color snow or confetti from being seen within the picture. It also keeps you from seeing color rainbows from around fine detail and edges of a black-and-white picture. This stage is also used to kill the color signal during weak or snowy signal conditions. Thus, the killer circuit must know the difference between the color burst signal and interference or noise conditions.

SANDCASTLE CIRCUIT OPERATION

Most modern color TVs have a sandcastle circuit located within an IC. The sandcastle circuit is a special device used by design engineers to inject three mixed signals into one pin of this IC. The IC separates these three signals and uses them for various internal functions. The three signals are the horizontal sweep pulses, a delayed horizontal sync pulse, and a vertical sweep pulse.

After separation inside the IC, the horizontal sweep, also called *flyback pulses*, provides horizontal blanking for the output signals of the chip, and the delayed sync pulses separates the color burst from the back porch of the horizontal sync pulse, and the vertical pulse provides vertical blanking. If you would look at this pulse on an oscilloscope that is developed from this chip, it appears as a sandcastle, thus the name given this circuit. If one of the input pulses is missing or the IC is defective, the TV screen will be blanked out.

FUNCTIONS OF THE SYNC CIRCUITS

The basic function of the TV sync circuits is to separate the horizontal and vertical sync pulses from the video signal. These separated pulses are then fed to the horizontal and vertical sweep stages to control and lock-in the color picture. These circuits need good noise immunity to maintain good, stable vertical and horizontal picture lock in.

Some color sets have the sync and AGC (automatic gain control) circuits combined. Normally the AGC circuit develops a bias in proportion to the sync pulses peak-to-peak level, which is then used as a dc voltage to control TV receiver gain. Keyed AGC circuits are used because they provide better noise immunity.

VERTICAL SWEEP DEFLECTION OPERATION

The vertical sweep oscillator stage receives a sync pulse from the vertical integrator stage, which forms and develops this pulse. This sync pulse keeps the vertical oscillator running at the vertical scanning rate. Some sets have digital countdown and divider circuits to per-form this task. The oscillator feeds the buffer and driver stages. The output current from the vertical power amplifier stage is then applied to the vertical winding of the deflection yoke, which is located around the neck of the picture tube. A pulse from the vertical out-put stage is used for picture tube screen blanking. Some of these pulses can also be used for convergence of the three color beams in the gun of the picture tube.

HORIZONTAL SWEEP DEFLECTION OPERATION

Older color TVs have a horizontal circuit that consists of a sawtooth generator that would drive the horizontal sweep and high-voltage generating transformer. This circuit is con-trolled by an AFC (automatic frequency control) circuit that compares the frequency of the oscillator with the sync pulse coming from the TV transmitter and then produces a correc-tion dc voltage for any oscillator frequency drift that might occur. The deflection current for the horizontal yoke coils, along with picture tube high voltage and focus voltage, plus other pulses, are generated by the horizontal sweep transformer. The horizontal sweep and HV stages need very good voltage regulation to produce a good sharp color picture.

Figure 4-3 shows various adjustment controls found on most TVs. These are horizontal hold (Horiz. Hld.), brightness level, vertical hold (Vert. Hld.), vertical height/vertical lin-earity, and sometimes a color killer and AGC adjustment.

Modern color TVs use a digital countdown divider system to generate and control pulses to drive the horizontal sweep stage. Figure 4-4 shows the horizontal sweep and

FIGURE 4-3   **Some of the adjustment control locations on TV receivers and some monitors.**

FIGURE 4-4   The high-voltage transformer found in a TV set.

high-voltage transformer section. Figure 4-5 shows the high-voltage lead and cup that plugs into the picture tube HV button. This lead will carry a voltage of 25,000 to 32,000 volts. Use caution because this voltage can still be present at this cap—even when the TV is turned off.

The horizontal output stage in these modern TVs are usually of a pulse-width design that not only sweeps the electron beam across the picture tube, but also develops other dc voltages to operate other circuits in the color TV chassis. This eliminates the heavy weight and costly price tag of a power transformer and also improves the efficiency of the ac power and current that the TV uses. A safety high voltage shutdown circuit is also used.

SOUND CONVERTER STAGE OPERATION

This stage converts the audio frequency in the video IF passband circuit (41.25 MHz) down to the sound IF frequency of (4.5 MHz) by heterodyning (beating) the picture and sound frequencies together and tuning to the different frequency. This stage is usually a diode detector and a 4.5-MHz tuned circuit.

SOUND IF AMPLIFIER OPERATION

This stage is sometimes called an audio IF amplifier. This stage is used for amplifying the 4.5-MHz sound IF signal to a level that is high enough to be detected by the sound detector.

THE SOUND (AUDIO) DETECTOR

The sound detector is also called the *audio detector*. This part of the circuit converts the FM-modulated 4.5-MHz sound IF frequency to an audible sound frequency that drives an audio amplifier stage(s).

AUDIO AMPLIFIER STAGE

This circuit is used to amplify the audio frequencies from the detector to power levels great enough to drive the speaker(s). If the TV receiver is MTS equipped, there might be two amplifiers (stereo) and the MTS-decoding circuits will precede the audio amplifiers.

TV POWER-SUPPLY OPERATION

As with any electronic device, the power supply is the heart of the device and makes all the other systems operate. If the TV is dead or not working properly, look at and check out the power-supply section first. The problem could be simple, such as an off/on switch that is defective, a blown fuse, or tripped "off" ac breaker for the wall socket. And check the condition of the power cord and be sure that it is plugged into the wall socket. Figure 4-6 shows the location of the main TVs power supply fuse located on the circuit board. Replace fuse with the same current rating if it has blown. If it blows again, suspect a short circuit or power-supply problems.

FIGURE 4-5   Location of the picture tube's high-voltage anode connection. Use caution when working around this section of a TV set or monitor.

FIGURE 4-6   **The main power clip-in fuse used in some TV sets.**

The transistor power supply regulator heatsink is shown in Fig. 4-7. If the regulator or sweep output transistors are defective (shorted), they might cause the main fuse in the power supply to blow.

This should now give you an overview of how these blocks and circuits in a color TV work together and what could go wrong. Take a closer look at the circuit operation within some of these blocks and see how they work and what to do when they don't.

SWEEP CIRCUITS AND PICTURE TUBE OPERATIONS

The following section of this chapter shows some advanced troubleshooting of color TV and computer monitor circuits. The horizontal and vertical sweep circuit operations and troubleshooting will be all worked in together.

Vertical sweep circuit operation   You can use the following information for color TV and computer monitor troubleshooting to isolate the vertical oscillator, driver, and sweep output stage problems. The vertical driver and output stages amplify the vertical oscillator signal, which provides the current drive needed for the vertical deflection of the yoke. A defective driver circuit, output stage, or yoke can cause loss of deflection, reduced height, or vertical linearity picture problems. Before you use signal injection to troubleshoot a vertical sweep problem, use a dc voltmeter (DVM) to confirm that you have proper bias voltages on the output stage components. The vertical stages are usually dc coupled to get good linearity. A wrong dc voltage affects all the components in the oscillator, driver, and output stages. A dc bias problem must be repaired before you can effectively use signal injection in the vertical stages. Use an analyzer, such as the Sencore VA62 or TVA92, to inject vertical and horizontal sweep signals into the circuit (Fig. 4-8).

Collapsed vertical raster   This problem will show up as a thin white horizontal line across the screen (Fig. 4-9). To isolate the trouble, inject the analyzer's vertical drive signal into the output of the vertical driver circuit (Fig. 4-10).

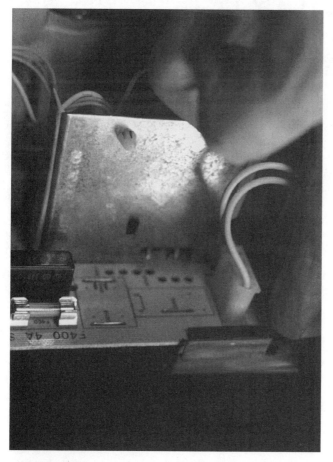

FIGURE 4-7   The heatsink for the regulator transistor and fuse that could blow if the transistor is shorted.

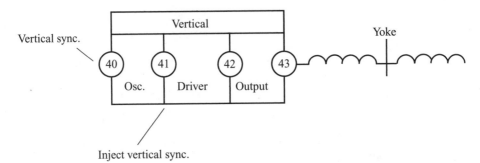

FIGURE 4-8   This block diagram shows where to inject the vertical sync pulse.

FIGURE 4-9   A thin, white horizontal line indicates a vertical sweep failure.

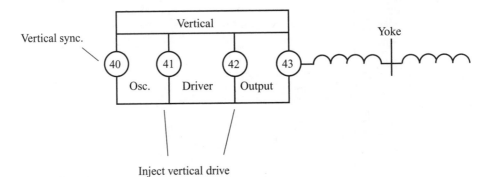

FIGURE 4-10   Arrows indicate where to inject the vertical drive test signal.

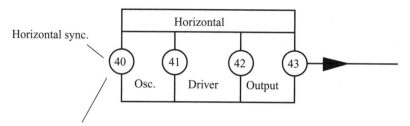

FIGURE 4-11   **The test signal is being injected into the horizontal oscillator stage.**

> Injecting a test signal into the vertical stages will not always produce full vertical picture deflection because most of the signals are uniquely shaped by feedback loops and waveshaping circuits.

   Look for the sweep (raster) to expand on the screen, but remember that it will probably not be a full screen. If the raster expands, either partially or fully, the circuits from the injection point to the output stage are good. If the sweep does not expand, check the output components or "ring test" the deflection yoke coils.

> The vertical drive test signal will not directly drive the vertical yoke coils.

Isolating horizontal sync problems   The horizontal sync pulses control the timing of the horizontal oscillator. Many computer monitors receive horizontal sync directly. TV receivers and some monitors have a composite sync, or "sync on video" input that requires the use of sync separators. Sync pulses that are low in amplitude, the wrong frequency, or are missing will cause the monitor to lose horizontal picture hold.

Loss of horizontal sync symptoms   Inject the video analyzer's horizontal sync drive signal into the input of the horizontal oscillator (Fig. 4-11). If the TV or monitor regains horizontal hold control and produces full horizontal deflection, the driver and output stages are operating properly. The next step is to troubleshoot the horizontal sync circuit path. If the TV or monitor displays the same symptoms with the drive signal from the analyzer applied, then troubleshoot the horizontal oscillator circuit.

DEFLECTION YOKE PROBLEMS

The changing current through the windings of the deflection yoke produces a magnetic field that scans the electron beam across the face of the picture tube. Yokes can develop shorted or open windings. An open or shorted winding might cause reduced vertical or horizontal size, or a complete loss of deflection.

   The analyzer ringer test can be used to find defective yokes—even if it has a single shorted turn. Readings of 10 rings or more are accompanied by a "good" display and shows that the winding does not have a shorted turn. "Bad" readings, less than 10 rings, indicate a shorted turn(S).

Collapsed raster symptom   For this symptom, picture would be pulled in and small; you need to ring the horizontal and vertical yoke windings. For this test, always disconnect the yoke from the circuit and unsolder any damping resistors (leave the yoke mounted on the picture tube neck).

If the horizontal and vertical yoke windings ring more than 10 rings, the yoke is good. If any of the windings ring less than 10, the yoke is defective and needs to be replaced.

KEY VOLTAGE READINGS

A lot of troubleshooting information can be revealed about a TV's operation by measuring the dc and peak-to-peak voltage at the collector of the horizontal output transistor. The Sencore analyzer has a dc and peak-to-peak voltmeter with the input protection required for measuring signals at this test point. The dc reading shows you if the B+ supply is working correctly, while the peak-to-peak reading shows if the output circuits are developing the needed high voltage.

INOPERATIVE COMPUTER MONITOR PROBLEM

With the analyzer or voltmeter, measure the dc voltage at the collector of the horizontal output transistor. If the B+ voltage is low or missing, unload the power supply by disconnecting the collector of the horizontal output transistor from the circuit. Measure the voltage at the output of the power supply regulator again. If the voltage is low or missing, troubleshoot the power supply. If the voltage is lower than that noted on the schematic, then something is loading down the supply. In this case, troubleshoot the output transistor, flyback transformer, or yoke (Fig. 4-12).

Testing sweep high-voltage transformers   The sweep or flyback transformer in a TV or computer monitor is used to develop the focus, high voltage, and other scan-derived power-supply voltages. The flyback is a high-failure component and it is also one of the most expensive parts in the TV or computer monitor.

Although an open transformer winding is easy to identify using an ohmmeter; the more common shorted transformer winding is nearly impossible to locate using the conventional testing methods. The Sencore analyzer has a patented ringer test that provides you with an easy-to-use, fail-safe method of finding opens and shorts in high-voltage sweep transformers.

ANOTHER MONITOR PROBLEM

For this test, connect the ringer across the flyback's primary winding and ring test the transformer. A "good" reading of 10 rings or more indicates that none of the windings in the flyback have shorts or opens. You do not need to ring any other winding. A shorted turn in any other winding will cause the primary to ring bad (Fig. 4-13).

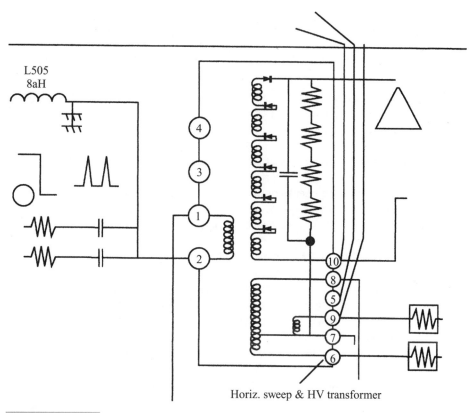

FIGURE 4-12   The Sencore's OVM meter has input protection to measure the P-P high-voltage pulses in this horizontal sweep and high voltage circuit.

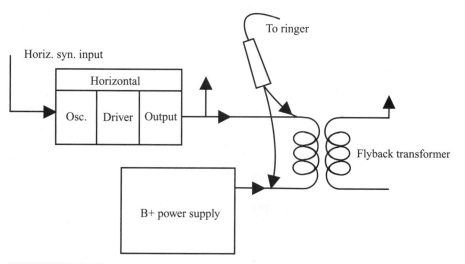

FIGURE 4-13   The ringer test hook-up for finding shorts and open conditions in a flyback transformer.

A "bad" reading, less than "10" rings, may be caused by a circuit connected to the flyback that is loading down the ringer test. Disconnect the most likely circuits in the following order:

1 Deflection yoke.
2 CRT (picture tube). Unplug the socket connection.
3 Horizontal output transistor collector.
4 Scan-derived supplies.

Retest the flyback after you have disconnected each circuit. If the flyback now rings "good," it does not have a shorted winding.

If the flyback still checks out bad after you have disconnected each circuit, unsolder it and completely remove it from the circuit. If the flyback primary still rings less than 10, the flyback is defective and must be replaced.

Testing the high-voltage diode multipliers During normal TV/monitor operation, a large pulse appears at the collector of the horizontal output transistor. The output connects to the primary of the flyback transformer and the pulses are induced into the flyback's secondary. The pulses are stepped up and rectified to produce the focus and high voltages. These voltage pulses are rectified by high-voltage diodes contained in the flyback package or in an outboard diode multiplier package.

Because these are high-voltage components, it is often difficult to determine dynamically if the diodes will break down under high-voltage conditions. The Sencore analyzer has a special test to determine if these diodes are good or bad.

HIGH-VOLTAGE PROBLEMS

It is only necessary to do this test if all of the following conditions are met:

1 The high voltage or focus voltage is low or missing.
2 The B+ and peak-to-peak voltages at the horizontal transistors are normal.
3 The horizontal sweep (flyback) transformer passes the ringer test.

With the analyzer, feed a 25-volt peak-to-peak horizontal sync drive signal into the primary winding of the flyback transformer. The step-up section of the transformer and the high-voltage diodes should develop a dc voltage between the second anode and high-voltage resupply pin on the flyback transformer. Measure this voltage with a dc voltmeter. Look up this voltage on the schematic to determine if the high-voltage diodes are good or bad.

HORIZONTAL OSCILLATOR, DRIVER, AND OUTPUT PROBLEMS

If the horizontal yoke, flyback, multiplier, horizontal output transistor, and B+ supply have tested good, but the TV still lacks deflection or high voltage, the horizontal driver circuit might be defective. A missing or reduced-amplitude horizontal drive signal could prevent the TV

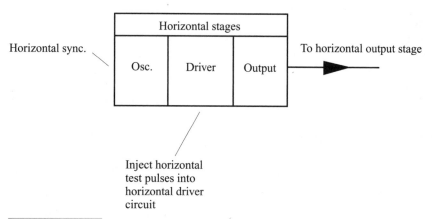

FIGURE 4-14   The injection points for test pulse injection into the horizontal driver circuit.

from starting and operating properly. Use the Sencore analyzer's horizontal drive signal to isolate problems in the horizontal drive circuit. Refer to the signal injection points in Fig. 4-14.

TV start-up problem

1 Before injecting into the horizontal drive circuit, test the flyback and yoke, the high-voltage multiplier, the horizontal output transistor, and the +3-V supply.
2 When injecting at the output transistor, disconnect the secondary winding of the driver transformer from the base.

Inject the horizontal drive signal into the driver circuit. Watch for horizontal deflection on the picture tube. If it returns, you are injecting after the defective stage. If nothing happens, inject the horizontal drive signal at the base of the horizontal output transistor. Refer to Fig. 4-15 for these injection points.

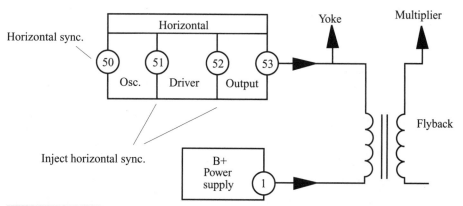

FIGURE 4-15   Horizontal drive signal test injection points. The base of the output driver transistor is a good injection point.

How to measure the TV's high voltage   The picture tube (CRT) requires a very high dc voltage to accelerate the electrons toward the screen of the CRT. This voltage is developed in the secondary winding of the flyback transformer and is amplified and rectified by the integrated diodes in the flyback, or by a separate multiplier circuit.

Measuring the high voltage at the second anode of the picture tube lets you know if the output circuit, sweep transformer, high-voltage multiplier, and power-supply regulation circuits are working properly. Additionally, some TVs and computer monitors have adjustments for setting the high voltage and focus voltage correctly.

Use extreme care when measuring and adjusting any voltage around the picture tube and high-voltage power supplies.

Blurred, out-of-focus picture symptom   For this problem, first measure the picture-tube high-voltage capacitor with a high-voltage probe. Compare these readings with the HV readings shown in the schematic. Also, if these voltages are OK, check the focus voltage and suspect that the CRT is weak.

Switching transformer checks   Switching transformers are used in power-supply circuits to step voltages up or down. They are one of the most common components to fail in switch-mode power supplies. Open windings are easy to find with an ohmmeter, but shorted turns are nearly impossible to locate with conventional test methods. The Sencore video analyzer's ringer test helps you to locate switching transformer with open or shorted windings.

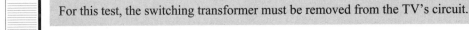

For this test, the switching transformer must be removed from the TV's circuit.

To perform this test, connect the Sencore analyzer ringer test leads across a winding on the switching transformer. A reading of 10 rings or more will show that the winding does not have a shorted turn. Perform this same test on all windings of the switching transformer.

THE VERTICAL SWEEP SYSTEM

In my feedback from many electronic technicians, most say that the vertical sweep systems are among the most difficult circuits in a monitor or TV to troubleshoot. Even the most small change in a component can cause reduced sweep deflection, nonlinear deflection, or picture fold-overs. These symptoms can be caused by a small circuit part or an expensive vertical yoke. Thus, you must think carefully of a strategy to take the guesswork out of isolating vertical sweep problems.

How vertical deflection works   Knowing how the vertical sweep deflection circuits operate requires an understanding of picture tube beam deflection. The electron beam travels to the face of the picture tube striking the phosphor surface coating to produce light on the front of the picture tube.

If the stream of electrons travels to the face plate of the tube without any control from any magnetic or electrostatic field, the electrons will strike the center of the screen and produce a white dot. To move this dot across the face of the picture tube screen requires that the electrons be influenced by an electrostatic or magnetic field.

In video display tubes, a magnetic field is produced by the vertical coils of a yoke mounted around the neck of the tube. The yoke is constructed with coils wound around a magnetic core material.

When current flows in the vertical yoke coil windings, a magnetic field is produced. The yoke's core concentrates the magnetic field inward through the neck of the picture tube. As the electrons pass through the magnetic field on the way to the tube's face plate, they are *deflected* (pulled upward or downward) by the yoke's changing magnetic field. This causes the electron stream to strike the picture tube face plate at points above and below the screen center.

To understand how electrons are deflected requires a review of the interaction of magnetic fields. As you refer to Fig. 4-16A and 4-16B, you might recall that an individual electron in motion is surrounded by a magnetic field. The magnetic field is in a circular motion surrounding the electron. As electrons travel through the magnetic field of the yoke, the magnetic fields interact. Magnetic lines of force in the same direction create a stronger field, but magnetic lines in opposite directions produce a weaker field. The electrons are then pulled toward the weaker field.

The direction of the current in the yoke coil determines the polarity of the yoke's magnetic field. This determines if the electron beam is deflected upward or downward.

How far the electrons are repelled when passing through the yoke's magnetic field is determined by the design of the yoke and the level of current flowing through the vertical coils. The higher the current, the stronger the magnetic field and resulting electron deflection.

A requirement of vertical sweep deflection in a TV or monitor is that the current in the coils of the vertical yoke increase an equal amount for specific time intervals. This linear

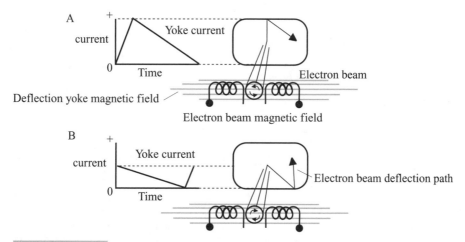

FIGURE 4-16   The yoke mounted on the CRT neck produces the magnetic field, resulting in electron beam deflection.

current change causes the deflection of the electron beam from the top to the center of the picture tube faceplate.

The waveforms shown in Fig. 4-16A and 4-16B represent a current increasing and decreasing in level, with respect to time. Figure 4-16A shows the current increasing quickly and then decreasing slowly back to zero. This would cause the electron beam to quickly jump to the top of the picture tube screen and then slowly drop back to the center.

Figure 4-16B shows the current increasing slowly in the opposite direction and then decreasing quickly back to zero. This would cause the electron beam to slowly move from the center to the bottom of the picture tube faceplate and then return quickly to the center.

During normal TV or monitor operation, the yoke current increases and decreases (Fig. 4-16A and 4-16B). The current changes directions alternating between the illustrations at approximately 60 times per second. The alternating current moves the electron beam from the top of the picture tube faceplate to the bottom and quickly back to the screen's uppermost area.

How the vertical drive signal is developed   The vertical circuit stages of the TV are responsible for developing the vertical drive signals. This signal is fed to an output amplifier, which produces alternating current in the vertical deflection yoke.

The vertical section consists of four basic circuits or blocks (Fig. 4-17). These include:

1 Oscillator or digital divider.
2 Buffer/pre-driver amplifier.
3 Driver amplifier.
4 Output amplifier.

The circuitry for these stages can be discrete components on the circuit board or might be included as part of one or more integrated circuits.

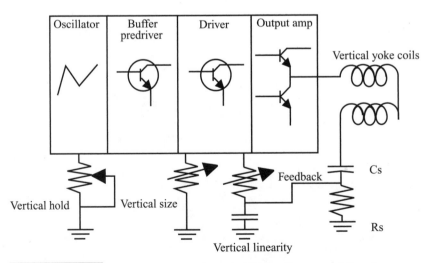

FIGURE 4-17   **The vertical section of a TV receiver consists of an oscillator, buffer, driver, and output amplifier.**

The vertical oscillator generates the vertical sweep signal. This signal is then fed to the amplifiers and drives the yoke to produce deflection. Vertical oscillators can be free running or the more modern digital divider generators.

These free-running oscillators use an amplifier with regenerative feedback to self generate a signal. More common types are RC (resistance-capacitance) oscillators associated with ICs or discrete multivibrator or blocking oscillator circuits.

A digital divider generator uses a crystal oscillator. The crystal produces a stable frequency at a multiple of the vertical frequency. Digital divider stages divide the signal down to the vertical frequency. You will usually find most of the digital divider oscillator circuitry located inside an integrated circuit.

The output of a vertical oscillator must be a sawtooth-shaped waveform. A ramp generator is often used to shape the output waveform of a free-running oscillator or digital divider. A ramp generator switches a transistor off and on, alternately charging and discharging a capacitor. When the transistor is off, the capacitor charges to the supply voltage via a resistor. When the transistor is switched on, the capacitor is discharged.

The vertical oscillator must then be synchronized with the video signal so that a locked-in picture can be viewed on the picture tube. The oscillator frequency is controlled in two ways.

1 A vertical hold control might be used to adjust the free-running oscillator close to the vertical frequency.
2 Vertical sync pulses, removed from the video signal, are applied to the vertical oscillator, locking it into the proper frequency and phase.

If the oscillator does not receive a vertical sync pulse, the picture will roll vertically. The picture will roll upward when the oscillator frequency is too low and downward when the frequency is too high.

Several intermediate amplifier stages are between the output of the vertical oscillator and output amplifier stage. Some common stages are the buffer, predriver, and/or driver. The purpose of the buffer amplifier stage is to prevent loading of the oscillator, which could cause frequency instability or waveshape changes.

The predriver and/or driver stages shape and amplify the signal to provide sufficient base drive current to the output amplifier stage. Feedback maintains the proper dc bias and waveshape to ensure that the current drive to the yoke remains constant as components, temperature, and power-supply voltages drift. These stages are dc coupled and use ac and dc feedback, similar to audio amplifier stages.

Notice that ac feedback in most vertical circuits is obtained by a voltage waveform derived from a resistor placed in series with the yoke. The small resistor is typically placed from one side of the yoke to ground. A sawtooth waveform is developed across the resistor as the yoke current alternates through it. This resistor provides feedback to widen the frequency response, reduce distortion, and stabilize the output current drive to the yoke. This vertical stage feedback is often adjusted with gain or shaping controls, referred to as the *vertical height* or *size* and *vertical linearity controls*.

The dc feedback is used to stabilize the dc voltages in the vertical output amplifiers. The dc voltage from the output amplifier stage is used as feedback to an earlier amplifier stage. Any slight increase or decrease in the balance of the output amplifiers is offset by slightly changing the bias. Because the amplifier's waveforms are slightly distorted, the bias change will shift the bias on the output transistors, somewhat, thus bringing the stage back into compliance.

Much of the difficulty in troubleshooting vertical stages is caused by the feedback and dc coupling between stages. A problem in any amplifier stage, yoke, or its series components alters all of the waveforms and/or dc voltages, making it difficult to trace the problem.

Vertical picture-tube scanning The vertical output stage produces yoke current that then pulls the electron beam up and down the face of the picture tube. The vertical yoke might require up to 500 mA of alternating current to produce full picture tube deflection. A power output stage is now required to produce this level of current.

A current output stage commonly consists of a complementary symmetry circuit with two matched power transistors (Fig. 4-18). The transistors conduct alternately in a push-pull arrangement. The top transistor conducts to produce current in one direction to scan the top half of the picture. The bottom transistor conducts to produce current in the opposite direction to scan the bottom part of the picture.

Most vertical output stages are now part of an IC package and are powered with a single positive power supply voltage. The voltage is applied to the collector on the top transistor.

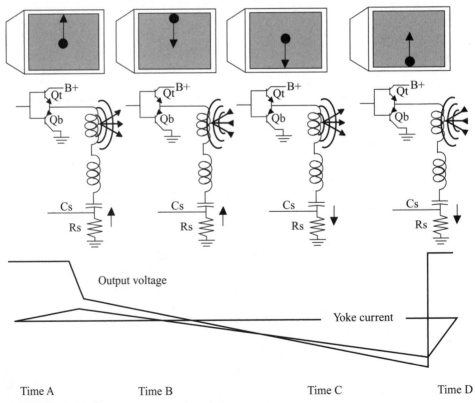

Time A   Time B   Time C   Time D

FIGURE 4-18   **The deflection currents and waveforms during four time periods of the vertical cycle.**

In this balanced arrangement, the emitter junction of the transistor should measure about one half of the supply voltage on this stage. In series with the vertical yoke coils is a large-value electrolytic capacitor. This capacitor passes the ac current to the yoke, but blocks dc current to maintain a balanced dc bias on the output amplifier transistors.

To better understand how a typical vertical output stage works, let's walk through the current paths at four points in time, during the vertical cycle illustrated in (Fig. 4-18). Starting with time A, the top transistor, Qt is turned on by the drive signal to its base. The transistor is biased on, resulting in a low conduction resistance from collector to emitter, which provides a high level of collector current. This puts a high plus (+) voltage potential at the top of the deflection yoke, resulting in a fast rising current in the yoke.

During time A, capacitor Cs charges toward a positive (+) voltage and current flows through the yoke and the top transistor, Qt. This pulls the picture tube's electron beam from the center of the picture tube up quickly to the top. During time A, an oscilloscope connected at the emitter junction displays a voltage peak, shown as the voltage output waveform in Fig. 4-18. The inductive voltage from the fast-changing current in the yoke and the retrace "speed-up" components cause the voltage peak to be higher than the positive (+) power supply voltage.

The current flowing in the deflection yoke during time A produces a waveform, as viewed from the bottom of the yoke to ground. This is the voltage drop across Rs, which is a reflection of the current flowing through the yoke.

During time B, the drive signal to Qt slowly increases the transistor's emitter-to-collector resistance. Current in the yoke steadily decreases as the emitter-to-collector (E-C) resistance increases and thus reduces the collector current. The voltage at the emitter junction falls during this time and capacitor Cs discharges. A decreasing current through the yoke causes the picture tube's electron beam to move from the top to the center of the screen.

To produce a linear fall in current through the yoke during time B demands a crucially shaped drive waveform to the base of Qt to meet its linear operating characteristics. The drive waveform must decrease the transistor's base current at a constant rate. Thus, the transistor must operate with linear base-to-collector current characteristics. These reductions in base current must result in proportional changes in collector current.

At the end of time B, transistor Qt's emitter-to-collector resistance is high and the transistor is approaching the same emitter-to-collector resistance as the bottom transistor, Qb. Capacitor Cs has been slowly discharging to the falling voltage at the emitter junction of the output transistors. Just as the voltage at the emitter junction is near one half of the positive (+) supply voltage, the bottom transistor begins to be biased ON to begin time C. This transition requires that the conduction of Qt and Qb at this point be balanced to eliminate any distortion at the center of the picture-tube screen.

During time "C", the resistance from the collector to emitter of transistor Qb is slowly decreasing because of the base drive signal and the increase of collector. The signal passes from capacitor Cs through the yoke and Qb. As Qb's resistance decreases and its collector current increases, the voltage at the emitter junction decreases. This can be seen on the voltage output waveform as it goes from one half positive (+) supply voltage toward ground during time C. The current increases at a linear rate through the yoke, as shown in the yoke current or voltage across Rs waveform (Fig. 4-18).

The resistance decrease of Qb must be the mirror opposite of transistor Qt's during time B. If not, the yoke current would be different in amplitude and/or rate, causing a difference in picture-tube beam deflection between the top trace and bottom trace times. At the end of time C, the emitter-to-collector resistance of Qb is low and Qb is slowly decreasing by the base increase of collector begins to discharge, producing current as the deflection yoke approaches a maximum level.

At the start of time D, the emitter-to-collector resistance of Qb is increased rapidly and collector current will decrease. This quickly slows the discharging current from capacitor Cs through the yoke and transistor. As the current is reduced, the trace is pulled quickly from the bottom of the screen back to its center. Time A begins again and the cycle is repeated again. This should now give you an overall view of how the horizontal and vertical sweep and scanning system produces a picture on your TV or computer monitor.

The basic inner workings of the color TV and PC monitor have now been covered. Another very important part of the color TV is the portion that you look at, the *picture tube* (*cathode-ray tube* or *CRT*).

The working of the color picture tube   The CRT works by producing (emitting) steady flow of electrons from the electron gun at the base (neck) of the CRT. These electrons are attracted to and strike the phosphor-coated screen of the CRT, causing the phosphors to emit light. Deflection circuits and a yoke outside the CRT produce a changing magnetic field that extends inside the CRT and deflects the beam of electrons to regularly scan across the entire face of the CRT, lighting the entire screen. The CRT can be divided into three functional parts (Fig. 4-19):

1 The electron gun cathode assembly.
2 The electron gun grids.
3 The phosphor screen and front plate.

The color picture tube is the last component in the video chain that lets you actually view a color picture on your TV or monitor. The major sections of a color set have previously been explained in this chapter, so now see how the CRT develops a color picture.

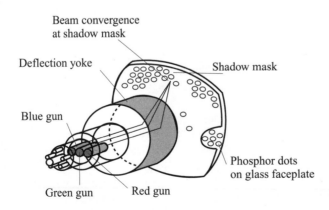

FIGURE 4-19   An inside view drawing of a picture tube that has an in-line gun assembly, metal dot mask, and phosphor dot triads on a glass faceplate.

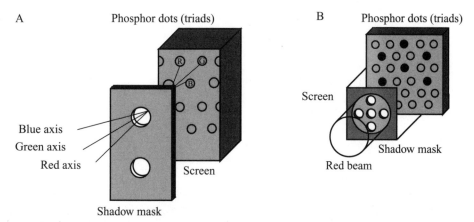

FIGURE 4-20   Shown in A is convergence of the blue, green, and red electron beams at the shadow mask. In B, each beam illuminates more than one phosphor triad.

Color CRTs use a metal shadow mask, phosphor screen, and three electron guns to produce red, blue, and green (RBG) colors that can produce a full color picture. These three colors are produced from phosphors that are excited by electron beams coming from three different guns, one gun for each of the (RBG) colors.

Figure 4-19 shows the relationship between shadow mask, electron guns, and the phosphors on the tube's faceplate. As you refer to Fig. 4-20A, notice that each beam converges through a hole in the shadow mask, while approaching the hole at a slightly different angle. Because of these different angles, the red beam hits the red phosphor, the blue beam the blue phosphor, and the green beam hits the green phosphor. However, each beam strikes more than one hole (Fig. 4-20B). With signals from the TV's red, green, and blue demodulators, these three electron beams are then mixed (matrixes) to different proportions to produce a very wide range of spectrum colors and intensities.

Over the years, TVs and monitors have used various types of picture-tube construction. The first color picture tubes used a delta gun arrangement with a dot shadow mask. As shown in Fig. 4-20A and 4-20B, the metal mask has evenly spaced holes with RGB phosphors clustered on the glass faceplate in groups of three. However, this triad arrangement had convergence problems because the three beams could not be made to meet at the shadow mask holes for certain areas of the faceplate.

How the electron CRT gun works   The electron gun consists of several different parts that together create, form, and control the electron beam. These parts are the filament (heater), cathode (K), the screen grid (G1), and the screen grid (G2). A monochrome (single color) CRT has just one electron gun, and a color CRT has three separate electron guns—one each for each color: red, green, and blue.

The cathode (K) is the source of the electrons, which are attracted to the screen. The cathode in most picture tubes look like a tiny tin can with one end cut out. It is coated with a material (such as barium or thorium) that emits large numbers of electrons when heated to a high temperature with the filament.

Several grids in front of the cathode attract the electrons away from the cathode toward the phosphor screen, control the rate of electron flow, and shape the cloud of electrons into a sharply focused beam.

The filament is mounted inside the cathode, and resembles the filament in a light bulb. The filament is used to heat the cathode. The filament is also called the *tube heater*. The filament is insulated from the cathode and does not make electrical contact.

The control grid is used to control the electrons. Without the control grid, the electrons would quickly leave in one big cloud with no control. The operation of the control grid can be compared to how a water faucet controls the flow of water.

GE in-line electron gun   General Electric developed the in-line gun with the slotted shadow mask in the mid 1970s. The metal mask has vertical slots instead of holes and the phosphors on the glass faceplate are RGB vertical strips, instead of dot triodes. The advantage of the in-line gun (Fig. 4-21) is simplification of convergence adjustments and a brighter picture level. When mated with properly designed yokes, the color convergence is considerably simplified. The Trinitron picture tube, invented and developed by the Sony Corporation, has a similar in-line design, except it has a three-beam electron gun and the shadow mask has a series of strips. The three common CRT gun patterns are in-line, delta, and Trinitron (Fig. 4-22).

In most cases, TV images are usually blobs of intensity and color. When a camera pans from one object to another, they are fuzzy because of bandwidth limitations of the video signal. In most cases, the images on computer displays consist of lines with sharp transi-

FIGURE 4-21   This photo shows the GE in-gun assembly and the adjustments used for convergence.

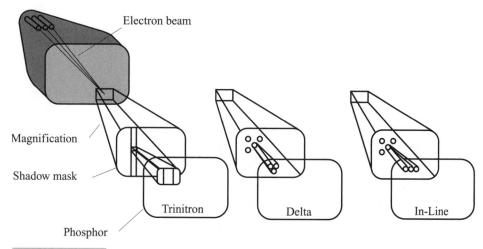

FIGURE 4-22   The three phosphor patterns for Trinitron (Sony), delta, and in-line picture tube formats.

tions of luminance. Usually the in-line/strip and slotted-mask CRT provides excellent pictures, but the in-line gun/dot mask-construction design displays text and graphics much better. A PC monitor color picture tube gun and socket assembly is shown in Fig. 4-23.

Color picture tube summation   A color TV contains all of the circuitry of the monochrome receiver, plus the added circuits needed to demodulate and display the color portion of the picture. To display the picture in color, three video signals are derived: the original red, green, and blue video signals.

The color CRT contains three color phosphors, each of which glows with one of the three primary colors when bombarded by electrons. These phosphors are placed on the inner surface of the picture tube faceplate as either triangular groups of the three colors (used in older models), alternating rectangles of the three colors, or alternating stripes of the three colors. Regardless of the version, all color tubes require three separate electron beams, each modulated with the video of one of the primary colors. All color tubes have some type of shadow mask placed behind the phosphors. This mask has a series of openings that allow each electron beam to strike only the correct color of phosphor.

The three beams must be precisely aligned to enable them to enter the opening in the mask at the correct angle and strike the correct phosphor. Stray magnetic fields could create enough error to cause the incorrect color to be displayed in parts of the picture. For this reason, color TVs have a coil mounted around the CRT faceplate and an automatic degaussing circuit to keep the picture tube and other nearby metal parts demagnetized.

Sometimes when a TV is moved to another location, the picture tube might have to be manually degaussed to clean up the color picture (Fig. 4-24).

LARGE-SCREEN PROJECTION TV OPERATION

Large-screen projection TVs are now produced in many screen sizes and price ranges. Most have provisions for "surround-sound" audio amplifier systems, audio/video, and cable TV

FIGURE 4-23   The picture tube socket and PC board assembly.

FIGURE 4-24   A degaussing coil being used to demagnetize a color TV picture tube faceplate.

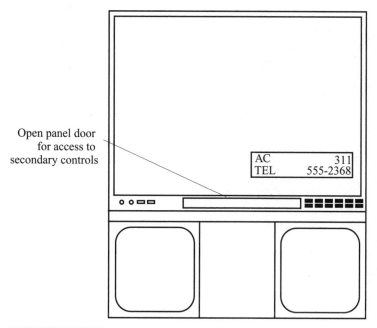

Open panel door
for access to
secondary controls

AC   311
TEL   555-2368

FIGURE 4-25   The front view of a typical projection color TV receiver.

and DBS dish input connections. A front view of a typical large-screen projection set is shown in Fig. 4-25. This type of TV projects the picture image onto the back of a translucent (Fresnel) screen that can then be viewed from the front. As shown in Fig. 4-26, the inside view these sets have three separate red, green, and blue (RGB) projection tubes to produce a bright picture.

A front-screen projection TV is illustrated in Fig. 4-27. These sets also use three separate red, green, and blue tubes to throw an image on a beaded projection screen, usually mounted on a wall.

In large-screen projection sets, high-definition, liquid-cooled projection tubes are used to provide a bright, high-resolution, self-converged picture display. Optical coupling is used between the projection tubes and the projection optics for display contrast enhancement. A screen with high-gain contrast and an extended viewer angle are now used on the newer-model projection receivers. Also, fault-mode sensing and electronic shutdown circuits are provided to protect the TV in the event of a circuit fault mode or picture tube arc.

Some projection TV system details   For their optics, some projection TVs use three U.S. precision lens (USPL) compact delta 7 lenses. This new lens, designed by USPL, incorporates a lightpath fold or bend within the lens assembly. This is accomplished with a front surface mirror that has a lightpath bend angle of 72 degrees. Because of this lightpath bend, the outward appearance of the lens resembles, somewhat, that of the upper section of a periscope. The lens elements and the mirror are mounted in a plastic housing. Optical focusing is accomplished by rotating a focus handle with wing lock-nut provisions. Rotation of the focus handle changes the longitudinal position of the lens' B element.

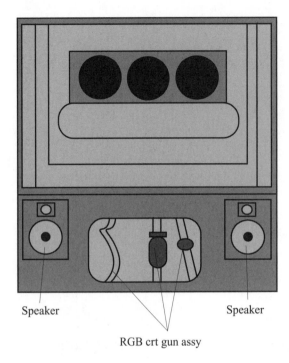

Speaker

Speaker

RGB crt gun assy

FIGURE 4-26   A front view
with viewing screen removed of a
rear projection color TV set,
showing component locations.

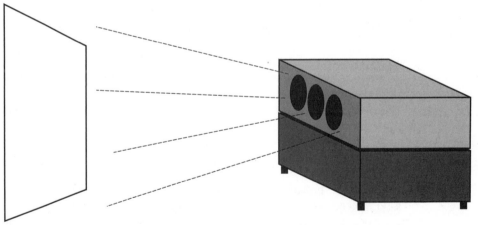

FIGURE 4-27   Drawing of a front screen projection TV set. This unit can set on
a table or be hung from the ceiling. Many of this units are used in home theater
installations.

Projection set lightpath profile   A side view of the TV lightpath is shown in Fig. 4-28. Note
the tight tuck of the lightpath provided by the Delta 7 compact optics. For comparison purposes, the lightpath profile of an earlier model projection set is shown in Fig. 4-29.

Liquid-cooled projection tubes   The rear-screen projection TVs use three projection
tubes (R, G, and B) arranged in a horizontal-in-line configuration. This type of config-

uration uses two (red and blue) slant-face tubes and one (green) straight-face tube. All tubes are fitted with a metal jacket housing with a clear glass window. The space between the clear glass window and the tubes faceplate is filled with an optical clear liquid. The liquid that is heat-linked to the outside world, prevents faceplate temperature rise and thermal gradient differentials from forming across the faceplate when under high-power drive signals. With liquid-cooled tubes, the actual safe power driving level can be essentially doubled over that of the older nonliquid-cooled tubes. This is highly desirable in terms of the large-screen picture brightness. The late-model sets use an 18-watt drive level to the picture tube, but the older-model projection sets had only an 8.5-watt drive level.

A side view of the jacket/tube assembly is shown in Fig. 4-30. The metal jacket shell extends back, well over the panel to the funnel seal and thereby functions as an effective x-ray

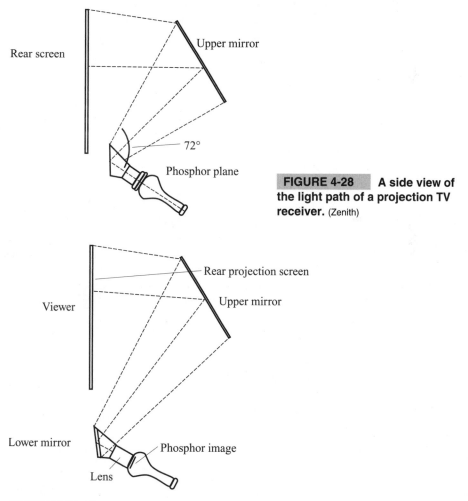

FIGURE 4-28   **A side view of the light path of a projection TV receiver.** (Zenith)

FIGURE 4-29   **A side view of the light path and mirrors of a projection TV.** (Zenith)

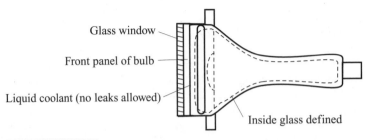

Glass window

Front panel of bulb

Liquid coolant (no leaks allowed)

Inside glass defined

FIGURE 4-30   **Drawing of a liquid-cooled CRT assembly.**

shield. The metal jacket also serves as the mechanical mounting and support for the picture tube assembly. The front of the metal jacket is elongated and the mounting holes are placed in the elongated sections. This is purposely done to permit the tightest possible tube-to-tube spacing for in-line tube placement.

Optical picture tube coupling A pliable optical silicone separator is mounted between the glass window on the liquid-cooled jacket assembly and the rear element of the Delta 7 lens. When under mounting pressure, the silicone separator makes close contact with these two lightpath interconnecting surfaces.

Self-convergence design Many large-screen projections have self-convergence and automatic convergence features. Final touch-up convergence can also be made with the remote control when in the service or set-up mode. This is accomplished in the receiver with the tilted faceplate of the red and blue tubes, in combination with shifted red and blue pointing angles, are image offsets that are used to provide for three-image convergence. This combination is required because of the shorter focal length in the Delta 7 lens design and its incompatibility with existing faceplate tilt angles. Because the receiver is a self-convergence system, registration of only the three images will be required. This is accomplished with special circuits located in the raster registration PC module.

Picture brightness and projection screen Usually, the projection screen for these projection sets is a two-piece assembly. The front (viewer side) piece will be a vertical lenticular black-striped section. The rear piece is a vertical off-centered Fresnel section. The black striping not only improves initial contrast, but also enhances picture brightness and quality for greater viewer enjoyment under typical room ambient lighting conditions.

The newer-large screen receivers demonstrate increased picture brightness over previous projection TVs. This is made possible by the use of liquid-cooled projection tubes and their ability to accommodate higher-power drive signals. The improvements will be substantial and some projection sets run almost twice the brightness level as the older models. Figure 4-31 shows the location of the circuit board modules and where the projection tubes are mounted in a late-model projection TV.

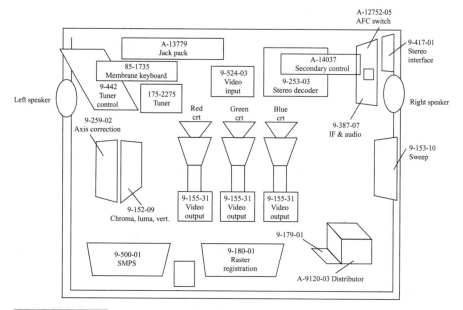

FIGURE 4-31   **Circuit modules and picture tube locations of a typical color TV projection set.**

What To Do When Your TV Has Problems

Some of the TV troubles were covered in the last portion of this chapter. Some of the trouble symptoms will be photos taken from the actual TVs with the problem. Some of the other problems are within the TV.

PROBLEMS AND WHAT ACTIONS YOU CAN TAKE

The symptom   The set will not operate (no sound or picture, dark screen).
  What to do:

■ Check the ac power outlet with an ac meter or plug in a known-working lamp. If no ac power is found, check and/or reset the circuit breaker to this outlet.

■ Check the ac line cord and plug from TV to the wall outlet. Some older TVs might have an interlock plug that removes ac power from the set when the back is removed. Be sure that this interlock plug is making a good connection.

■ Check and/or reset the circuit breaker on back of a TV. Other sets will have a main power fuse located on the chassis. Check fuse with a ohmmeter. Replace any blown fuse with same current (amp) rating as the blown (open) fuse. If the fuse blows again, the set probably has a shorted rectifier diode in the power supply or some other circuit is shorted or drawing too much current.

■ Check the on/off switch for proper mechanical operation. Use an ohmmeter to see if the switch is working (on and off contact) electrically.

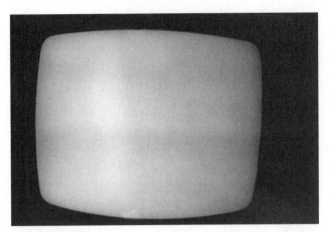

FIGURE 4-32   **The symptom for this TV set's problem is a blank (white) screen and no sound.**

The symptom   The TV has no sound or picture. The set produces a smooth white picture (Fig. 4-32).
What to do:

■ Check the TV cable, antenna lead in, cable lead from DBS antenna, and be sure that all of these cable connections are good and tight. Replace the coax cable and connections if it is found defective.

■ Inside the TV is a separate tuner box that will have a shielded cable that plugs into the main chassis. Check this cable for clean, tight connections.

■ For older TVs, the tuner knob will turn and click. This indicates that it is a mechanical tuner with switch contacts. Dirty or corroded contacts can cause a loss of picture and sound. Remove the tuner cover and spray the contacts with a tuner cleaner and lubricant.

 When working inside a TV, always be very careful because high voltage is present.

■ Some TVs have a control, usually on the back, labeled *AGC (Automatic Gain Control)*. If this control is misadjusted, the picture and sound will be missing. Try readjusting the AGC.

The symptom   Picture width reduced (pulled in from the sides, as shown in Fig. 4-33).
What to do:

■ Check the dc voltage from power supply. If not correct, readjust the B+ level control if the set has one.

■ A shorted coil winding in the horizontal sweep transformer or deflection yoke could cause this problem.

The symptom   Very bright narrow horizontal line across the screen. This problem is caused by the loss of vertical sweep.

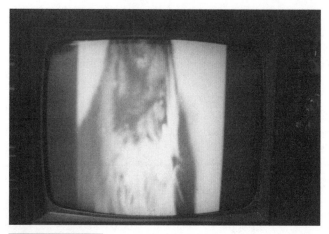

FIGURE 4-33   **The picture (raster) is pulled in from both sides of the screen.**

What to do:

- Check and adjust the vertical hold control.
- Check, clean, and/or adjust vertical height and linearity controls.
- Check vertical oscillator and output transistors and or IC stages.
- Check lead-wire plugs or solder connections to the deflection yoke.
- The loss of vertical sweep could also be caused by an open vertical coil winding in the deflection yoke, which is mounted on the neck of the picture tube.

The symptom   The picture is reduced at top and bottom (Fig. 4-34). This is also a vertical sweep problem.

FIGURE 4-34   **The picture pulled down from top and bottom. This is usually a vertical sweep circuit problem.**

What to do:

■ Check the vertical sweep output stage components.
■ It could also be a shorted winding in the vertical coils of the deflection yoke. This might show up as keystone raster shape.
■ Check and adjust the vertical hold control.
■ Check and adjust vertical size and linearity controls.
■ Some sets have a vertical centering control. If your set has one, check and adjust it because a defective centering control will cause the picture to shrink down in size.
■ Check for low dc voltages in the set's power supply and in the vertical sweep stages.
■ Check out any of the large (electrolytic) capacitors in the vertical sweep stage or that couple this stage to the deflection yoke. To make a quick check, just bridge another good capacitor across the suspected one and see if the picture fills out.
■ A large black bar at the top or bottom of the picture tube could be caused by some type of RF noise interference. Change channels and if this black bar disappears, that is your problem. A problem in the cable system could cause this same symptom.

The symptom   Small horizontal black lines appear across the picture and it might tend to weave (Fig. 4-35).
What to do:

■ The power supply might have poor low voltage regulation or faulty filter capacitors. Check B+ voltage with a meter and adjust the voltage level if your set has an B+ adjustment control. This symptom could also be caused by some type of signal interference.

FIGURE 4-35   **Small narrow black lines appear across the screen, and the picture might bend or weave.**

■ The degaussing circuit might not be turning off after the TV warms up. Check it by unplugging the degaussing coil that goes around inside the picture tube faceplate. The thermal resistor or diode in the power supply might be defective.

The symptom   An arcing or popping sound. This is usually around the large red HV lead and rubber cup on the picture tube. Also, in and around the HV sweep transformer stage.
What to do:

■ This will usually be some type of high-voltage arc. Use caution when checking out this problem. Check the large high-voltage lead (usually red in color) that goes to anode of the picture tube. Clean the rubber cup that snaps onto the CRT.
■ Check the amount of high voltage because it might be too high. You will need a special HV meter probe. Check that all ground straps around the picture tube are making good connections.
■ A blue arc in the guns (neck) of the CRT could indicate loose particles in the gun assembly or a defective tube. To clear the gun short you can carefully place the face of the tube on a flat, soft pad and gently tap the neck of the tube. This can remove any particles in the gun and clear the arc.

The symptom   The screen is blank except for small white horizontal lines (Fig. 4-36). The set has good sound.
What to do:

■ These symptoms usually indicate a video amplifier problem. The power supply voltage and high voltage to the CRT are probably OK. Most TVs and monitors have the video

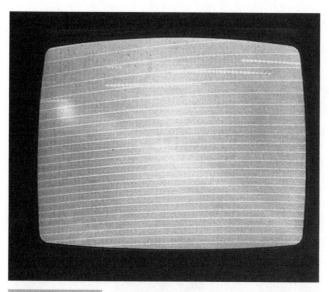

FIGURE 4-36   A blank (white) screen symptom, with small white lines going across the screen. The sound is good.

board and CRT socket in one unit. This PC board will be plugged into the picture socket. Check out this video board and clean the CRT socket assembly.

■ The blank picture could also indicate a picture tube failure. A short in the CRT guns could cause this problem. The blank screen might be all one color, such as red, green, or blue.

■ In some cases, a blanking problem might cause this symptom.

The symptom   The picture is not clear and has poor focus (Fig. 4-37).
What to do

■ Check and adjust the focus control. The control might also be defective.
■ Check the focus lead wire (large in size) and the pin on the picture tube socket.
■ Clean all pins on the picture tube socket.
■ The focus circuit could be defective and be supplying improper focus voltage.
■ The picture tube be defective.

The symptom   The TV has no picture or sound. Only snow and sparkles are seen on the screen. Only a hissing sound heard in the speakers.
What to do:

■ A snowy picture is shown in Fig. 4-38. The problem could be within the TV tuner. The RF amplifier stage or input balun coils could be damaged from lightning coming into the coax cable or antenna lead wire.

■ If you are using an outside antenna, the antenna or coax cable could be open or a connection could be loose or faulty.

■ If you have a cable splitter and or amplifier in your home, it might have failed. These devices are used if you operate two or more TVs from the same cable or antenna.

■ If you are using a DBS satellite receiver, it might not be working properly.

■ If you have an older TV with a mechanical tuner, the contacts might have become dirty. You can clean them with tuner spray.

FIGURE 4-37   **An out-of-focus or blurred picture symptom.**

FIGURE 4-38   **Picture has snow and sparkles. The sound is just a hissing noise.**

The symptom   Figure 4-39 shows a TV picture that rolls around and will not lock in.
What to do:

■ Try to adjust the vertical and horizontal controls to lock the picture in. If it will not lock in, the problem is in the sync or AGC circuits. In this case, you will need a professional to repair your set.

The symptom   The TV has a good picture, but no sound or distorted sound.
What to do:

■ The speaker voice coil might be open. Check it with an ohmmeter or substitute a known-good speaker.

■ Be sure that the set's volume level is turned up and it is not in the Mute mode. This can easily be overlooked on remote TVs with screen readouts.

■ Check all wiring and plug connections that go from the TV's main chassis to the speaker. If the leads plug into the speaker, be sure that they are clean and tight. If they are soldered, the connections might have a cold solder joint. Resolder these connections, if necessary.

■ Also, check to see if any external speakers might have shorted wiring, which would cause a loss of sound or distortion.

■ For distorted audio (sound), check the speaker cone for damage or warpage, or a voice coil that might be rubbing. Replace speaker with the same impedance (ohms) as the original one.

Conclusion   For some of the TV symptoms and problems just covered, you will need a professional TV technician to solve or correct them. You can take a small TV into the service shop. However, the large-screen or projection TVs will need to be repaired by a professional servicer in your home.

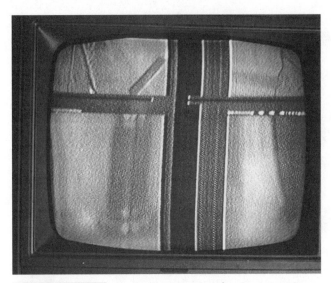

FIGURE 4-39   The picture is unstable, moves up and down, and tends to slip sideways. Picture cannot be locked in with horizontal or vertical hold controls. Newer sets with advanced sync circuits will not have a vertical hold control. These picture symptoms usually indicate a fault in the sync and/or clipper stages.

For any TV problem, you need to find out if the TV is defective or if the signal coming into your home is either missing, substandard, or weak. A good test is to disconnect the receiver in question and connect a known-good TV. If the test set has the same symptoms, then you know that the signal into your home has a problem. You will need to call the cable company, check the outside antenna, or check the DBS system, if you are using one. If the picture is OK on the test set, you know you have a problem with your primary and/or large-screen receiver.

Digital/HDTV Operation Overview

Let's now review the operation of the digital HDTV system that will now replace the analog TV system that has been the standard used in the United States. The digital HDTV system format was developed in the United States by the Advanced Television System Committee (ATSC) and was then approved by the FCC.

The complete conversion to an all-digital HDTV system will take more than 10 years to implement, but the FCC has already assigned all U.S. television broadcast stations a new digital transmitter frequency. The new ATSC format allows terrestrial transmission of digitally coded program material that will have a higher video resolution as well as CD-quality audio. This digital format also has a wide-screen format (16×9 aspect ratio), as opposed to the standard 4×3 aspect ratio of today. The highest-resolution program material or picture content is referred to as *high-definition television* or *HDTV*.

The ATSC format also provides the capability of broadcasting multiple lower resolution programs simultaneously, should the program material not be broadcast in "high definition." These multiple programs are transmitted on the same RF carrier channel used for one HDTV program. This is referred to as *multicasting*. The capability to broadcast several digital "channels" simultaneously involves the use of compression technology, which is not possible with the present analog system. The standard-definition signal will be noise-free, similar to the picture quality viewed on a digital satellite system, and quite an improvement over the present NTSC broadcast standard.

With this digital technology, broadcasters can supplement DTV programs with other data. Broadcasters can use the unused or "opportunistic" bandwidth to deliver computer information or data directly to a computer or the TV receiver. Digital broadcasting will allow new services to be created and let the broadcaster provide multiple channels of digital programming in different resolutions, while providing data, information, and/or other interactive services.

HDTV PICTURE QUALITY

HDTV produces much better picture quality. Lines of resolution go from the 525 interlaced to 720 lines and on up to 1080 lines. Then the ratio between picture width and height increases to 16:9. Conventional TV aspect ratio is 4:3. This change in picture quality opens up a lot of new options.

Digital television refers to any TV system that operates on a digital signal format. DTV is classed into two categories: HDTV and SDTV.

The SDTV (standard-definition TV) refers to DTV systems that operate off the 525-line interlaced or progressive standard. This will not produce the higher-quality video that HDTV does.

In addition to the higher-quality picture that HDTV delivers, it also has an advanced sound system. This audio system is supported by Dolby digital audio compression and also includes surround sound provisions.

SET-TOP CONVERTER BOX

The set-top converter box is used for receiving many different signals, including high-definition digital, standard digital, satellite digital, analog cable, and the standard UHF/VHF signals. For several years TV stations will be transmitting more analog than digital signals, and these options will be most useful.

The set-top converter decodes an 8-level vestigial sideband (VSB) digital signal that is transmitted by the TV station. VSB is the digital broadcasting system now being used in the United States.

The set-top converter can decode the digital signal for a standard TV receiver; however, the picture quality will not be improved much. Without a high-resolution screen, about the only useful feature will be that the box converts a digital signal to analog that can be viewed on a standard set.

In the future, built-in digital systems within the HDTV sets will be included in new TV sets. With so many standard TV sets in use, it will take a while for built-in digital TV receivers to become the majority units.

Most sets now marketed use terms such as HDTV-ready, digital-ready, or HD-compatible. This term does not indicate the TV set can produce a digital signal, only that it has a jack available to plug in a set-top decoder. Most of these types of sets do have enhanced screen resolution.

DIGITAL VIDEO FORMATS

There are several video formats; however, the most common formats are the 720 and 1080. Combinations of interlaced, progressive, and various frames per second, which are mated to these two resolutions make up the majority of these formats. HDTV and set-top manufacturers will supply the units that will read these formats. In addition, they will also supply equipment that will decode the complex audio signals.

DIGITAL TELEVISION SIGNAL

Terrestrial HDTV transmission is accomplished on an 8-level vestigial sideband, or 8-VSB. It is derived from a 4-level AM VSB and then trellis-coded into a scrambled 8-level signal (cable will use an accelerated data rate of 16 VSB). A small pilot carrier is then added and placed in such a way that it will not interfere with other analog signals. A flowchart illustrating these events of the data stream is shown in Fig. 4-40.

Satellite systems already are transmitting digital HDTV signals. Direct TV has two HDTV channels now and plans more in the near future. Digital satellite systems will have a head start in sending out high-definition TV over the conventional TV stations.

NOTES ON COMPATIBILITY

Electronics service technicians and TV reviewers are concerned about these new digital TV systems and whether they will be compatible with standard VCRs, camcorders, DVDs, and other entertainment products. In most cases it appears that they will be. Just about all equipment manufacturers are providing their products with composite video and analog inputs on these digital HDTV receivers.

INTRODUCTION TO THE DTV DELIVERY SYSTEMS

Some of the first DTV programs will appear on terrestrial broadcast stations. And DTV is also available via the DBS (Direct Broadcast Satellite) dish systems. You will be seeing more HDTV programs on the cable as more cable companies convert to a wideband digital system.

Digital broadcasting (DTV) provides many new challenges and opportunities for the professional electronics technician. Usually an outdoor antenna will be required to receive HDTV station programs.

It's very important to remember that the reception characteristics of a digital signal and an analog signal are quite different. A DTV receiver does not behave like a standard NTSC analog television receiver. When an analog NTSC broadcast is received, as the signal strength decreases, the amount of noise in the picture increases, and eventually, the picture is full of snow or will be blanked out. In contrast, a digital broadcast is completely noise-free until the signal level is too low for the receiver to decode. Once the digital signal threshold is reached, the picture will freeze and/or blank out. As discussed in Chap. 10, precise pointing of the

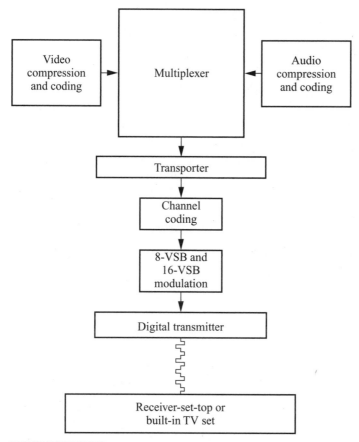

FIGURE 4-40 **Flowchart illustrates the data stream used in the HDTV system.**

HDTV antenna is very important. The antenna will have to be positioned to receive the best average sum of all digital signals within the viewing area. In some instances, the HDTV viewer may need more than one antenna because of different locations of the transmitter towers. A signal strength indicator will be built into many receivers to help position the antenna.

The digital signal is transmitted using a standard 6-MHz bandwidth. This is the same bandwidth that NTSC analog TV broadcasters use. The DTV signals are also broadcast in the same spectrum, or range of frequencies, that NTSC is broadcast in (primarily UHF). In most applications, the same antenna can be used for both HDTV and NTSC reception. Some new antenna designs are in the works and they will blend in with their surroundings and be less noticeable than traditional rooftop antennas.

The new satellites that broadcast the DTV signal do not use the same satellites currently used for Direct TV and USSB. However, the DTV satellite will be close enough to the others to allow the same dish to receive both regular DBS programming and DTV programming. A new dish and receiver will be needed for these dual-purpose systems. In addition to reception (antenna), consideration must be given to signal distribution. Again, it is necessary to remember that the signal does not become noisy as the signal strength weakens.

The signal levels (and picture quality) that are tolerable to the viewer with the present system may have too much noise to operate with digital signals. The installation of a low-loss, high-quality signal distribution system may be required.

HDTV FORMATS AND MODES

Generally, the 1080p, 1080i, and 720p (p = progressive, i = interlaced) formats are considered high-definition formats. However, limitations in current receiver technology prevent these formats from being available in current consumer TV devices. No doubt, in future years, as technology improves, these higher resolutions (above 1080) may become available. Although the format of the broadcast material might change, the receiver will use the same digital processing to convert the different formats.

WILL NTSC BROADCASTS DISAPPEAR?

The transition to digital TV is expected to take up to 10 years and possibly longer. At some point in the future, no analog broadcast stations will be on the air. Once that point is reached, all of the analog channels will be reallocated to other user spectrum services.

During the transition period, there will be available set-top converter boxes that will decode the digital signal and allow it to be displayed on a standard analog NTSC receiver. However, with this setup there will be a decrease in picture resolution. Set viewers will be able to use their current NTSC analog TV sets until the NTSC transmitter has to shut down. At this point in time, in order to watch TV programs, the viewer will have to purchase a new digital TV or a converter box that receives the digital signal and converts it to analog.

STANDARD-DEFINITION AND HIGH-DEFINITION BASICS

Picture resolution can be specified in *pixels* or *lines*. Resolution is the maximum number of transitions possible on the screen in a horizontal and vertical direction. The maximum resolution that a CRT (picture tube) can display is determined at the time the CRT is produced. The greater the amount of horizontal and vertical pixels, the greater the resolution capability. The resolution of computer monitors is most often given as the number of pixels it can display. This is given in both horizontal and vertical directions, i.e., 1920(h) × 1080(v). Pixels are also used to describe the resolution capabilities of the new digital ATSC and HDTV format.

In broadcast television, the resolution of the studio camera that captures the video is what determines the highest resolution possible. The picture resolution produced by the camera is given in pixels, as it is for a CRT. This is the current resolution limitation as the transition to high-definition digital TV takes place. In NTSC, the ability of a signal voltage (analog) to quickly change from low to high, in order to produce a dark to white transition, is comparable to a pixel.

The number of lines transmitted in the current NTSC analog format is 525.[*] This is considered standard-definition (SD) transmission. A standard-definition transmission of a 525-line NTSC signal can be transmitted in the analog or digital (ATSC) mode. A higher-definition (HD) transmission can be transmitted only in the digital television (DTV) mode.

---

*In SDTV, or standard-definition television 525 lines of resolution are transmitted, but only 480 of these lines are viewable. SD can be sent as an NTSC analog or digital (DTV) transmission.

DIGITAL TELEVISION QUESTIONS AND ANSWERS

Q: What is "Digital Television" (DTV)? And what's the status of high-definition television (HDTV)? Are HDTV and DTV the same thing?

A: The FCC, its Advisory Committee on Advanced Television Service, and the Advanced Television Systems Committee (a consortium of companies, research labs, and standards organizations) have defined 18 different transmission formats within the scope of what it broadly calls the *digital television standard.*

   DTV is the umbrella term for all 18 formats.

   Six of these formats are considered "high definition" because they constitute a significant improvement over the resolution quality of current TV, referred to as NTSC format set up nearly 50 years ago. Most TV set viewers will see a great improvement in image quality even with the other 12 formats because of digital transmission. The TV viewers will also benefit from DTV formats such as wide-screen theater-like displays, enhanced audio quality, and new data services.

Q: Besides better resolution, audio performance, and data services, are there any more reasons to have an HDTV set?

A: One of the basic improvements with HDTV is the way it is transmitted. Digital transmission can deliver a near-perfect signal, free of ghosts, interference, and picture noise.

Q: Will you be able to view the new HDTV broadcasts on a current conventional TV set?

A: Yes. You can watch HDTV broadcast programs by using a special HDTV decoder box device. These set-top boxes will receive digital transmissions and convert all 18 formats to standard TV, and can be connected as easily as a VCR up to your TV set.

Q: Will people be able to watch high-definition TV using this set-top box and a standard TV set?

A: Not high definition, but a big improvement that will provide many different solutions for better images and sound using A/V equipment that you may already have. The decoder box will supply output HDTV broadcasts with Dolby digital audio giving more precise localization of sounds and a more convincing, realistic ambience. You may already have a multichannel, multispeaker audio system and can take advantage of digital TV's enhanced sound quality.

   Many HDTV decoders will also provide three high-quality connections for monitors. Component video outputs will allow you to connect the box to most home theater LCD projectors and direct-view sets with component inputs to provide optimum image quality. Many large-screen TVs can be connected via S-Video, which maintains high image quality by separating the luminance and chrominance signals. You can even connect this box to a standard VGA computer monitor, which provides a more "crisp" and detailed picture than the conventional TV.

Q: How are these HDTV signals received?

A: In most locations you should be able to receive HDTV with any standard UHF antenna. The exact style of antenna that is required for optimal reception may vary depending on your geographic location and distance from the TV tower. Consult your electronic distributor for advice for selecting the optimal antenna for your location.

5

FLAT PANEL MONITOR/LARGE SCREEN PROJECTION SET AND HDTV DIGITAL TV SYSTEM OPERATION

CONTENTS AT A GLANCE

Introduction to Flat Screen HDTV and Monitor Displays

The viewing screen used since the inception of television is actually a vacuum tube with a faceplate and neck (gun) assembly that is often called the cathode-ray tube (CRT). The CRT has three electron beams (for red, green, and blue) that sweep across the phosphor strips or dots to produce a color picture. The CRT creates pixels by illuminating these RGB phosphors. As the beams sweep they are controlled or modulated for various brightness levels to produce a viewable color picture. This CRT operation is illustrated in Fig. 5-1. One of these optical gun assemblies is shown in Fig. 5-2.

Rear screen HDTV projection receiver sets use three smaller, very high brightness CRTs to project an image on the back of the viewing screen. And, of course, the three projection set CRTs scan continuously across the screen columns. As shown in Fig. 5-3, the projection CRTs expand smoothly along rows of phosphor dots. In this same figure you will note that the fixed pixel displays sample the image in both directions, thus response capabilities are much faster. Figure 5-4 lets you compare a high-definition monitor VGA to XGA, with 4×3 resolution and 16×9 resolution pictures, respectively.

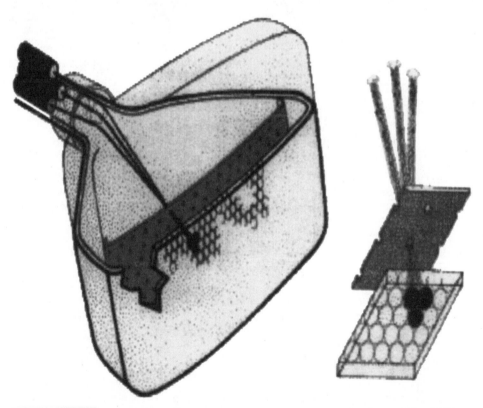

FIGURE 5-1   Illustration of how the electron beam scans the face of the CRT.

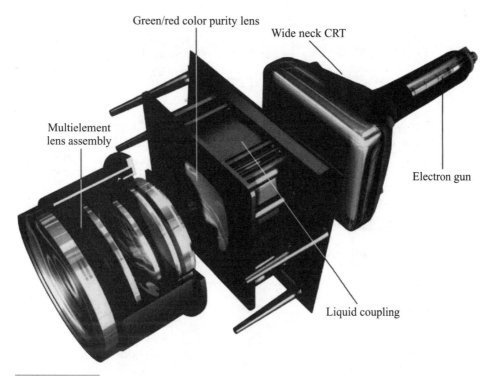

Green/red color purity lens

Wide neck CRT

Multielement
lens assembly

Electron gun

Liquid coupling

FIGURE 5-2   **Drawing of a CRT optical gun assembly used in a projection TV receiver.**

Let's now look at the benefits of the *plasma panel* for a high-definition monitor as used for TV viewing:

- Unlimited installation possibilities
- A flat, thin screen that is cool and convenient
- Can be easily hung on the wall
- Swing it out from the wall for viewing comfort
- Extremely high performance
- XGA level computer monitor
- Perfect picture through matrix technology
- Matrix style display has a perfect geometry

Figure 5-5 shows two views of HDTV plasma monitor wall installation arrangements. The features of a plasma HDTV monitor are as follows:

- Encased cell structures
- Black striping
- An automatic format converter
- 8× processing
- PureCinema

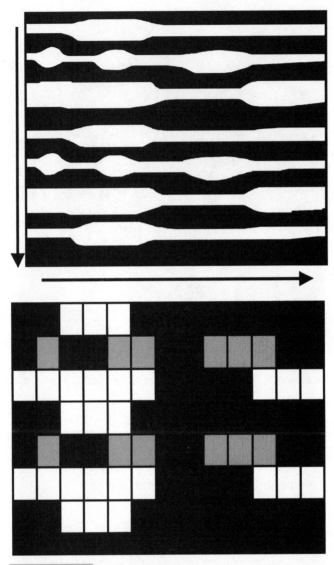

FIGURE 5-3   In the top drawing, a projection set's CRT is scanned continuously across columns on the screen. In the bottom drawing a fixed pixel (LCD flat screen) display will sample the image in both directions, thus response capabilities are much faster.

- HD progressive scan processing
- New pixel driving sequence
- New menu system
- More remote control features

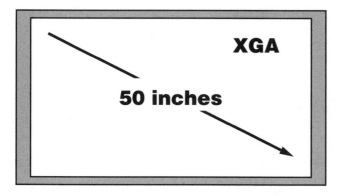

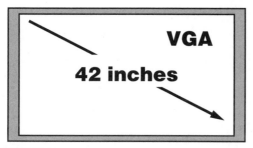

FIGURE 5-4   Comparison of the HGTV monitor VGA to the XGA with 4 × 3 resolution format.

Current Plasma Panel Technology

The current generation of plasma display panels uses a row-type configuration for the construction of the elements of each cell. Refer to Fig. 5-6, and you will note that each element of the pixel is individually illuminated. The current design problem with some types of plasma panels is that they suffer from light leakage from element to element within the vertical color column. This light leakage is illustrated in Fig. 5-7.

Newer model plasma monitors use a new encased cell structure that prevents light from leaking from cell to cell. This encased cell structure technique is illustrated in Fig. 5-8. Figure 5-9 is a close-up view of this encased structure. In addition, the encased cell structure has been able to increase the overall light output and efficiency. To obtain this added light output, the additional top and bottom walls are now coated with phosphor and also emit light.

The advanced HDTV plasma monitor screen uses a black stripe coating design technique. The *black stripe* coating helps produce deep solid blacks by absorbing external light and reducing light reflections. Producing black striping at XGA resolution requires extreme precision during plasma screen production. The black stripe coating is illustrated in Fig. 5-10.

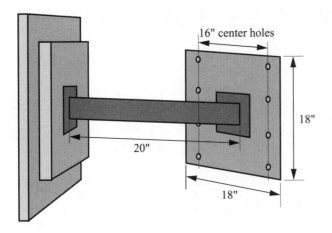

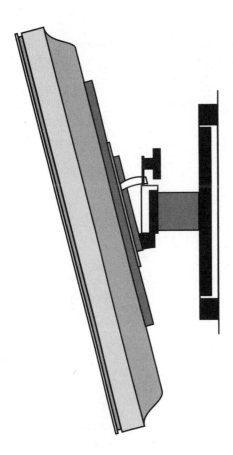

FIGURE 5-5 Two views of the HGTV plasma monitor.

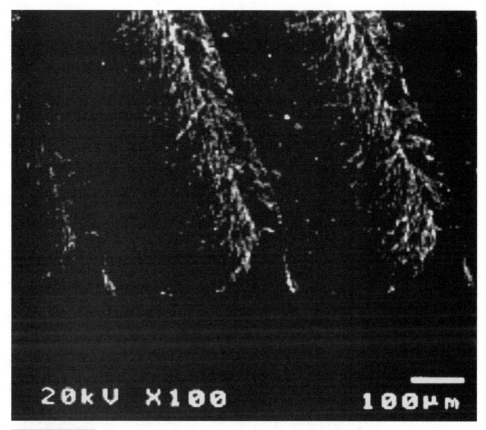

FIGURE 5-6   The current generation of plasma display panels.

PLASMA PANEL PROGRAMMING

Some models of flat plasma screens can be achieved in three different ways: by accessing the manual controls at the monitor, using the remote control hand unit, or using your PC as a controller. Now let's take a brief look at these adjustments.

PLASMA MONITOR ADJUSTMENTS

Generally, plasma units have three main operating modes: *normal,* which allows setting of the screen-size switching and full auto-zoom; *menu,* which is used for setting the picture quality and image positioning; and *integrator,* which mainly adjusts white balance (a mode that enables adjustment by using your PC). Although the plasma unit is preset at the factory, ambient conditions where the plasma panel is installed may require some fine-tuning changes. Should this adjustment process not go as planned, you may perform an initialization (reboot) and the system will return to the factory preset conditions.

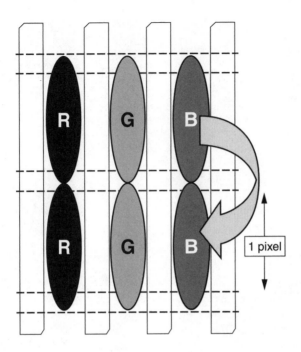

FIGURE 5-7 Current plasma panels suffer from light leakage that infiltrates from one element to another.

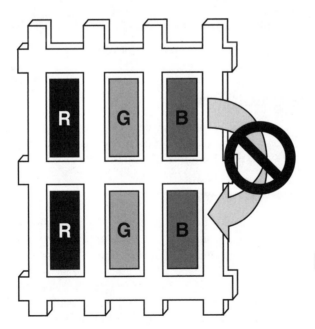

FIGURE 5-8 The improved encased cell structure prevents light from leaking from cell to cell.

FIGURE 5-9   Close-up view of an encased cell structure for more light output.

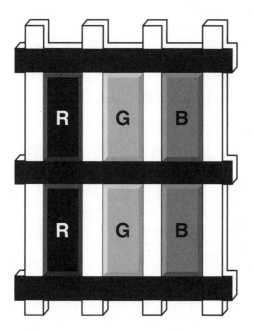

FIGURE 5-10   Black stripe coating absorbs external light and reduces light reflections.

HDTV DIGITAL VIDEO PROCESSING

Let's now see how video signal digital processing is used on some plasma HDTV systems. The automatic format converter (AFC) technique achieves 8 times the normal NTSC signal density for the ultimate in high picture quality from conventional video input sources. The original signal and 2× and 4× conversion formats are illustrated in Fig. 5-11. Figure 5-12 gives you more details on how AFC obtains the ultimate in high-quality pictures from conventional video sources. The illustrations in Fig. 5-13 give some comparisons of various brands of plasma screens for digital video processing.

PURECINEMA VIDEO PROCESSING

A Pioneer model features a PureCinema circuit that detects a film-based source and converts it to a progressive format with precise processing of the smoothest presentation. Note in Fig. 5-14 that each original still film frame is recreated and displayed alternately 2 or 3 times for incredibly pure images.

Another digital AFC technique is now used to convert 1080i HDTV signals to 1080p (the p stands for progressive) for more efficient processing and a sharper picture. Figure 5-15 illustrates this 2× signal processing conversion.

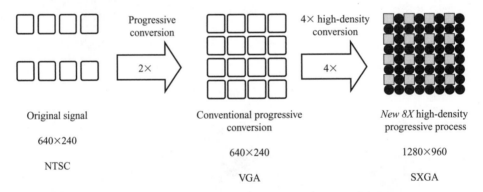

FIGURE 5-11 Automatic format converter (AFC) gives 8 times the normal NTSC signal density.

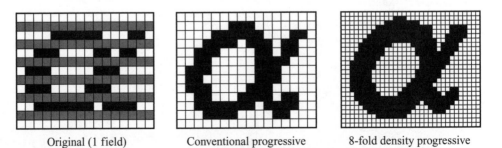

Original (1 field)   Conventional progressive   8-fold density progressive

FIGURE 5-12 When AFC circuitry is used, more picture detail is achieved.

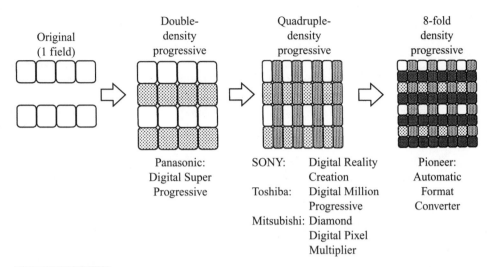

FIGURE 5-13 Comparison of various brands of plasma flat screen displays.

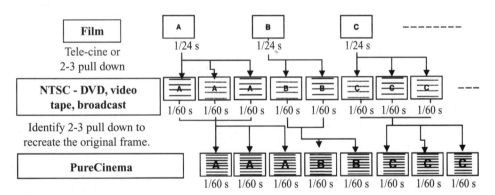

FIGURE 5-14 PureCinema circuit detects a film-based source and converts it to a progressive format.

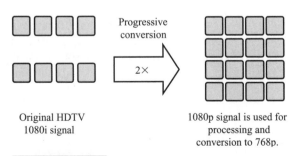

FIGURE 5-15 AFC converts 1080i HDTV signals to 1080p for improved video processing and a sharper picture.

TIPS FOR PLASMA PANEL INSTALLATION

When it comes time to install the plasma panel HDTV unit on a wall or ceiling, you must make some structural inspections. For a wall or ceiling mounting, there needs to be some sturdy frame or studding behind the panel mounting surface. Stud finders, bubble levels, and careful measurement calculations will take the guesswork out and make a professional looking installation.

These plasma panel monitors have variable-speed cooling fans located at the panel top. And there are many air vents located all around the sides of the panel frame. During installation of these units, make sure that these fans and vents have a large enough area clearance for good airflow.

Some plasma monitor screens may have a problem referred to as *pseudo-contour*. This is a pattern of striped shadows that may accompany a moving image that contains certain colors or different levels of brightness. In later plasma models, system designers used different video drivers and tweaked the various circuits and have kept this effect to a minimum.

PLASMA HDTV MAINTENANCE TIPS

Let's now take a brief look at some maintenance tips that should be performed on plasma monitors. Some of the same precautions that are performed for conventional CRT receivers can be used for the plasma display panels also. Still images affect the plasma screen the same way they affect a CRT screen, since phosphors are used in both applications. Blue and green phosphors degrade faster than red, so it is a good idea to adjust the white balance every 1000 hours of operation. Note that when certain plasma monitors (older models with screen savers) are programmed, the screen saver function is not usually activated.

The plasma display screen is usually coated with an antiglare material that can easily be damaged. When cleaning the screen surface, use caution while gently wiping the surface with a soft cloth.

Note:

Some types of cleaning solutions may discolor the monitor screen surface, or cause it to become opaque.

As noted previously, some plasma models require forced-air cooling because of a restricted air space of the slim cabinet enclosure. Dirt and dust should be removed from the vents to keep the internal temperature cool. For dirt and dust removal, you should use a low-suction vacuum cleaner with a soft brush attachment. Also, check and make sure all of the fans are operating properly.

Flat Panel LCD Displays

The liquid-crystal display (LCD) is now being used in many products for video viewing and has replaced the CRT in some products. The early model LCDs were produced on glass panels, but later versions are fabricated on quartz. This quartz process is akin to semiconductor electronics production. The CRT and LCD have generally been based on analog technology and have light-level modulation, which means they have analog signal limitations.

The LCD panel has been found in laptop computers for many years and is now being used in monitors. The LCD panel consists of components encased as a sandwich of glass plates. Between the glass plates are the liquid crystals (which contain tiny molecules that are influenced by magnetic fields) and the small chambers that control them. In a working matrix display, these cells will have a thin-film transistor (TFT) that will turn ON and OFF and produce an intermittent magnetic field.

The glass layer panel of an LCD has polarizing filters. Light from behind the plates shines through the filter and is twisted by each individual liquid crystal molecules. These molecules are twisted by the magnetic field, which influences them as illustrated in Fig. 5-16.

When these molecules are lined up with a polarizing filter, the molecule blocks out light, causing a dark spot. Now, when turned 90 degrees out-of-phase with the filter, the molecule will let light pass through. Thus, any angle in between will result in various brightness levels.

Later-model color LCDs, shown in Fig. 5-17 in a laptop computer, use multiple liquid-crystal elements to produce a variety of colors. Each pixel or dot of the screen now contains three or more elements that can be combined to produce color. These elements are each mounted against a color filter. The varying light intensities are then combined in each bunched group of elements and cause the viewer's eyes to see a single-color dot. As an example, 100 percent blue, 100 percent red, and no green elements will combine to produce a purple color dot or pixel.

Unlike CRT screens, which lines go across many times a second, each pixel in an active-matrix LCD screen can be turned off and on individually and at the same time. This is why LCD screens are flicker free and much easier on your eyesight. As an example, after using a laptop with an LCD panel for a while, it's almost like looking at a bright piece of paper.

The light weight and top quality of LCDs makes them very popular in many more devices than the laptop computer. As an example, LCD technology made it possible to have color hand-held computers (HPCs) and Palm Pilots, as shown in Fig. 5-18.

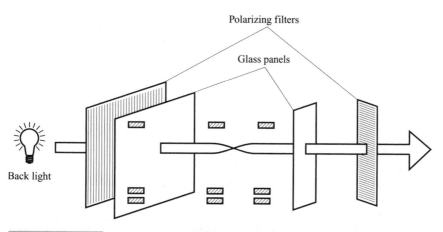

FIGURE 5-16   **Passing polarized light through liquid crystals makes it possible to control the intensity of light via a filter at the other end. By a magnetic field influence, liquid crystal molecules can be made to twist the light to varying degrees in and out of phase.**

FIGURE 5-17   This young lady is using a laptop computer, which has an LCD screen.

The viewable size of the LCD monitor is a little smaller than the CRT version. A 17-inch CRT monitor has about the same size viewable area as a 15-inch LCD monitor. Pricewise, a 17-inch CRT monitor costs approximately half as much as the LCD flat panel monitor. Thus, you have to consider the space-saving feature and other advantages to justify the cost differences. And the LCD panels are going down in price almost every month.

The type of interface used in the LCD display will also have a bearing on its cost. Some computers are equipped with an output that matches certain digital inputs on LCD monitors. With this type input you can send the digital signal from your PC's video card right into the monitor's display circuitry. With no analog conversion, the signal and picture are much cleaner and sharper. With no converter required in the flat screen LCD monitor, the cost is less than for an analog LCD monitor.

LCD PANEL VIEWING ANGLES

Most LCD panels today offer a pretty good viewing angle. This wide-angle design generally uses enhancements to its backlighting techniques and grooves within the polarizing filter located on the front of the LCD panel. Check on wide-angle enhancements and view

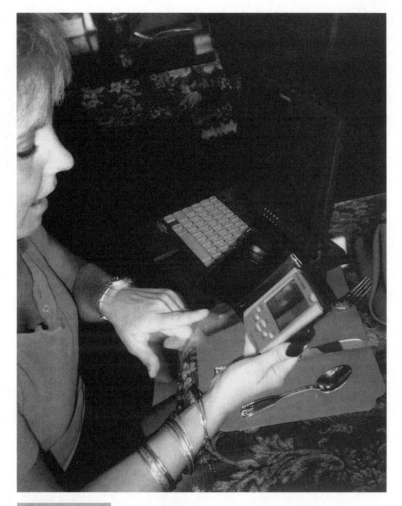

FIGURE 5-18   A Palm Pilot equipped with an LCD screen is shown in operation.

a demo model before purchasing the LCD monitor. Even though the LCD monitor is more expensive than the CRT unit, it is becoming a good option as the price drops and the technology improves.

Digital Chip TV Projection System

There is now available a bright projection TV using silicon-based digital technology combined with new materials and processes that allows the monolithic integration of an efficient digital light switch controlled by a digital address chip to produce a fast digital projection display. This technology, invented and developed by Texas Instruments, is called the digital micromirror device (DMD). The digital light processing (DLP) projection system based on the DMD has great image fidelity, inherent digital stability, and noise immunity. This unique DMD technology has several advantages over other projection systems used for home theater viewing. The digital advantage of DLP enables precise image quality with digital gray scale and color reproduction. The DLP system has no convergence or CRT burn-in problems to deal with. By nature, a CRT is an analog device and drifts with time, but DLP devices will not. Just about the only DLP upkeep is replacing a burnt-out light bulb.

The makeup of a DLP projection set is the DLP chip, signal processing at the input to convert all signals to the chip's own resolution, a light source, color filter system, cooling fan, and projection optics.

The DMD is a solid-state light switch with thousands of tiny mirrors, built on hinges on top of a static random access memory (SRAM). Each mirror is programmed to switch one pixel of the 1,280,720-pixel image. The hinges let the mirrors be tilted between two states; 10 degrees for ON and −10 degrees for OFF. These square mirrors on the DMD are 16 micrometers, separated by 1-micrometer gaps, giving a fill factor of up to 90 percent. This indicates that 90 percent of the pixel/mirror area can actively reflect light to create a projected image.

Because the DLP chip will accept only digital signals, analog signals must be converted to digital signals at the chip's input. An interlaced video signal has to be converted to progressive scan with interpolative processing. At this point, the signal goes via video processing and is converted to red, green, and blue (RGB) data. This progressive RGB data is then formatted into binary bit planes of data. With the video information now in digital format, it is then fed into the DMD. Now each pixel of data is mapped directly to its own mirror in a 1:1 ratio, giving precise, digital control. If the signal is 640 by 480 pixels, the central 640 × 480 mirrors on the DMD would be active. The other mirrors outside this area would be shut to the OFF position. For this reason, the video processor will upconvert the incoming signal to match the DLP ICs display resolution, which will then fill the complete screen.

By electronically "writing" the memory cell below each mirror with the binary bit plane signal, each mirror on the DMD array will be electrostatically tilted to either the ON or OFF position. How long each mirror tilts in either direction determined by pulsewidth modulation (PWM), a technique similar to that used in modern TV power supplies. These mirrors are capable of switching more than 1000 times per second. This rapid speed allows for digital gray scale and color reproduction. Illustrated in Fig. 5-19 is a view of a single DMD mirror and how it is hinged.

The DLP system is very like an optical projector, with light from the projection lamp passing through a condensing lens and a color filter system and focusing on the DMD device. With the mirrors in the ON position, they reflect light through the projection lens and onto the theater screen to produce a digital, square-pixel picture. The DLP system provides an accurate reproduction of the gray scale and color levels. Each of the video frames is generated by a digital, 8 to 10 bits per color gray scale; thus the exact digital picture can be re-created again and again.

With DLP, the human eye will see more visual information and perceives a higher-resolution picture. Most home theater systems feature several screen formats or sizes. And with DLP, there is no uneven CRT deterioration of brightness or picture quality or adjustment compensations.

Texas Instruments licenses its DLP technology to various TV receiver manufacturers and many other types of imaging devices; thus you will be seeing more of this type of display in the near future as the costs decrease.

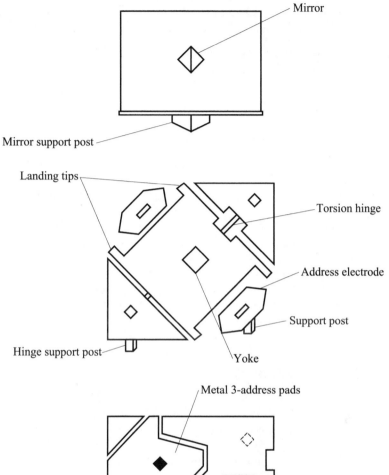

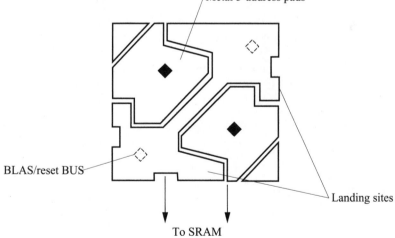

FIGURE 5-19 A drawing of a single DMD mirror and hinge arrangement.

Large-Screen Projection TV Systems

Projection TV receivers have special added components that are not found in a conventional direct view TV set. They are as follows:

1 Three CRT image tubes (red, green, and blue) that display the television picture on a screen.
2 The optics to project and magnify the image from the display tubes.
3 A screen upon which the magnified image is focused. The image can be projected on the front or rear of a screen, thus the terms *front-screen* and *rear-screen* projection TV sets.

Figure 5-20 illustrates a projection TV system in its most basic form. A small screen, direct-view television receiver is optically coupled to a viewing screen. The only change required in the normal receiver's electronics is a reversal of the vertical sweep. This sweep reversal is required because of the image inversion by the optics of the lens system.

Referring to Fig. 5-21, you will see a three-tube, in-line, front-screen projection system setup. For a rear-screen projection system the same technique is used, but there have to be added optical folds, and mirror bends to the optical path. For rear-screen projection, all CRTs and their respective optical axes must be perpendicular to the screen's vertical axis. For this reason, optical distortions of the vertical plane, such as vertical nonlinearity and keystone, are not generated.

THE BASIC TV PROJECTION SYSTEM

Most projection TV receivers (except the DLP chip system covered earlier in this chapter) use a projection system that is referred to as a *three-tube, in-line refractive system.* This in-line projection system consists of three direct-view display tubes—one red, one green, and one blue special projection tube.

The three display tubes are mounted in line on the horizontal axis. The red, green, and blue images on the three display tubes are then optically projected with lenses and mirrors

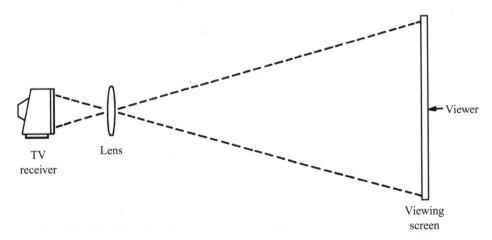

TV receiver

Lens

Viewer

Viewing screen

FIGURE 5-20   Basic rear screen projection of a video image.

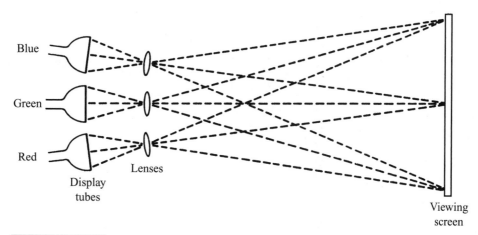

FIGURE 5-21 Drawing of a three-tube (CRT) in-line, front screen TV projection system.

to a viewing screen. This is shown in the simplified illustration of Fig. 5-22. At the viewing screen, all three projected tube images are properly registered and converged to produce the correct color picture rendition.

THE OPTICAL LIGHT PATH

The projection TV system is a combined electronic, optical, and mechanical system arrangement. The three individual electronically formed images are combined optically on the projection viewing screen. The original images are optically magnified, approximately 10 times, and aimed through two mirrors in a folded light path to the viewing screen.

The basic elements in the light path consist of a projection screen, an upper or second mirror, a lower or first mirror, projection lenses, and the red, green, and blue CRTs that form the three individual images.

THE PROJECTION LENS SYSTEM

Many of the production projection TV receivers use the U.S. Precision Lens (USPL) compact delta 7 lens. This lens, designed by USPL, incorporates a light-path fold, or bend, within the lens assembly. For a better understanding of the USPL CRT system optical compound assembly, refer to Fig. 5-23. The light path is established with a front mirror surface that has a bend angle of 72 degrees. Because of this light-path bend, the outward appearance of the lens resembles, somewhat, that of the upper section of a periscope. The lens elements and the mirror are mounted in a plastic housing. Optical focusing is accomplished by rotating a focus handle with wing nut locking provisions. Rotation of the focus handle changes the longitudinal position of the lens element.

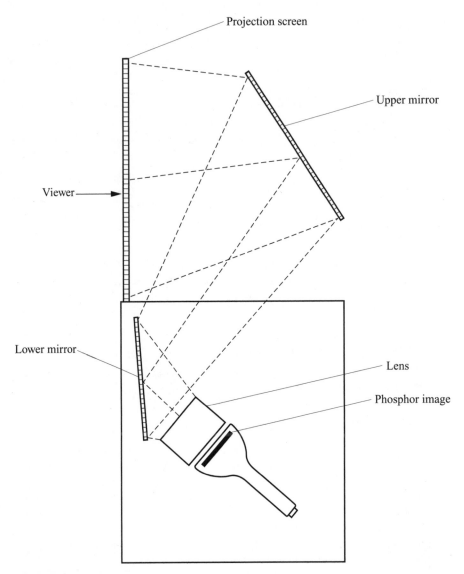

FIGURE 5-22   Simplified component placement of a rear screen projection TV receiver.

LIQUID-COOLED PROJECTION CRTS

The basic consumer TV projection sets use three CRTs (red, green, and blue) placed in a horizontal in-line configuration. There are two (red and blue) slant-face CRTs and one (green) straight-face CRT. The tubes are fitted with a metal jacket housing that has a clear glass window. The space between the clear glass window and the tube's faceplate is filled with a clear optical liquid. This liquid, which is insulated and self-contained, prevents

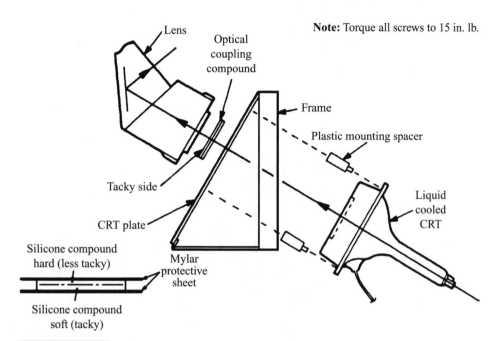

Note: Torque all screws to 15 in. lb.

Lens

Optical coupling compound

Frame

Plastic mounting spacer

Tacky side

Liquid cooled CRT

CRT plate

Silicone compound hard (less tacky)

Mylar protective sheet

Silicone compound soft (tacky)

FIGURE 5-23 Drawing of the U.S. Precision Lens (USPL) assembly with a light-path fold.

faceplate temperature rise and thermal gradient differentials from forming across it when under high-power drive signals. With these liquid-cooled tubes, the actual safe power driving level can almost be doubled compared to that of non-liquid-cooled CRTs. This technique increases the overall system's screen brightness, as the drive level wattage can be increased twofold.

SPECIAL PROJECTION SCREEN DETAILS

Most TV projection screens are constructed of a two-piece assembly. The front (viewing side) section will have a vertical lenticular black-striped section. The rear portion is a vertical, off-center Fresnel construction. The black striping not only improves initial contrast but also enhances picture brightness and quality for more viewing pleasure under typical room ambient conditions found in the home theater setting.

The Fresnel lens consists of many concentric rings, as shown in Fig. 5-24. Each ring is made to reflect light rays by the desired amount, resulting in a lens that can be formed into thin sheets.

If the surface of this sheet is divided into a large number of rings, each ring face may be flat and tilted at a slightly different angle. The resulting cross section of the lens resembles a series of trapezoids.

As you view the details of the Fresnel lens in Fig. 5-24, you will note that the lens is incorporated onto the back (projection side) of the set's TV projection screen.

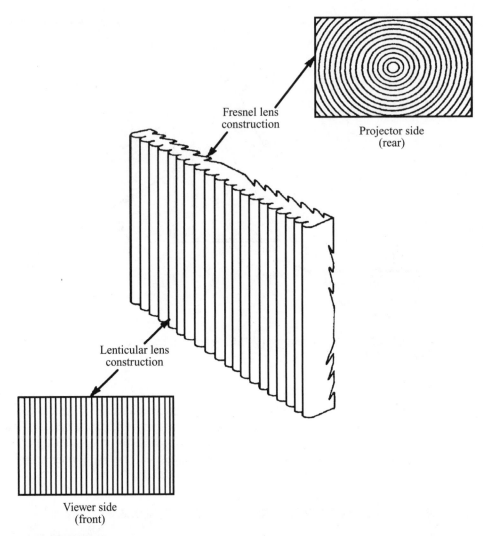

Fresnel lens
construction

Projector side
(rear)

Lenticular lens
construction

Viewer side
(front)

FIGURE 5-24   Fresnel lens on front of a projection TV receiver that illustrates its construction in detail.

PROJECTION SET DIGITAL CONVERGENCE

These large-screen projection receivers require a convergence circuit to compensate for mis-convergence caused by any difference of the red, green, and blue beam's mechanical align-ment. The digital convergence circuit can adjust the convergence accurately by generating a crosshatch pattern for adjusting and moving the cursor, displaying the points of beam adjustments.

Simplified digital convergence   A digital convergence circuit block diagram is shown in Fig. 5-25. A simplified operation of the digital convergence circuitry section is as follows:

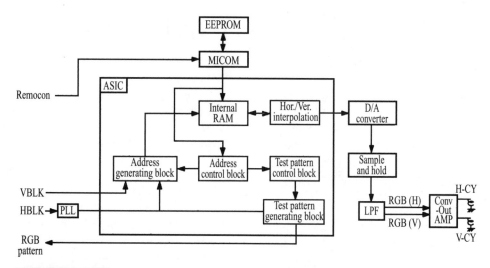

FIGURE 5-25 Block diagram of a digital convergence system that is used in some projection sets.

The EEPROM memory chip has the convergence data for all of the adjustment points. The average number of points for most big screen projection sets is 45.

The micron controls the convergence data to send from an electronically erasable and programmable read-only memory (EEPROM) to an application-specific integrated circuit (ASIC) when powering the set ON and OFF after adjustment. The PLL generates the main clock for the system by synchronizing to the horizontal blanking signal.

The address generating block generates the number (position) of scanning lines by synchronizing to the vertical and horizontal blanking (BLK) signals.

The horizontal/vertical interpolation block calculates convergence interpolation data of the actual scanning position in real time and then reconstructs it to fit the digital-to-analog (D/A) converter, and then sends it onto the D/A converter.

The test pattern control and test pattern generating blocks generate the test pattern and cursor during the convergence adjustment mode.

The D/A converter converts digital convergence adjustment data from the ASIC into analog data. It uses a 16-bit D/A converter circuit for this task.

The sample and hold block demultiplexes convergence data from the D/A converter into horizontal/vertical values. In addition, to avoid glitches caused by setting the time of the D/A converter, this block samples stabilized output from the D/A converter after a constant time frame.

The LPF block interpolates among adjusting points horizontally. This means that this block connects adjusting convergence points smoothly from the stair-like output data by a filtering process.

For the final convergence adjustment, there is a compensation of the magnetic field by a flowing amplitude convergence compensation waveform through coil CY generated by the successive operation that is used to compensate any misconvergence.

Digital Television (HDTV) System Overview

We will now give you the simplified overview of the basic digital TV/HDTV operational system. The digital TV format was developed by the Advanced Television Systems Committee (ATSC) for compatibility with the existing NTSC and future digital TV transmission endeavors. These digital (DTV) broadcasts occupy the same 6-MHz channels that have been used for the conventional NTSC system. However, instead of a single analog program, the digital system can provide a full range of programs and options. Now we will review this system operation and see how it all fits together.

You will find that the ATSC format provides the capability of broadcasting multiple, lower-resolution programs simultaneously, should the program material not be broadcast in HDTV. These multiple programs are transmitted on the same RF carrier channel used for only one HDTV program. This technique is referred to as *multicasting*. Compression technology allows the simultaneous broadcast of several digital channels. The present NTSC analog system is unable to do this. The standard-definition signal (DTV) will be noise-free, with quality similar to the picture quality you would view on a digital satellite system, and a much sharper picture than the present NTSC TV broadcasts.

With this digital technology, broadcasters can insert DTV programs with additional data. With these unused bandwidth slots, TV stations can deliver computer information or data directly to a computer or TV receiver. In addition to new services, digital broadcasting allows the TV station provider to have multiple channels of digital programming in different resolutions, while providing data, information, and/or interactive services.

HDTV PICTURE IMPROVEMENT

As stated before, HDTV broadcasts produce a much improved picture quality as compared to conventional TV broadcasts. Lines of resolution are increased from 525 interlaced to 720 lines, and up to 1080 lines. Also, the ratio between picture width and height increases to 16:9, as compared to NTSC's conventional aspect ratio of 4:3. Not only are picture quality and sharpness improved, but many new options are possible.

Digital television refers to any TV system that operates on a digital signal format. DTV is classified under two categories: HDTV and SDTV.

Standard-definition TV (SDTV) refers to DTV systems that operate off the 525-line interlaced or progressive sweep scan line format. This format will not produce as high a quality video as HDTV is capable of.

Another feature, in addition to the high-quality video picture that HDTV delivers, is the advanced sound system.

ANALOG/DIGITAL SET-TOP CONVERSION BOX

A converter, or set-top box, is used to receive and process many different signals including high-definition digital, standard digital, satellite digital, analog cable, and the conventional NTSC VHF/UHF TV station signals. For some future years TV stations will be transmitting analog signals (some will broadcast both analog and digital), since viewers will continue to use their analog TV receivers because of the cost of purchasing a new HDTV receiver.

These converter set-top boxes can decode an 8-level vestigial sideband (VSB) digital signal that is being transmitted by some TV stations. VSB is the digital system being used in the United States at this time.

The set-top converter box decodes the digital signal for a standard TV receiver; however, the picture quality will be improved only slightly. Without a high-resolution screen (with more scan lines etc.) to detect the digital signal and process it, the VSB-to-analog conversion is the only function the set-top box can perform.

Many of the HDTV receivers now have digital systems built into the units. More and more production TV receivers have these built-in digital features.

You will now find TV sets for sale that advertise the term HDTV ready, digital ready, or HD compatible. These terms do not indicate that the TV set can produce a digital signal, only that they have a jack available in which to plug in a set-top decoder. Most of these sets do, however, have an enhanced screen resolution.

HDTV VIDEO FORMATS

You will find several video formats available; however, the most common are those with 720 or 1080 lines of resolution. A majority of formats use either interlaced or progressive resolution and vary the number of frames per second. Cable and other sources have HDTV set-top boxes available that will read these various formats. And of course, there is equipment available to decode the complex audio signals.

OVER-THE-AIR TELEVISION SIGNALS

Local terrestrial HDTV broadcast transmission is accomplished on an 8-level vestigial sideband, or 8-VSB. It is derived from a 4-level AM VSB and then trellis coded into a scrambled B-level signal (cable will use an accelerated data rate of 16-VSB). A small pilot carrier is then added and placed in such a way that it will not interfere with other analog signals. A flow chart that illustrates these data stream events will be found in Fig. 5-26.

Digital satellite systems have been transmitting digital HDTV signals for several years. DirecTV and the Dish Network have several channels operational and plan more in the near future. Digital satellite systems thus have a head start on conventional TV stations for delivering high-definition programs.

THE COMPATIBILITY QUESTION

Consumers ask quite frequently if these new digital TV receivers will be compatible with the VCRs, camcorders, DVD players, and other electronic devices that they now have. In most cases the answer is yes. In almost all cases the equipment manufacturers are designing their electronic devices with composite video and analog inputs for their digital HDTV receivers.

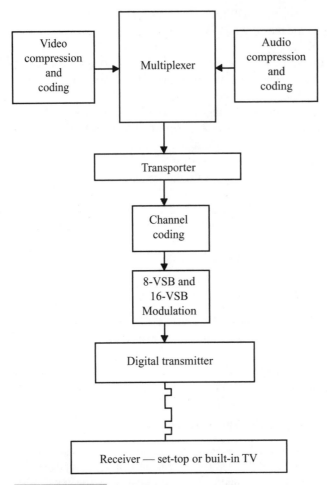

FIGURE 5-26 Simplified block diagram of a digital over-the-air TV broadcast system.

RECEIVING THE DIGITAL SIGNAL

Some of the first DTV programs will be transmitted on the commercial and public broadcast stations. And digital HDTV is also available on DirecTV and the Dish Network satellite systems. You will also be seeing more HDTV programs on cable as more companies convert to the wideband digital cable system.

Typically, an outdoor antenna will be required to receive HDTV program signals. Keep in mind that the reception characteristics of a digital signal as compared to an analog signal are quite different. A DTV receiver does not behave like a standard NTSC analog television receiver. When receiving an analog NTSC broadcast, as the signal strength decreases, the amount of noise (snow) in the picture increases. This noise may come and go, or the picture will stay full of snow or become blanked out. However, a digital broadcast signal will be completely noise-free until the signal is too low for the receiver to decode. Once the digital signal threshold is reached, the picture will either freeze, fall apart in blocks, or blank out.

The antenna needs to be positioned to receive the best average sum of all digital signals within your viewing area. In some cases, the HDTV viewer may need more than one antenna due to the varied locations of the transmitter towers. A signal strength indicator built into some HDTV receivers will help you to position the antenna for best reception.

The digital signal is transmitted on the same 6-MHz bandwidth that the conventional NTSC analog system uses. The DTV signals are also broadcast in the same spectrum or range of frequencies that the NTSC uses, which is primarily UHF. In most applications, the same antenna can be used for both HDTV and NTSC reception. However, some new antenna designs are currently being developed. These antennas will blend in with their surroundings and be less noticeable than the older rooftop antennas.

The new satellites that broadcast DTV signals are not the same satellites currently used for DirecTV or the Dish Network. However, the DTV satellite broadcasts will be close enough in specs to allow the same dish to receive DBS programs and the DTV service. However, a new dual dish and receiver is available for the dual-purpose applications. In addition to reception (antenna and dish), consideration must be given to signal distribution. Keep in mind that the signal does not become noisy as the DTV signal weakens. The signal levels and picture quality that you have been accustomed to with the analog system may have too much noise for proper operation of the digital system. Thus, the installation of a low-loss, high-quality signal distribution system may be required.

VARIOUS HDTV FORMATS

There are more than a dozen formats and possible standards for the transmission of digital television video. These cover the number of pixels per line, the number of lines per picture, the aspect ratio, the frame rate, and the scan type. Some of these formats, at this time, have not been put into practice, and not all of these formats qualify as high definition. However, this digital technology will result in a vast improvement of video and audio quantity and quality.

It's usually considered that the 1080p, 1080i, and 720p formats are high-definition formats. But keep in mind, the limitations in current TV receiver technology prevent these formats from being included in TV models now in the showrooms. However, with the advanced technology some models are now available with the high-resolution (1080p) formats. It is possible that the broadcast material may change, but the receivers will use the same digital processing to convert the various formats.

Future NTSC TV Reception

It's been predicted that the transition to digital TV will have a time frame of approximately 10 years. At some future time, there will be no analog TV stations on the air. When this point is reached, all analog spectrum space will be reallocated by the FCC to other radio services.

During this transition period, set-top converter boxes can be used to decode the digital signal and allow this output signal to display a picture on a conventional NTSC receiver. Of course, with this set-up there will be a decrease in picture resolution. Today's conventional TV set owners can continue to use their analog TV sets until the NTSC broadcast transmitters go off the air. When this time is reached, if you want to view TV video program channel you will need to purchase an HDTV set or converter box that changes the digital signal to an analog NTSC signal for your old TV receiver.

HDTV and NTSC Transmission Basics

TV picture resolution can be specified in pixels or lines. Resolution is the maximum number of transition periods (changes) possible on the screen in a horizontal and vertical direction. The maximum resolution that a CRT, or picture tube, can display, is determined by its specs when it is produced. The greater the amount of horizontal and vertical pixels, the greater the CRT's resolution capability. The resolution of a computer monitor screen is generally specified by the number of pixels it can display. This is listed in both horizontal and vertical directions, for example, 1920 horizontal and 1080 vertical. Pixels are also used to rate the resolution of the new ATSC and HDTV screen formats.

In broadcast television, the resolution of the studio camera that captures the video is what determines the highest resolution possible. The picture resolution produced by the camera, given in pixels, is very similar to that of a CRT screen. This is the current resolution limitation as the transition to high-definition digital TV takes place. With NTSC, the ability of an analog signal to quickly make a transition from low to high levels is comparable to a pixel channeling from black to white.

The number of lines transmitted in the current NTSC analog format is 525. This is considered standard-definition (SD) transmission. A standard-definition transmission of a 525-line NTSC signal can be transmitted in the analog or digital (ATSC) mode. A high-definition transmission can be transmitted only in the digital television mode.

Simplified HDTV Transmitter Operation

Many years ago, when I was a lad, the National Television Systems Committee created the analog television specifications and standards known as NTSC. The new digital standard, for HDTV/DTV, was developed by the Advanced Television System Committee (ATSC). The primary objective of ATSC was to develop a digital transmission format that would fit within a 6-MHz bandwidth. Another major goal in developing the ATSC format was to allow expansion and versatility in the transmission of additional content such as electronic program guide (EPG) information and digital data such as text content. Using this new digital transmission technique, a broadcast TV station can transmit multiple digital programs simultaneously within a 6-MHz bandwidth. However, in some situations, and in order to broadcast multiple digital programs within the allotted 6-MHz bandwidth, the maximum picture resolution may have to be compromised.

To better understand how a high-resolution digital picture is developed for transmission within a 6-MHz bandwidth, a simple overview of the digital encoder/transmitter (Fig. 5-27) should be useful to you. The HDTV transmitter block diagram consists of two parts. The packet generation section multiplexes compressed video and audio, along with additional services data, into a single digital bit stream. The vestigial sideband (VSB) transmission sec-

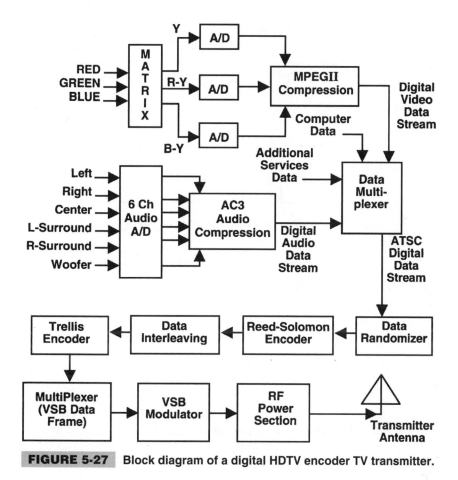

FIGURE 5-27   Block diagram of a digital HDTV encoder TV transmitter.

tion then scrambles this digital data to allow for error correction during decoding and reconstruction of the signal. The VSB transmission section also adds the data sync and transmits the data via the RF power amplifiers and antenna.

THE HDTV BASIC AUDIO SYSTEM

The HDTV digital audio system has built-in provisions to transmit six channels of high-fidelity audio for a full theater stereo surround sound experience. This digital audio system consists of the complete audio path, from the point where it enters the audio encoders at the HDTV transmitter, to the audio decoder output in the HDTV receiver.

Digital audio signal processing   In the digital (DTV) system, the audio portion is also transmitted digitally. Of course, the original audio sound is analog, and the human ear picks up the sound in an analog format. Thus, to make the complete HDTV system work, the audio portion needs some type of conversion process. This process is called analog-to-digital (A/D) conversion. The complimentary process in the receiver is referred to as digital-to-analog (D/A) conversion. At this time we will give you a simple explanation of the audio digital signal processing (DSP).

The digital signal processing converts analog signals to digital form for computer processing. Regardless of the type of analog signal, the basic blocks of the system are the same. For a better understanding of this system refer to the basic circuit blocks shown in Fig. 5-28.

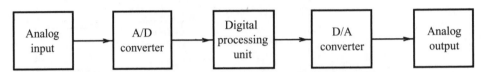

FIGURE 5-28   Block diagram for conversion of analog signals to digital signals for computer processing.

Digital audio processing   The degree of digital signal processing will vary from simple EQ operations to the complex audio operations used in HDTV devices. To start the process, the analog signal is first A/D converted. The A/D conversion has three subprocesses:

- Sampling
- Quantization
- Binary notation

The sampling process   The audio analog signal is sampled with a frequency that is approximately 2 times the maximum frequency found in the audio signal. This sampling technique produces a more accurate conversion of the audio signal.

Quantized binary sampling   For this audio signal process, the sampled signal is divided into certain number levels, and each level of binary is coded. The amplitude of the audio signal is quantized into eight levels, each corresponding to a 3-bit binary code between 000 and 111. If the maximum analog amplitude level is 7 volts, then eight voltage levels can be expressed in binary code (0 to 7). However, errors occur in quantization because number values between the whole numbers are considered as the numbers above or below them. As an example, 6.2 volts and 5.7 volts would be rounded off to 6 volts. Quantization errors are reduced by increasing the number of bits for quantization. In practice, a sampling frequency of 44.1 kHz is used. This frequency is just above 2×20 kHz, as 20 kHz is considered to be the theoretical highest frequency of the human hearing range. In addition, this frequency solves the problem of aliasing frequencies. The aliasing frequencies are those lower than the sampled frequency and are created when the sampling frequency is less than 2 times the highest frequency that is sampled. In the audio spectrum sampling, the aliasing frequencies are audible.

Audio signal coding   With each audio analog voltage level sampled and given a binary code, a serial data stream is formed. As many as 20 bits are used in a digital audio HDTV system to produce over 1 million levels. Parity bits are added to this data for error checking and correction. A *parity bit* is a binary digit that is added to an array of bits to make the sum of the bits always odd, or always even, to ensure accuracy. This process is referred to as *ECC encoding.*

The audio signal is finally modulated in a format called EFM (8- to 14-bit modulation). EFM is a system in which 8-bit data is converted to 14-bit data for the purpose of avoiding continuous ones and zeros in the data stream during the audio signal transmission.

The above information is a simplified explanation and does not necessarily represent the way an actual HDTV audio system operates. It was used for an easier-to-understand digital audio system concept. Actual HDTV audio systems will likely differ from this simplified version. You will find some audio data system processors that are divided into two blocks: one is the PES packetization, and the other is the transport packetization. Also, some of the functions of the transport subsystem may be included in the audio coder or the transmission subsystem.

Note that some HDTV audio systems may contain six audio channels dedicated to audio programming. These six channels, also referred to as 5.1 channels, are as follows:

1 Left channel
2 Center channel
3 Right channel
4 Left surround channel
5 Right surround channel
6 Low-frequency enhancement (LFE)

When certain conditions are required, the transport subsystem can actually transmit more than one of these elementary audio bit streams. Note that the bandwidth of the LFE channel is usually limited to the range of 3 to 120 kHz.

Using audio compression   Because of the huge amount of audio to be conveyed, and to keep the digital HDTV video signal within the 6-MHz TV channel bandwidth, the video signal must be compressed. And the same technique is also used for the audio system. The compression of the audio portion of the HDTV system is desired for two reasons:

1 To make the channel bandwidth efficient and able to carry more audio information
2 To reduce the bit memory and bandwidth space required to store the program material

The purpose of audio compression is to reproduce an audio signal faithfully, with as few bits as possible, while maintaining a high level of stereo sound quality.

Recovering digital audio   In the audio digital recovery process, the audio signal is demodulated to produce the ECC coded signal. Following a stage where the parity bits are removed, the digital signal representing the original audio signal is produced. This digital signal is then D/A converted and filtered to reproduce the original analog audio signal faithfully.

SOME HDTV QUESTIONS AND ANSWERS

Let's now review some HDTV questions that you, the TV viewer, may ask and the answers to them.

Q: How are various DTV and HDTV signals received?

A: In most locations you should be able to receive HDTV reception with a standard UHF outside antenna. Inside antennas are not very effective. The type or model of the antenna needed for best reception will vary due to your location and distance from the TV transmitting station towers. You should consult your electronics service center for the proper TV antenna needed for your situation and location.

Q: Will you be able to watch high-definition TV using a set-top box and a standard NTSC TV receiver?

A: No, not high definition, but a much better picture will be viewed. The video and sound will be improved on the equipment you now have. The decoder box will output HDTV broadcasts with Dolby digital audio, providing more precise localization of sounds and a more convincing, realistic ambiance. You may already have a multichannel, multispeaker audio system allowing you to take advantage of digital TV's enhanced sound quality.

  Most HDTV decoders will also provide three high-quality connections for monitors. Component video outputs will allow you to connect the box to most home theater LCD and DLP format projectors and direct-view sets with component inputs to provide optimum image quality. Many large-screen TVs can be connected via S video, which maintains high image quality by separating the luminance and chrominance signals. You can even connect this box to a standard VGA computer monitor, which provides a more crisp and detailed picture than the conventional NTSC TV receiver.

Q: Will I be able to view the new HDTV broadcasts on a conventional NTSC TV receiver?

A: You will be able to watch HDTV broadcasts by using a special HDTV decoder box device. These set-top boxes, which connect as easily as VCR or DVD players, will receive digital signals and convert all formats to standard NTSC reception. This will let you view the HDTV digital programming but not in actual sharp, clear HDTV format.

Q: Besides better resolution, audio performance, and data services are there any more reasons to purchase a new HDTV receiver?

A: A good reason to invest in an HDTV set is the way the signal is transmitted. Digital transmissions can deliver a near perfect signal—free of ghosts, interference, and picture (snow) noise.

Q: What is digital television? And what is the current status of high-definition HDTV? Are HDTV and DTV the same thing?

A: The FCC, its Advisory Committee on Advanced Television Service, and the Advanced Television Systems Committee, a consortium of companies, research labs, and standards organizations, have defined 18 different transmissions formats within the scope of what is broadly called the Digital Television Standard. DTV is the umbrella term for all 18 formats.

  Six of these are considered high definition because they constitute a significant improvement over the resolution quality of the current NTSC format. The current

NTSC TV system was established over 50 years ago. If you have not viewed an HDTV program yet, I am certain you will see a great improvement in image quality even with the other 12 formats because of the digital transmission concept. You will also reap benefits from the DTV formats such as wide-screen theater-type displays, enhanced audio quality, and new data services.

Q: When will HDTV direct-view sets, HDTV decoders, and home theater projection receivers become available?

A: Many models have been on the market for the past few years. As of this writing, early 2002, a good selection of all models is in the dealer showrooms at prices that are becoming lower monthly. The large flat screen HDTV panels are now becoming more affordable to the general TV customer, also. In the past year or so, many set manufacturers have prepared the set buyer in advance by marketing HDTV-ready projection TV receivers with their multiple high-quality video inputs and direct-view large-screen TVs with component video inputs.

Q: What is the scoop on digital signals from cable systems and satellite TV systems? Aren't some cable and satellite systems already transmitting digital signals at this time? Will digital HDTV sets display signals from these systems?

A: All of the above is correct. Some cable and satellite systems already use digital technology to transmit their TV programming. These systems require the viewer to use a converter box for that service. Many of the digital standards are not compatible with each other or the ATSC format.

Recap of the Digital TV and HDTV Systems

The HDTV (either 1080i interlaced or 720 progressive scan) and SDTV (480 interlaced) will offer the exciting experience of clearer, more detailed digital video and audio than today's NTSC signal. The TV broadcasters can choose the type and number of signals they transmit within their allotted bandwidth (6 MHz) and transmission rate (19.3 Mbps). In the future, as new DTV products are developed with advanced features, the broadcasters will be rolling out new services. A few of these possibilities are as follows:

■ Up to four SDTV programs broadcast from one TV station simultaneously, where you currently only receive one. These pictures will be clearer than NTSC, free from interference like snow and ghosts, but will not have as much resolution as an HDTV picture.

■ On-screen data, such as educational material, or team statistics presented during a game in progress.

■ Pay-per-view movies and premium channels, as on present satellite channels and cable TV systems. Access to Web sites related to the program you are viewing.

■ Home shopping and purchasing using your remote control to make your choices or ask for more details. And much, much more.

6

DIRECT BROADCAST SATELLITE (DBS) SYSTEM OPERATION

Introduction to Satellite TV

At this time, several direct digital satellite TV systems are in operation around the world. This chapter shows how these systems work and gives you information on various items you can check if the receiver and dish do not work. You might also want to obtain another of my books from "McGraw-Hill" that has complete instructions for installing one of these 18 inch direct TV dishes and various troubleshooting information.

FIGURE 6-1   **The satellite dish is shown mounted on a mast below a conventional TV antenna .**

Introduction to Satellite TV

These TV satellites or "birds," as they are often called, revolve around the earth at over 22,000 miles in a geosynchronous orbit which makes it appear that they are not moving. These TV satellites pick up signals with their receivers and then send the video signals via onboard high-power 120-watt transmitters back down to earth in a pattern that covers all of the 48 main land states. The signal is strong enough to be picked up with a small 18-inch dish that is shown in Fig. 6-1. These TV satellites operate like an amateur radio repeater.

In the geosynchronous orbit, the satellite is placed over the equator at approximately 22,300 miles above the earth. A satellite in this type of orbit will not wander north or south and will have an earth-day rotation. This satellite in the sky will appear to stand still in a fixed position because its speed and direction matches that of the earth's rotation.

The uplink transmitter station pointing at the satellite in a geostationary orbit, and the downlink to your dish will not require tracking equipment because the earth's rotation matches that of the satellite.

KEEPING THE SATELLITE ON TRACK

Because the earth's gravitational pull is not the same at all places as the satellite rotates around it and the moon also affects its position in space, the satellite is always being pulled off course and must be corrected.

Position and attitude controls are used to counter these gravity pulls and keep the satellite in its proper slot. These adjustments are accomplished by on-board rocket thrusters that are fired to obtain course corrections. In fact the lifespan of the satellite is determined by how fast these thrusters use up the fuel for stabilization. Once the fuel is used up, the satellite will wander off course and become unusable.

In the early days of satellites, the spacing between them was four degrees. Now, with much improved antenna directivity, the satellites can be placed at 2-degree spacing.

POWERING THE SATELLITES

Because the satellite is not a passive device, it has to have the ability to collect and store electrical energy.

Solar cells are used to power the DBS satellites, but there are times when the satellite is in darkness. At these times, nickel-hydrogen batteries are used and then they are recharged by the solar panels. Over the years, the solar panels are hit by particles in space and the batteries lose efficiency, which is the main reason that the satellite becomes inoperative. The DBS satellite transmits compressed digital video signals, which produces very high-definition picture quality.

DBS Satellite Overview

All communication services, from military, police, radio and television, and even communications satellites are assigned special bands of frequencies in a certain electromagnetic spectrum in which they are to operate.

To receive signals from the earth and relay them back again, satellites use very high frequency radio waves that operate in the microwave frequency bands. These are referred to as the *C band* or *KU band*. C-band satellites generally transmit in the frequency band of 3.7 to 4.2 Gigahertz (GHz), and is called the *Fixed Satellite Service band (FSS)*. However, these are the same frequencies occupied by ground-based point-to-point communications, making C-band satellite reception more susceptible to various types of interference.

The KU-band satellites are classified into two groups. The first include the low- and medium-power KU-band satellites, transmitting signals in the 11.7- to 12.2-GHz FSS band. And the new high-power KU-band satellites that transmit in the 12.2-GHz to 12.7-GHz Direct Broadcast Satellite service (DBS) band.

Unlike the C-band satellites, these newer KU-band DBS satellites have exclusive rights to the frequencies they use, and therefore have no microwave interference problems. The RCA system receives programming from high-power KU-band satellites operating in the DBS band.

The C-band satellites are spaced closed together at locations of 2 degrees. The high-power KU-band DBS satellites are spaced at 9 degrees, with a transmitter power of 120 watts or more.

Because of their lower frequency and transmitting power, C-band satellites require a larger receiving dish, anywhere from 6 to 10 feet in diameter. These whopper platters are at times referred to as "BUDs" or "Big Ugly Dish." The higher power of KU-band satellites enables them to broadcast to a compact 18-inch diameter dish.

How the Satellite System Works

A satellite system is comprised of three basic elements:

1 An uplink facility, which beams programming signal to satellites orbiting over the earth's equator at more than 22,000 miles.

2 A satellite that receives the signals and retransmits them back down to earth.

3 A receiving station, which includes the satellite dish. An RCA satellite receiver is shown in Fig. 6-2.

The picture and sound data information originating from a studio or broadcast facility is first sent to an uplink site, where it is processed and combined with other signals for transmission on microwave frequencies. Next, a large uplink dish concentrates these outgoing microwave signals and beams them up to a satellite located 22,247 miles above the equator. The satellite's receiving antenna captures the incoming signals and sends them to a receiver for further processing. These signals, which contain the original picture and sound

FIGURE 6-2   Front view of the DBS satellite receiver.

information, are converted to another group of microwave frequencies, then sent up to an amplifier for transmission back to earth. This complete receiver/transmitter is called a *transponder*. The outgoing signals from the transponder are then reflected off a transmitting antenna, which focuses the microwaves into a beam of energy that is directed toward the earth. A satellite dish on the ground collects the microwave energy containing the original picture and sound information, and focuses that energy into a *low-noise block converter (LNB)*. The LNB amplifies and converts the microwave signals to yet another lower group of frequencies that can be sent via conventional coaxial cable to a satellite receiver-decoder inside your home. The receiver tunes each of the individual transponders and converts the original picture and sound information into video and audio signals that can be viewed and listened to on your conventional television receiver and stereo system.

HOW THE RCA SYSTEM WORKS

The RCA DSS system is a DBS system. The complete system transports digital data, video, and audio to your home via high-powered KU-band satellites. The program provider beams its program information to an uplink site, where the signal is digitally encoded. The uplink site compresses the video and audio, encrypts the video and formats the information into data "packets." The signal is transmitted to DBS satellites orbiting thousands of miles above the equator at 101 degrees West longitude. The signal is then relayed back to earth and decoded by your DSS receiver system. The DSS receiver is connected to your phone line and communicates with the subscription service computer providing billing information on pay-per-view movies, etc. Figure 6-3 illustrates the overall operation of the DSS satellite system.

Now, here's a technical overview at how the total DSS system transports the digital signals from the ground stations via satellites into your home.

Ground station uplink The program provider sends its program material to the uplink site, where the signal is then digitally encoded. The "uplink" is the portion of the signal transmitted from the earth to the satellite. The uplink site compresses the video and audio, encrypts the video, and formats the information into data "packets" that are then transmitted with large dishes up to the satellite. After this signal is received by the satellite, it is relayed back to earth and received by a small dish and decoded by your receiver.

MPEG2 video compression The video and audio signals are transmitted as digital signals, instead of conventional analog signals. The amount of data required to code all of the video and audio information would require a transfer rate well into the hundreds of *Mbps (megabits per second)*. This would be too large and impractical a data rate to be processed in a cost-effective way with current equipment. To minimize the data-transfer rate, the data is compressed using *MPEG2 (Moving Picture Experts Group)*, a specification for transportation of moving images over communication data networks. Fundamentally, the system is based on the principle that images contain a lot of redundancy from one frame of video to another as the background stays the same for many frames at a time. Compression is accomplished by predicting motion that occurs from one frame of video to another and transmitting motion data and background information. By coding only the motion and background difference, instead of the entire frame of video information, the effective video data rate can be reduced

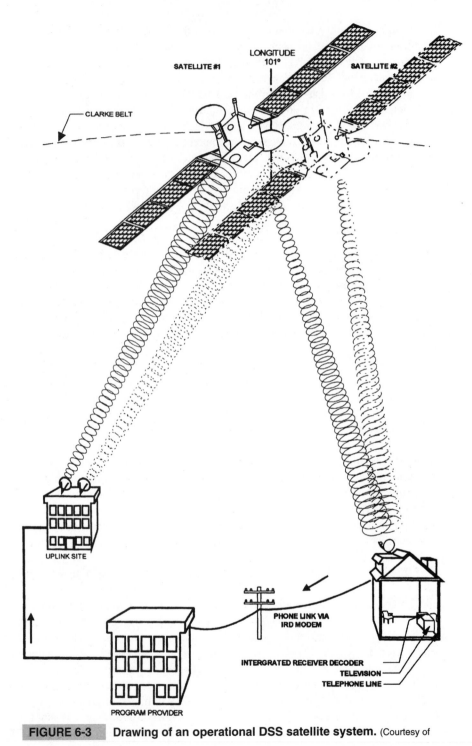

FIGURE 6-3   **Drawing of an operational DSS satellite system.** (Courtesy of
Thomson Multimedia.)

from hundreds of Mbps to an average of 3 to 6 Mbps. This data rate is dynamic and will change, depending on the amount of motion occurring in the video picture.

In addition to MPEG video compression, MPEG audio compression is also used to reduce the audio data rate. Audio compression is accomplished by eliminating soft sounds that are near the loud sounds in the frequency domain. The compressed audio data rate can vary from 56 *Kbs (kilobits per second)* on mono signals to 384 Kbps on stereo signals.

Data encryption   To prevent unauthorized signal reception, the video signal is encrypted (scrambled) at the uplink site. A secure encryption "algorithm" or formula, known as the *Digital Encryption Standard (DES)* is used to encode the video information. The keys for decoding the data are transmitted in the data packets. Your customer Access Card decrypts the keys, which allows your receiver to decode the data. When an Access Card is activated in a receiver for the first time, the serial number of the receiver is encoded on the Access Card. This prevents the Access Card from activating any other receiver, except the one in which it was initially authorized. The receiver will not function when the Access Card has been removed. At various times, the encryption will be changed and new cards will be issued to you to protect any unauthorized viewing.

Digital data packets   The video program information is completely digital and is transmitted in data "packets." This concept is very similar to data transferred by a computer over a modem. The five types of data packets used are Video, Audio, CA, PC compatible serial data, and Program Guide. The video and audio packets contain the visual and audio information of the program. The CA (Conditional Access) packet contains information that is addressed to each individual receiver. This includes customer E-Mail, Access Card activation information, and which channels the receiver is authorized to decode. PC compatible serial data packets can contain any form of data the program provider wants to transmit, such as stock market reports or software. The Program Guide maps the channel numbers to transponders and also gives you TV program listing information.

Figure 6-4 shows a typical uplink block diagram for one transponder. In the past, a single transponder was used for a single satellite channel. With digital signals, more than one satellite channel can be sent out over the same transponder. Figure 6-4 illustrates how one transponder handles three video channels, five stereo audio channels (one for each video channel plus two extra for other services, such as second language), and a PC-compatible data channel. Audio and video signals from the program provider are encoded and converted to data packets. The configurations can vary, depending on the type of programming to be put on stream. The data packets are then multiplexed into serial data and sent to the transmitter.

Each data packet contains 147 bytes. The first two bytes (remember, a byte consists of 8 bits) of information contained in the SCID (Service Channel ID). The SCID is a unique 12-bit number from 0 to 4095 that uniquely identifies the packet's data channel. The Flags consist of 4-bit numbers, used primarily to control whether or not the packet is encrypted and which key to use. The third byte of information is made up of a 4-bit Packet-Type indicator and a 4-bit Continuity Counter. The Packet Type identifies the packet as one of four data types. When combined with the SCID, the Packet Type determines how the packet is to be used. The Continuity Counter increments once for each Packet Type and SCID. The next 127 bytes of information consists of the "payload" data, which is the actual usable information sent from the program provider. The complete Data Packet is illustrated in Fig. 6-5.

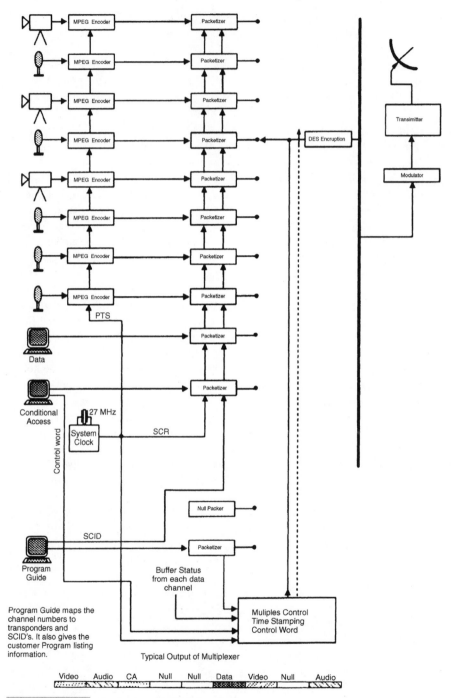

FIGURE 6-4   **Typical uplink DBS configuration.** (Courtesy of Thomson Multimedia.)

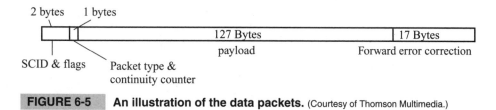

2 bytes   1 bytes

| | | 127 Bytes | 17 Bytes |

SCID & flags   Packet type &
                    continuity counter

payload   Forward error correction

FIGURE 6-5   **An illustration of the data packets.** (Courtesy of Thomson Multimedia.)

Right hand circularly polarized wave   Left hand circularly polarized wave

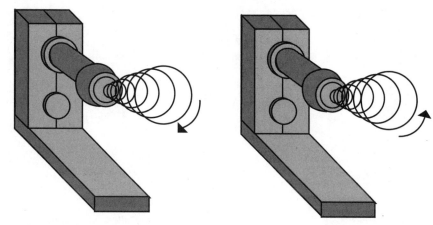

FIGURE 6-6   **The left- and right-hand circularly polarized signal transmitted from the satellite.** (Courtesy of Thomson Multimedia.)

THE DIRECTV SATELLITES

Two high-power KU-band satellites provide the DBS signal for the receiver. The satellites are located in a geostationary orbit in the Clarke belt, more than 22,000 miles above the equator. They are positioned less than $\frac{1}{2}$ degrees apart from each other with the center between them at 101 degrees West Longitude. This permits a fixed antenna to be pointed at the 101-degree slot and you are able to receive signals from both satellites. The downlink frequency is in the K4 part of the KU-band at a bandwidth of 12.2 GHz to 12.7 GHz. The total transponder channel frequency bandwidth is 24 MHz per channel, with the channel spacing at 14.58 MHz. Each satellite has 16 different 120-watt transponders. The satellites are designed to have a life expectancy of 12 or more years.

Unlike C-band satellites that use horizontal and vertical polarization, the DBS satellites use circular polarization. The microwave energy is transmitted in a spiral-like pattern. The direction of rotation determines the type of circular polarization (Fig. 6-6). In the DBS system, one satellite is configured for only right-hand circular-polarized transponders and the other one is configured for only left-hand circular polarized transponders. This results in a total of 32 transponders between the two satellites.

Although each satellite has only 16 transponders, the channel capabilities are far greater. Using data compression and multiplexing, the two satellites working together have the possibility of carrying over 150 conventional (non-HDTV) audio and video channels via 32 transponders.

FIGURE 6-7   **A roof-mounted DBS dish being installed and adjusted.**

Dish operation   The "dish" is an 18-inch, slightly oval-shaped KU-band antenna. The slight oval shape is caused by the 22.5-degree offset feed of the LNB (Low Noise Block converter), which is depicted in Fig. 6-7. The offset feed positions the LNB out of the way so that it does not block any surface area of the dish, preventing attenuation of the incoming microwave signal. Figure 6-7 shows the DBS dish being installed on a roof.

Low-noise block (LNB)   The LNB converts the 12.2-GHz to 12.7-GHz downlink signal from the satellites to the 950-MHz signal required by the receiver. Two types of LNBs are available: dual and single output. The single-output LNB has only one RF connector, but the dual-output LNB has two (Fig. 6-8). The dual-output LNB can be used to feed a second receiver or other form of distribution system. Figure 6-9 illustrates how the signal path is received from the satellite. The basic package comes with a single-output LNB. The deluxe receiver system has the dual-output LNB installed in the dish.

   Both types of LNBs can receive both left and right-hand polarized signals. Polarization is selected electrically with a dc voltage fed onto the center connector and shield of the coax cable from the receiver. The right-hand polarization is selected with +13 volts while the lefthand polarization mode is selected with +17 Vdc. If you suspect coax or connector trouble, you can check for this dc voltage at the dish and at the antenna terminal on the back of the receiver. If you have proper dc voltage at the receiver antenna connection, but very low or no voltage at the dish coax connection, the cable is bad and needs to be replaced. Use a volt/ohmmeter for this check.

Receiver circuit operation   The DBS receiver is a very complex digital signal processor. The amount of speed and data that the receiver processes rivals even the fastest personal computers (PCs) on the market at this time. The information received from the satellite is a digital signal that is decoded and digitally processed. No analog signals are found, except for those exiting the NTSC video encoder and the audio *DAC (digital-to-analog converter)*. A block diagram of the DBS receiver is shown in Fig. 6-10.

The downlink signal from the satellite is downconverted from the 12.7- to 12.2-GHz range to the 950- to 1450-MHz range by the LNB converter. The tuner then isolates a single digitally modulated 24-MHz transponder. The demodulator converts the modulated data to a digital data stream.

The data is encoded at the transmitter site by a process that enables the decoder to reassemble the data and verify and correct errors that might have occurred during the transmission. This process is called *forward error correction (FEC)*. The error-corrected data is output to the transport IC via an 8-bit parallel interface.

Single output LNB   Dual output LNB

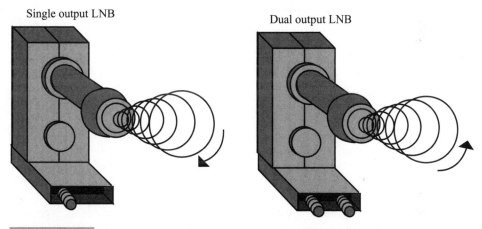

FIGURE 6-8   **The single- and dual-output LNB located on the dish.** (Courtesy of Thomson Multimedia.)

FIGURE 6-9   **An illustration of how the satellite signal is received by the dish antenna.** (Courtesy of Thomson Multimedia.)

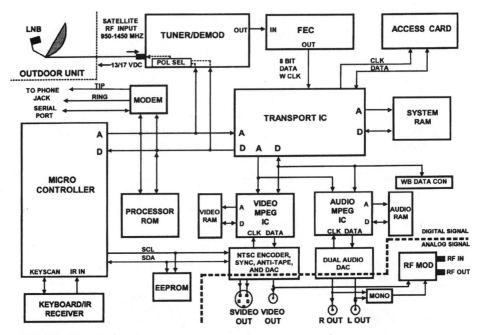

FIGURE 6-10   **A block diagram of the DBS receiver.** (Courtesy of Thomson Multimedia.)

The transport IC is the heart of the receiver data-processing circuitry. Data from the FEC block is processed by the transport IC and sent to respective audio and video decoders. The microprocessor communicates with the audio and video decoders through the transport IC. The Access-Card interface is also processed through the transport IC and is used to turn on or validate the receiver.

The Access Card receives the encrypted keys for decoding a scrambled channel from the transport IC. The Access Card decrypts the keys and stores them in a register in the transport IC. The transport IC uses the keys to decode the data. The Access Card also handles the tracking and billing for these services.

Video data is processed by the MPEG video decoder. This IC decodes the compressed video data and sends it to the NTSC encoder. The encoder converts the digital video information into NTSC analog video that is then made available to the S-Video and the standard composite video output jacks.

Audio data is also decoded by the MPEG audio decoder. The decoded 16-bit stereo audio data is sent to the dual DAC, where the left and right audio-channel data are separated and converted back into stereo analog audio. The audio is the fed to the left and right audio jacks and is also mixed together to provide a mono audio source for the RF converter.

The microprocessor receives and decodes IR remote commands and front-panel keyboard commands. Its program software is contained in the processor ROM (Read Only Memory). The microprocessor controls the other digital devices of the receiver via the address and data lines. It is responsible for turning on the green LED on the on/off button.

The receiver modem The modem in the receiver connects to your phone line and calls the program provider and transmits the pay-per-view programs purchased and reports them for billing purposes. The modem operates at 1200 bps and is controlled by the microprocessor.

When the modem first attempts to dial, it sends the first number as touch-tone. If the dial tone continues after the first number, the modems switches to pulse dialing and redials the entire number. If the dial tone stops after the first number, the modem continues to dial the rest of the number as a touch-tone number. The modem also automatically releases the phone line if you pick up another phone on the same home extension.

Diagnostic test menus The DBS receiver contains two diagnostic test menus. The first test is a customer-controlled menu that checks the signal, tuning, phone connections and the access card. The second test menu is servicer controlled. It checks out the majority of the receiver for problems.

Customer-controlled diagnostics The customer controlled test helps you, the customer, during installation or any time the receiver appears not to function properly.

- *Signal test* Checks the value of error bit number and the error rate to determine if the antenna connections are properly installed.
- *Tuning test* Checks to ensure that a transponder can be tuned. The test is considered successful and this part of the test is halted if proper tuning occurs on 1 of the 32 transponders.
- *Phone test* The phone test checks for dial tone and performs an internal loopback test. In Fig. 6-11, the system test indicates a phone connection problem. Some checks you can make is to plug a working phone into the phone line plugged into the back of the DBS receiver. If the phone works OK, the receiver modem could be defective. You will need to take it to a repair station for service. If the test phone does not work, check all plug-in connections or replace the phone cable to the unit.

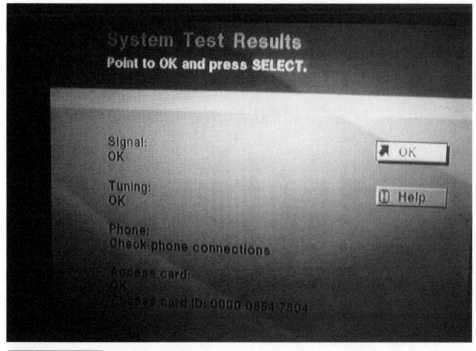

FIGURE 6-11   How the system test results appear on your TV receiver screen.

■ *Access card test* This test sends a message to the Access Card and checks for a valid reply.

The response for all tests will be an "OK" display or an appropriate message informing you of the general nature of the problem.

To enter the system test feature:

■ Select "Options" from the "DBS Main Menu." See Fig. 6-12.
■ Next, select "Setup" from the "Options" menu (Fig. 6-13).
■ Now select "System Test" from the "Setup" menu. Your TV screen will appear as in Fig. 6-14. Next, select "Test" from the System Test menu (Fig. 6-15).

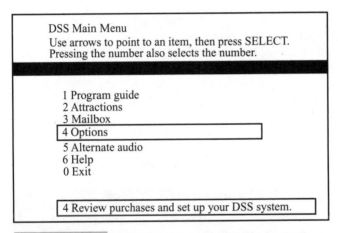

FIGURE 6-12   **A DBS main menu that is now in the options mode.** (Courtesy of Thomson Multimedia.)

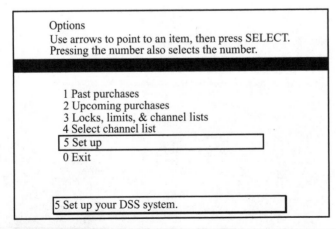

FIGURE 6-13   **The options menu in the setup mode.**

(Courtesy of Thomson Multimedia.)

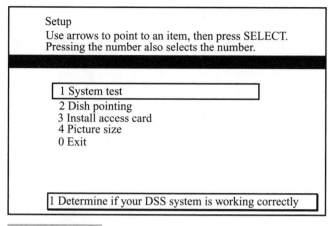

FIGURE 6-14   **The setup menu for the system test.**
(Courtesy of Thomson Multimedia.)

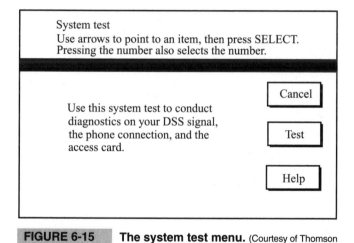

FIGURE 6-15   **The system test menu.** (Courtesy of Thomson
Multimedia.)

The system test results are displayed automatically when the test is completed. The following two screens (Fig. 6-16) show whether the receiver passed or failed the test. If the Access Card passes the test, the Access Card ID number will be displayed in the window of the menu.

Controlled Diagnostics for Troubleshooting

The servicer-controlled test provides a more in-depth analysis of the receiver and overall system operation for proper operation. The test pattern checks all possible connections between components as a troubleshooting aid. The following information is provided for system diagnostics.

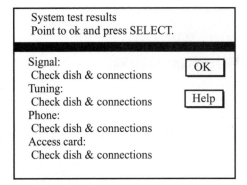

FIGURE 6-16   **The menu for the system test results.** (Courtesy of Thomson Multimedia.)

1 IRD (receiver) serial number
2 Demodulator vendor and version number
3 Signal strength
4 ROM checksum results
5 SRAM test results
6 VRAM test results
7 Telco (phone) callback results
8 Verifier Version
9 Access Card Test and Serial Number
10 IRD ROM version
11 EEPROM test results

The response for all of these tests will indicate the test was or was not successful.

In addition, this menu will allow entry into the phone prefix menu so that the installer can set up a one-digit phone prefix.

SERVICE TEST

To enter the service test mode feature of the DBS system, use the front-panel buttons of the receiver, not the remote control unit. For service test, simultaneously press the front-panel "TV/DBS" and the "Down" arrow buttons. The following screen menus will come up on your TV (Fig. 6-17).

The test results are automatically displayed after the test is completed. You or the service technician are given the option to exit the test or run the diagnosis again.

Front-panel control buttons Also included in the Service Test Menu are provisions for testing the modem and setting a single-digit phone number. During the service test, the modem will dial the phone number that appears in the boxes at the top of the test menu. The phone number can be changed by using the "Down" arrow keys on the remote control or receiver to move the cursor past the "Prefix" prompt to the number boxes. Once the boxes are selected, the number can be entered or changed with the number keys on the remote or by us-

ing the "Up"/"Down" keys on the remote or the receiver. The prefix can be changed by selecting the "Phone Prefix" on the display and changing the number with the number keys on
the remote control or by using the arrow keys on the remote-control hand unit or the receiver
front panel. The receiver front-panel control buttons are shown in Fig. 6-18.

Pointing the dish When you are installing your dish, you have to consider where to locate
the dish so as not to have any trees or buildings blocking the signal from the satellite. Figure
6-19 shows the dish pointing from sky-high view. You first have to determine the satellite's
position in the sky. You determine the side to side (azimuth) and the up/down (elevation)
bearings from your location to the satellite. These change with different locations across the
United States. For example, the azimuth and elevation for Minneapolis are different from
those in Houston, Texas. These changes are caused by the satellite's position in the geostationary orbit. Also, the azimuth of the dish changes as you move either east or west.

A World View of the DSS System

One way to better understand the DBS system is to look at the different parts of the system—from the studio down to the DSS receiver and the remote-control unit that you are
using in your home. Refer to the "bird's eye" view of the DSS system (Fig. 6-20).

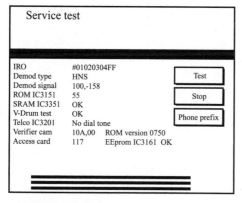 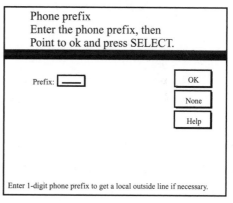

FIGURE 6-17   **The service test mode.** (Courtesy of Thomson Multimedia.)

FIGURE 6-18   **DBS receiver front-panel control button
locations.** (Courtesy of Thomson Multimedia.)

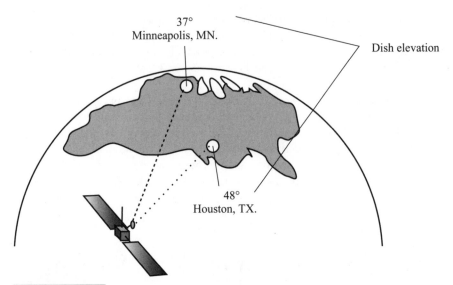

FIGURE 6-19   **An earth view of the DBS satellite reception elevations.**

(Courtesy of Thomson Multimedia.)

- *Uplink control center*   This building houses the equipment that transmits the programming via very large dishes up to the orbiting satellites.
- *Satellite*   The satellite relays the programming signals back to your satellite dish. The satellite is parked above the equator, in a geostationary orbit 22,300 miles above the earth.
- *Dish antenna*   The small dish receives the satellite signals. Because the satellite is so powerful, the dish only needs to have a diameter of 18 inches.
- *Program provider authorization center*   This center processes your billing statements. Your system is linked to the customer service center through the phone jack located on the back of your receiver.
- *DSS system home view*   Figure 6-21 illustrates the parts of the system required inside your home and satellite dish outside the house.
- *Receiver*   The receiver receives the TV digital program information and sends it to your TV for viewing or to a VCR for recording.

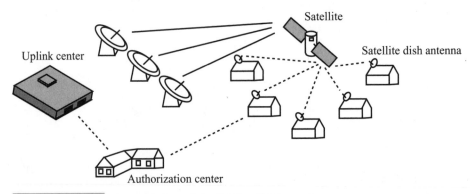

FIGURE 6-20   **A world view of the DBS satellite system.** (Courtesy of Thomson Multimedia.)

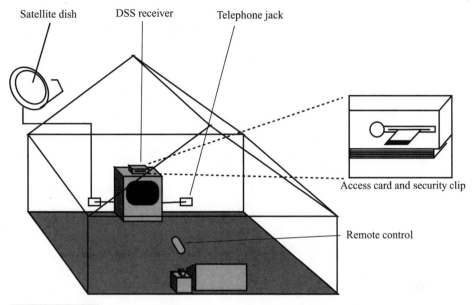

Satellite dish   DSS receiver   Telephone jack

Access card and security clip

Remote control

FIGURE 6-21   **A typical DBS home installation setup plan.** (Courtesy of Thomson Multimedia.)

- *Telephone jack*   A cable from this jack connects to back of the receiver and is plugged into the wall phone jack. The receiver uses a toll-free number once per month to update your Access Card. This update only takes a few seconds and ensures that you will have continuous service. The system automatically hangs up if you pick up the phone when the receiver is calling out the update information.
- *Television*   If your TV is remote controllable, you can program the Universal TV remote to change channels and the volume level.
- *TV universal remote*   The Universal remote is included with the receiver. This unit not only controls the system, but most other remote controllable TVs. Just point the remote at the set you want to control.
- *Access card*   Each receiver must have an Access Card. The card must be inserted before you can use the system. The card provides system security and authorization of DSS services. Do not remove the card, except when issued a new card as a replacement for the original.
- *Security card clip*   The clip is installed (Fig. 6-21). This clip fits over the Access Card and helps prevent the card from being inadvertently removed. To remove the clip, squeeze the top and bottom together and slide the clip off of the Access Card.

FRONT-PANEL RECEIVER CONTROLS

The following is information on front-panel receiver operation and functions.

- *On/off/message control*   This control turns the receiver on and off. When the receiver is turned off, a flashing light indicates that a message has been sent by the customer service center. The receiver never actually has the power turned off, but is put into a Standby mode.

- *TV/DSS switch*   This button switches the "OUT TO TV" connection from DSS programming to the normal TV antenna or cable input. This is similar to the TV/VCR button on many VCRs.
- *Arrow keys*   These keys allow you to move around the program guide and menu to make your selections. Use these arrows to point before selecting an item on the menu. When you are not in the program guide or the menu system, the up/down arrows can be used to change channels.
- *Menu button*   This button brings up the menu on the screen of your TV for program selections.
- *Select/display button*   Push this button to select an item you have pointed to when using the program guide or menus. Also brings up a channel marker showing the time, channel, and other program details when you are viewing a program or previewing a coming attraction.

The front-panel receiver buttons can be used to control the receiver when the remote control is not close by. See Fig. 6-22 for these front panel control locations.

- *Access Card slot*   Insert the Access Card in the receiver with the arrow face up and pointing toward the unit. The receiver is shipped with the Access Card inserted into the slot. Do not remove the Access Card, except to install a new card issued as a replacement for the original card.

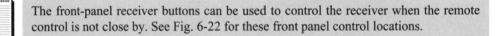

Do not stack electronic components or other objects on top of the receiver. The slots on top of the receiver must be left uncovered to allow for proper airflow circulation to the unit. Blocking airflow to the receiver could degrade performance or cause damage to your receiver or to other components.

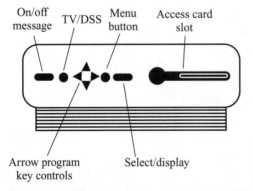

FIGURE 6-22   Callout locations of a typical DBS receiver's front control panel. (Courtesy of Thomson Multimedia.)

Connecting the Receiver

The following four drawings give you some examples of some hookups that are generally used to connect the receiver to a TV or other components. You can also refer to your TV and VCR owner's manuals for more specific information in regards to connecting your own components.

CONNECTION A

This hookup (Fig. 6-23) provides the best possible picture and stereo audio quality. To use connection A, you must have:

- TV with S-Video input, plus separate RF and audio/video inputs (jacks).
- VCR with RF input and output.
- S-Video, coaxial, and audio/video cables.

CONNECTION B

This hookup (Fig. 6-24) provides good picture and stereo audio quality. To use connection B, you must have:

- TV with separate RF and audio/video inputs (jacks).
- VCR with RF input and output.
- Coaxial and audio/video cables.

CONNECTION C

This hookup (Fig. 6-25) provides good picture and mono audio quality. To use connection C, you must have:

- TV with RF input (jack).
- VCR with RF input and output.
- Coaxial cables.

CONNECTION D

These connections (Fig. 6-26) provide a good picture and mono audio quality. To use connection D, you must have the following items:

- TV with an RF input (jack).
- Coaxial cables for the connections.

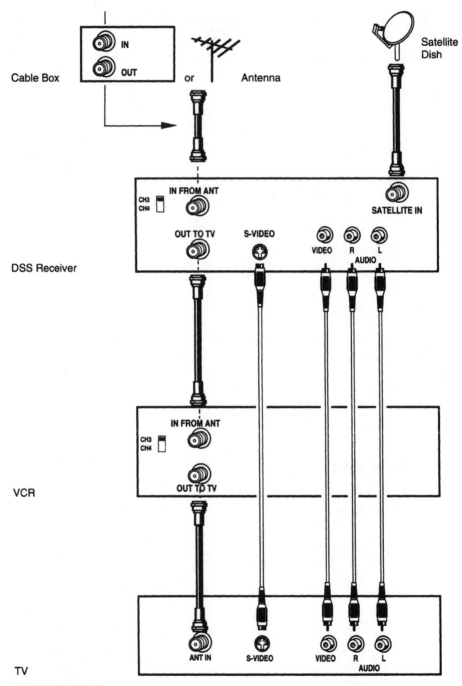

FIGURE 6-23   **Cable connections for the "A" hook-up of a DBS system.**

(Courtesy of Thomson Multimedia.)

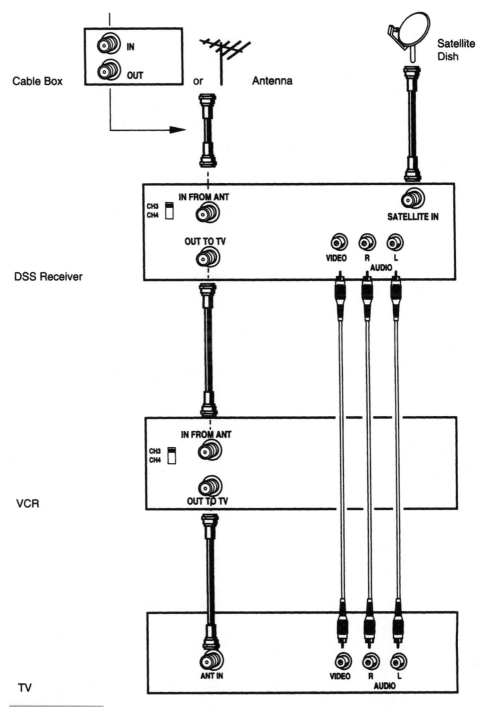

FIGURE 6-24 The "B" connection hook-up for the DBS receiver. (Courtesy of Thomson Multimedia.)

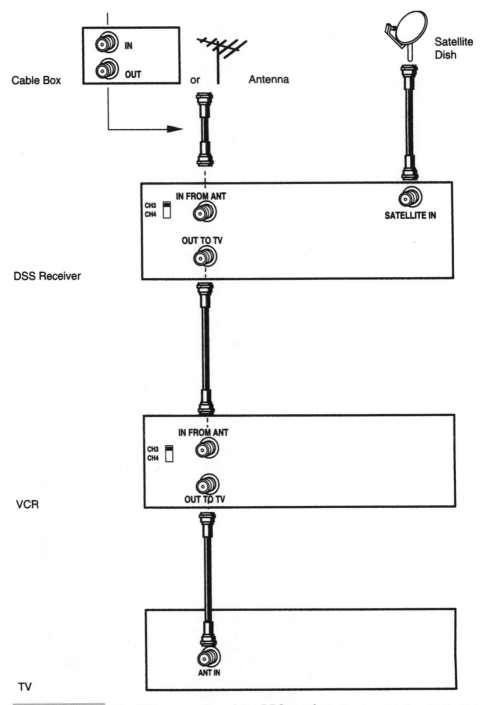

FIGURE 6-25   The "C" connection of the DBS receiver. (Courtesy of Thomson Multimedia.)

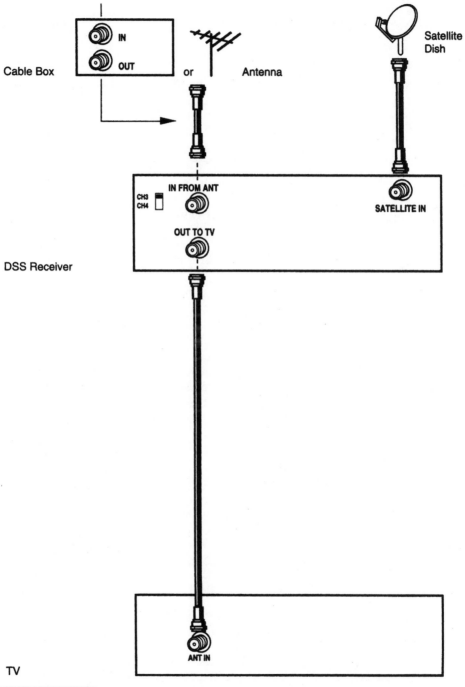

FIGURE 6-26   The DBS receiver and cable box "D" hook-up connections.

(Courtesy of Thomson Multimedia.)

Readjusting and Fine-Tuning the Dish Position

If you find the received signal is weak, it's possible the dish has moved and will need readjustment. You can do this by using the receiver's alignment system to fine-tune the position of the dish.

For this fine-tuning we will be using the TV receiver's on-screen display. Refer to Fig. 6-27 and you will see the difference between the early DSS and later-model DSS screens. The numeric display range is from 0 to 100. One hundred is a strong signal while zero is a very weak signal. The bar display extends across the screen—the stronger the signal, the farther the bar reaches across the screen. When you are aligning the dish, both the tone and on-screen display can be used to peak the dish's position.

The receiver's signal strength screen uses two systems to help you to fine-tune the position of the dish. The first of these is an audio tone. To listen to this tone, connect a TV set, headphones, or amplifier to the audio jacks on the rear of the receiver. There are two methods of audio depending upon the system you may have. With the early DSS models, when the dish is not pointing at the satellite and not receiving the signal, you will hear short bursts of tone. When the signal is being received, you will hear a continuous tone. Later DSS models use a pulsed low-frequency tone when not receiving a signal. As soon as signal is captured, it changes to a continuous higher-frequency tone. As the signal strength increases, the tone changes to a continuous higher frequency. If the signal weakens, the audio tone decreases in frequency. Refer to Fig. 6-28 for these tone explanations.

When the signal strength screen is activated, the receiver uses a search routine to lock the tuner to a transponder on the satellite. This search routine requires several seconds. The transponder that the receiver searches for is displayed on the alignment screen and is set at the factory. If this transponder is not active, another one can be manually entered via the signal strength menu.

The early DSS units used a search routine. When aligning the dish, move the dish only after the receiver is finished with one complete search routine. Both the on-screen display and audio tone signal the end of a search routine. The video display will flash "Rotate the

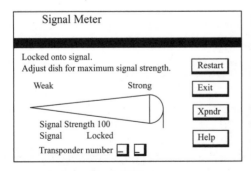

Early DSS

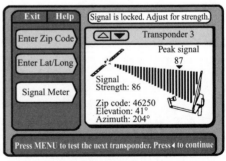

Later DSS

FIGURE 6-27   **The signal strength menu as it appears on the TV screen.**

(Courtesy of Thomson Multimedia.)

TONE BURST	RECEIVER TUNING	CONTINUOUS TONE		REC TU
NOT RECEIVING SIGNAL		**Early DSS**	SIGNAL RECEIVED	

INTERMITTENT LOW TONE	RECEIVER TUNING	STEADY HIGHER TONE	STEADY HIGHEST TONE	STEADY LOWER TONE	STEADY HIGHEST TONE
NOT RECEIVING SIGNAL		SIGNAL RECEIVED	OPTIMUM SIGNAL	WEAKER SIGNAL	OPTIMUM SIGNAL

Later DSS

FIGURE 6-28 What you can expect when aligning the dish with the audio tone.

dish 3 degrees" every other search routine. This ensures that the receiver has had the chance to complete one complete search operation before the dish is moved again. The receiver outputs a tone burst at the end of every search routine. When aligning the dish, move it every other tone burst. This ensures that the receiver has completed one search routine before the dish is moved. Keep in mind that once the DSS signal is received, the audio tone switches to a continuous tone.

Although later DSS receivers use a continuously updated display, it still requires some time to capture the transponder signal. Care should be taken to rotate the dish very slowly to avoid "skipping" through the signal.

VIDEO DISPLAY DISH ALIGNMENT

To use the video display to align the dish, a TV set must be positioned so that you can view it. You may be able to look through a door or window or run an RF cable from the OUT TV connector on the satellite receiver to a portable TV set. You will have to be able to see the TV set from the dish as you make these adjustments. The displays on early- and later-model DSS screens are different, but they give the same information. Refer to Fig. 6-29 to note these differences.

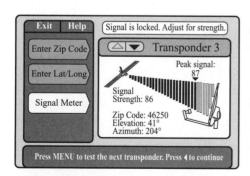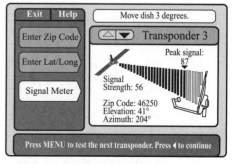

FIGURE 6-29 How the later-model DBS system's signal strength meter appears on the TV screen.

Aligning the dish with the video display

1 Position a portable TV in view of the DSS dish.
2 Turn the TV set on and tune to channel 3 or 4.
3 Very carefully turn the dish left or right on a tick mark of the alignment tape and pause. See Fig. 6-30.
4 Watch the signal strength display shown on the TV screen while the receiver goes through one complete cycle. Once the "Rotate the dish" display appears, move the dish another tick mark of the alignment tape and wait for one complete tuning cycle.

Note:

The tuning cycle on later model DSS IRD's is much faster, but the dish should still be moved very slowly. Moving the dish more than 3 degrees at a time may result in passing the correct position without acquiring a signal. The alignment tape on the mast of the DSS dish is marked in 3 degree increments. When you rotate the dish one tick mark, you are moving the dish 3 degrees.

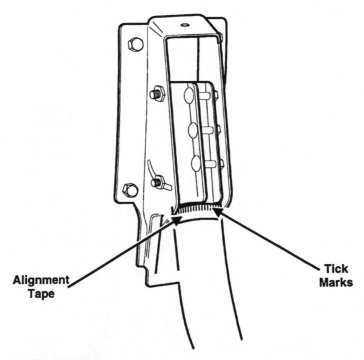

Alignment Tape

Tick Marks

FIGURE 6-30   How you adjust the azimuth with the alignment tape markings.

5 Continue this sequence and stop at the signal strength display's largest number or the longest bar display. If you turn the dish all the way to one direction without locking onto a good signal, move back to the starting point and repeat the same process in the opposite direction. If you do not receive a signal, then some troubleshooting may be required. Refer to "Some Possible DBS System Problems and Solutions" later in this chapter.

6 Once the azimuth is aligned, fine-tune the elevation adjustments. Start this adjustment by loosening the elevation bolts.

7 While watching the TV screen display, move the dish up or down 3 degrees at a time. Once the elevation has been moved 3 degrees, wait for the "Rotate" display to appear before moving the dish again.

8 Continue this sequence until the largest number or the longest bar display appears on the signal strength display. Then, tighten both of the azimuth and elevation bolts to secure the dish into a locked-down position.

ALIGNING DISH WITH AN AUDIO TONE

1 To listen to the audio tone, use a TV set, stereo unit, or wireless headphones. Connect the satellite receiver's audio output connector to the device you are using.

2 Now turn the dish either right or left one tick mark of the alignment tape and pause. If the receiver outputs a continuous tone, mark that position on the mast. If the receiver continues to emit tone bursts, wait for one complete tuning cycle (listen for the audio tone to beep twice) and move the dish (in the same direction) for 3 more degrees.

Note:

Later-model DSS IRD units use intermittent low-frequency tones when there is no signal. Once the signal is captured, it changes to a continuous low frequency. Continue this sequence until a continuous tone is heard, mark that position on the mast. If you move the dish up to 30 degrees in one direction, return the dish to the starting point and begin this same sequence again, moving the dish in the opposite direction.

3 Once a continuous tone is heard and the mast marked, continue rotating the dish 3 degrees (in the same direction) until the tone switches back to the burst mode. Mark the mast at that point.

Note:

It is possible that a continuous tone is received immediately after you enter the signal strength menu. If so, move the dish in 3-degree increments until the tone switches back to the burst mode (lower frequency tone for later model DSS). Mark that position on the mast and return the dish position to the starting point. Now, rotate the dish in the opposite direction in the 3-degree increments. Stop when the tone switches back to the burst mode (lower-frequency tone for later DSS) and mark that position. Once this is completed, continue to step 4.

Note:

For later-model DSS units, once the mast has been marked, rotate the dish back toward the original starting point (in the opposite direction). The frequency should immediately switch to the higher tone. Continue to rotate the dish until the frequency again lowers. Mark the mast at that point. This marks the range of acceptable signal strength of the dish.

4 Physically center the dish between the two marks. This should be the optimum position of the dish. Tighten the bolts securing the LNB support arm to the mast.

5 Next, fine-tune the elevation adjustment. To do this, record the current position of the dish, then loosen the bolts securing the elevation adjustment.

6 While listening to the audio tone, move the dish up in 3-degree increments. Stop moving the dish when the tone switches to the burst mode (lower-frequency tone for later-model DSS IRDs). Mark that position on the elevation scale.

7 Now lower the dish to the starting point.

8 While listening to the tone, lower the dish in 3-degree increments. Stop lowering the dish when the tone switches to the burst mode (lower-frequency tone for later model DSS IRDs). Mark that position on the elevation scale. Refer to drawing in Fig. 6-31.

9 Center the elevation of the dish between the upper and lower marks. This is the optimum elevation for the dish. Tighten the bolts securing the elevation adjustments. Your dish should now be set for maximum signal strength reception.

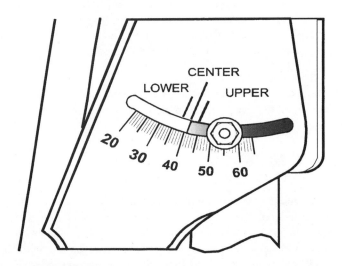

FIGURE 6-31   **Illustration of how to make elevation adjustments with an audio tone.**

Some Possible DBS System Problems and Solutions

You might lose your DBS signal if the dish becomes covered in snow (Fig. 6-32) or frozen with ice. Clean the snow off or melt the ice. Also, some dishes have a built-in de-icer for cold weather conditions and some heater kits can also be installed. The ice or snow on the dish cuts the microwave signal down very low and the TV will display a "hunting for signal" message. Also, recheck the position of the dish because it might have been turned a bit from ice and windy winter storm conditions.

Other loss-of-signal conditions or intermittent receiver operation could be due to loose cable connections or defective coax cables. Check the cables for an open center lead or shield or a short between the cable center lead and the outside shield. This can be checked with a very inexpensive ohmmeter. Refer to Chapter 1 for how to use the ohm and voltmeters. Be sure that all cable connections are crimped tight. Sometimes the center copper

FIGURE 6-32   Snow or ice buildup on the dish can prevent DBS signal reception. A dish heater is available that will melt any ice or snow if you live in a northern clime.

connector lead wire will bend or will be too short to make a good contact. As explained earlier, just follow the TV screen troubleshooting menu.

For remote-control problems, check out or replace the batteries. Be sure that the battery contacts are clean and tight. If you see any green looking corrosion on the battery or the battery contacts, clean them so that they are clean and bright. Should the remote unit become wet or get dropped in water, remove the batteries and wipe dry as soon as possible. Then use a hair dryer to completely dry it out. If you can remove the case, you can dry it much easier with the hair dryer. After the remote is dried out, replace the batteries. There's a good chance that the remote will operate.

DBS Glossary

Access Card   Identifies the DBS service providers and is required for your DSS system to work. Do not remove the Access Card, except when a new card has been issued to replace the original one.

Alternate audio   Refers to the different audio channels that can be broadcast in conjunction with a video program. A foreign-language translation is an example.

Attractions   Previews of special programs broadcast by your program provider.

Azimuth   Refers to the left-to-right positioning of your dish antenna. When you enter your zip code (or latitude and longitude), the display screen provides the number corresponding to an azimuth setting for your location.

Channel limit   Allows you to select which channels that can be viewed when the system is locked.

Receiver   Receives, processes, and converts the digitally compressed satellite signals into audio and video.

Elevation   Refers to the up and down positioning of your DBS dish. When you enter your zip code (or latitude and longitude), the display screen provides the number corresponding to the elevation setting for your location.

Key   The user-defined four-digit password that allows you to limit access to certain features of your DBS system.

Limits   There are three kinds of limits. The *Ratings Limit* allows you to control program viewing of rated programs by ratings level. The *Spending Limit* controls spending on a cost-per-program basis. The *Channel Limit* allows you to select which channel can be viewed when the system is locked.

Locks   The locks are a means of restricting access to certain features of the DBS system. The lock is controlled by a four-digit key that acts as a password. The closed or open lock icon in the channel marker indicates whether your system is locked or is unlocked.

Mailbox   Stores incoming electronic messages sent to you by your program providers. The mailbox is accessed through the on-screen menu system, and can store as many as 10 messages of 40 characters each.

Main menu   This is the first list of choices in the on-screen menu system. Press the Menu button on the remote-control unit or buttons on the receiver front panel to bring up the Main menu.

Some information in this chapter is courtesy of Thomson Consumer Electronics Company (RCA).

HOW VIDEO CAMERAS
AND CAMCORDERS WORK

Camcorder Features and Selections

Now, a large selection of video camcorders are available. The older, large camcorder (Fig. 7-1) takes the standard up to 6-hour recording tape. These are still very popular. The much lighter weight and easier-to-carry models are the 8-mm and VHS-C camcorders. The VHS-C small tape can be put into an adapter and played back on a standard VHS machine. The standard VHS and VHS-C camcorders are easier to use for playback.

If you have a 8-mm VCR, then you would want an 8-mm camcorder to play back your tapes. One advantage of an 8-mm camcorder (Fig. 7-2) is that you can place two hours of recording on one 8-mm tape. With a VHS-C cassette camcorder, you have only 30 minutes (or up to 90 minutes with a much poorer quality).

FIGURE 7-1   A full-size VHS cassette Zenith camcorder.

FIGURE 7-2   An 8-mm dual-battery system camcorder.

The JVC model GR-AX900 is VHS-C and it has the following features. It does nice time-lapse video, creates still shots and can then mix these videos into programs for editing functions when copying a tape to your VCR. The Sony model CCD-TRV40 has stabilization and a large zoom lens. The Sony also has a built-in speaker and a 3-inch LCD color monitor screen. This is great for reviewing the videos you just recorded and it can be used as a large viewfinder for recording.

There is also the S-VHS-C, a high-end format comparable to Hi8 cassette format. At this writing it appears that the 8-mm cassette camcorder will be the most popular along with digital recorders.

Unless you have an 8-mm VCR, you will have to use an 8-mm camcorder as a VCR; you will need to hook up a cable to your TV and standard VCR to view or copy the tapes. This is easy to do and is quite simple if your VCR and TV set have jacks on the front of the units. You do not need any cable connections with the VHS-C device, but seating the cassette into the adaptor may not be easy and you need to keep the tape from going slack when loading the cassette into the adapter or camera.

Digital Video Images

Digital TV video is now on the horizon and digital camcorder technology has been used for a few years. Camcorder digital techniques let you do special effects, such as to merge one shot into another scene, enlarge a picture, or make still frames very sharp and jitter free. Camcorders now use computer-type digital coding with images consisting of zeros and ones. This also allows you to put your video pictures into your computer and perform all sorts of picture manipulations.

You will find that digital camcorders have the same size of cassettes, which eliminates different formats. The tiny MiniDV video cassette is smaller than an 8-mm or VHS-C cassette, but holds 60 minutes of recordings.

These tiny cassettes lets the camcorder companies build small, light-weight models that you can easily carry anywhere.

The digital Sharp model VL-DC1U is a little larger than other models in the View-Cam line and features a large color viewing screen of four inches.

The JVC model GR-DV1U weighs in around 18 ounces with the battery and tape installed. It is about the size of a paper back book and has great special recording effects.

The Sony digital camcorder weighs about 22 ounces and sports a $2\frac{1}{2}$-inch color "pop out" viewfinder screen and has a built-in speaker.

To playback a digital tape, you will need to connect a cable from your camcorder and plug it into a TV. Another way is to connect the digital camcorder to a VCR to make an analog video tape, and then play it back through your TV.

Digital recording with camcorders has been used for special effects, manipulating picture size, and merging various video scenes into one for a good many years. The camcorders are now using computer-style coding to render the image into ones and zeros. The home camcorder uses only one image sensor instead of the three that professional cameras use, but the picture quality is still very good.

Digital camcorders use the same size cassette, thus you do not have any incompatibility problems. These small-size cassettes allow the cameras to be small in size and light in weight, which lets you take them everywhere you want to go.

The MiniDV videocassette is smaller than an 8-mm or VHS-C cassette, yet can record for 60 minutes. An optional slower tape speed on a Sony unit allows 90 minutes taping.

Let's now check out the specs of some digital camcorders that are small, lightweight, and give you very crisp video images. However, their cost is a good bit more than that of the camcorders you have been pricing.

Sharp VL-DC1U   This unit is larger than some of the JVC and Sony models we have used. This is a newer Sharp digital camcorder and is called the View Cam line. The VL-DC1U features a large viewscreen, which makes recording easier and more accurate. It is smaller in physical size than other viewcams but has a large 4-inch color screen.

JVC GR-DV1U   This JVC compact unit weighs in at only 18 ounces with the battery and tape installed and is about the size of a thick paperback book. After using it for some time, I felt it could have a better shape for holding it more comfortably. However, the special effects during recording or playback were excellent.

Sony DCR-PC7   This 22-ounce, hand-size Sony has a 2½-inch color LCD screen that pops out from the size of the camera body and can serve as a large viewfinder for taping shots or can be used to look back at the videos that have been previously recorded. It also has a small built-in speaker to check out the recorded audio. The prices I have seen, after it has been on the market a while, are between $2500 and $2850.

Video Camera/Camcorder Basics

There are actually only two types of video cameras, which are determined by the type of pick-up device they have to convert light to electronic signals. One camera type uses a vacuum-tube pick-up device and the other uses a solid-state pickup. The two types of solid-state pick ups are *CCD (charge-coupled devices)* and *MOS (metal-oxide semiconductor)*, although the CCD is more popular. More on the CCD chip later in this chapter.
In the mid-1980s, all video cameras used the tube-type pick-up device for imaging. As solid-state imaging chips become available, they were quickly used in portable consumer cameras because of their small size and weight advantages.

As the cost of solid-state chips have decreased and their resolution and light sensitivity has increased, consumer, industrial, and even TV broadcasters now use solid-state CCD pick-up for cameras and camcorders. Other advantages of CCD pick-ups are their increased ruggedness, decreased image lag, better sensitivity, less power-consuming drive circuits, higher-level output signals, and a lot less circuitry that a vacuum pick-up tube requires.

How Video Cameras Work

It really does not matter what type of pick-up device is used, the operating circuits are very similar from one video camera to another. Figure 7-3 shows the major circuits and signal flow of most basic camera types. Besides the lens and pick-up device, the signal processing

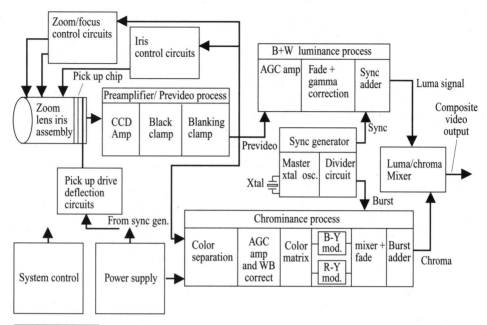

FIGURE 7-3   A simplified video camera block diagram.

and control circuits are very similar to circuits found in many other video products covered in this book.

WHAT IS A CAMCORDER?

Camcorders are in demand for consumers and for TV stations and networks for outside news gathering, etc. Camcorders combine a camera, a VCR record/playback section, and a viewfinder that is also used for looking at the video playback (Fig. 7-4). The camera in a camcorder also shares some of its electronics, such as the power supply, control system, and video circuits, with the VCR portion.

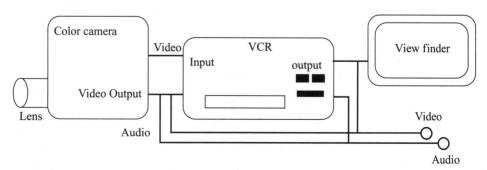

FIGURE 7-4   This block diagram illustrates that the camcorder is a combination of a camera, VCR, and viewfinder.

Usual camcorder faults   The most common failures with a camcorder are caused by the mechanical nature of the VCR transport section and the camera lens, plus the handheld, portable nature for the way the camcorder is used. The same mechanical failures that occur in standard home VCRs also occur in the VCR section of camcorders, but usually not as often because they are not used as much. Worn rubber and broken gears are common failures with camcorders. The camera lens assembly, including the iris, focus, and zoom control motors and gears also have a high failure rate.

The lens problems, as well as broken circuit boards and poor/broken solder connections are usually caused by rough handling and dropping, which will occur with a handheld portable device. If you drop the camcorder on its lens, it could cause lens damage, motor problems, or stripped gears. These type of mechanical failures with the camera section are usually quite easy to diagnose and you may be able to repair your self.

Other portions of the camcorder that develops troubles in either the camera or VCR section is of an electronics nature. In the VCR section, this would be servo, cylinder head, preamp, chroma, luminance or black-and-white, power supply, and system-control stages. In the camera section, electronic failures include sync generator, CCD imager, chrome, luma, power supply, and control problems.

Determining which camcorder section is faulty   Now look at ways to localize camcorder problems:

- *Localizing the problem area*   You need to determine whether the camcorder failure is related to the VCR, camera, or electronic viewfinder, and if the failure is mechanical or electrical. Then, see if you can correct the problem yourself or should you take the unit in for professional work.
- *Mechanical troubleshooting*   Do this to isolate the worn or damaged mechanical parts, which are causing improper VCR or camera operation.
- *Electronic troubleshooting*   This is performed to isolate the defective component that is causing the VCR or camera to operate incorrectly.
- *Alignment information*   Use this to determine if your camcorder needs alignment or adjustments caused by wear, drift, normal usage, or parts that have been replaced. The alignment for camcorders requires special equipment, jigs, and technical skills.

Performance check out   After you have performed any repairs or had your camcorder repaired at a service center, you should make some recordings and use all of the control functions to be sure that it is functioning properly.

Video camera functional blocks   The following is a brief description of the operational blocks that make up a typical video camera. Notice that, depending on individual camera design, the layout order for some of the blocks might be a little different for various brands of cameras.

Lens/iris/motors   The lens assembly focuses light from the scene you're viewing onto the light-sensitive surface of the pick-up device. The auto-iris circuit controls the amount of light that passes through the lens by operating a motor to open and close the iris diaphragm (Fig. 7-5). Under bright lighting conditions, the iris controls the amount of light falling on the pick-up device and thus the amplitude (strength) of the prevideo output signal.

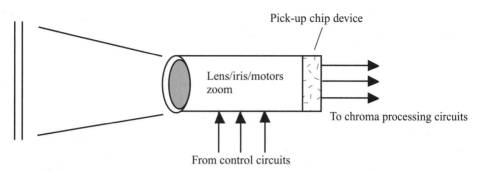

FIGURE 7-5   The lens, iris, and control motors adjust the light that passes through to the pick-up device.

Proper operation of the auto-iris circuit is crucial for video output because the iris diaphragm is spring-loaded closed; a failure in the iris control or drive circuit prevents light from reaching the camera pick-up device. The focus drive circuit generates the signals necessary to operate the focus motor. In cameras with auto-focus, the control circuit reacts to high-frequency information in the prevideo signal, or to an infrared or LED sensor. The zoom drive circuit generates the signals necessary to operate the zoom motor by reacting to input from the camera zoom control-button contacts.

Sync generator circuitry   The sync generator provides synchronization for all the other camera circuits. The output signals are developed by dividing down the signals from a master crystal-controlled oscillator. The master oscillator typically operates at two, four, or eight times the 3.58-MHz chroma burst frequency. The sync generator provides horizontal and vertical drive signals to the pickup device, composite sync and burst for the video output, and 3.58-MHz subcarrier reference signals for the R-Y (red) and B-Y (blue) modulators. The block diagram for the sync generator operation is shown in Fig. 7-6.

Camera pick-up devices   Presently, three types of image pick-up devices are used in consumer, broadcast, and industrial video cameras. These are vacuum tube, MOS, and CCD (charge-coupled devices) devices. The CCD devices are solid-state pick-ups, made of a large number of photodiodes arranged horizontally and vertically in rows and columns (Fig. 7-7). CCDs are now the most commonly used image pick-up devices.

Tube pick-up devices (Vidicon, Saticon, and Newvicon are common types) use magnetic yoke deflection and a high-voltage supply to scan an electron beam across a light-sensitive surface. These tube pick-ups suffer the same scanning irregularities that television picture tubes have, plus more, and require many scan-correction circuits to produce an acceptable output signal. Also, the very low-level output signal from the tube pick-ups (200 μV or less) requires an extremely high gain, with a very low-noise preamplifier as the first signal stage. Tube pickups have been replaced by solid-state CCD and MOS image devices in consumer cameras/camcorders and are being phased out of most broadcast and industrial cameras.

Solid-state MOS and CCD pick-up devices are very similar to each other in operation and performance, with only a few significant differences. Conversion of light to electrical energy occurs at each of the individual photodiodes, which produce a small electrical

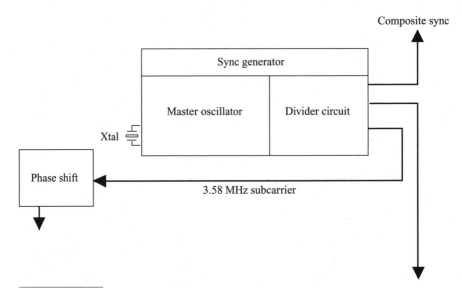

FIGURE 7-6   The sync generator provides timing signals for the remaining camera stages.

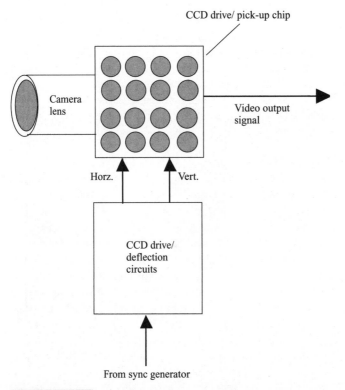

FIGURE 7-7   The pick-up CCD converts reflected scene lighting into electrical signals.

charge when light from the scene is focused on their exposed surface. A method of matrix scanning is used to repeatedly collect each of these charges and assemble them into a video signal. The scanning method used to collect these charges is one of the major differences between MOS and CCD pick-up devices.

MOS devices use a scanning method that results in three or four signal output lines. These lines carry white, yellow, cyan, and green color signals. (no green for the older three-line devices). One disadvantage of MOS devices is that the output signals are at a fairly low level (40 to 50 mV) and require low-noise preamps to bring the signals up to a usable level for standard signal color-processing circuits.

CCD devices use a scanning method that results in a single video output line. This signal contains all of the necessary luminance and chrominance information required to generate NTSC composite video. Also, the level of the output signal is high enough that no preamp is required. An advantage of CCD devices is that they have been more reliable than MOS devices.

With all types of pick-up devices, when color is desired, a multicolored filter is placed in front of the pick-up device's light-sensitive surface. This, along with the scanning of the device, results in the production of an extra high-frequency signal that carries information about color in the scene to be viewed. The location of the CCD chip is being pointed out in Fig. 7-8 photo of an 8-mm camcorder that has its side cover removed for service.

— CCD Chip

FIGURE 7-8   The CCD chip location in an 8-mm camcorder.

Just about all cameras and camcorders now use the *charge-coupled chip device* (CCD) for the image sensor, or pickup, device. In the majority of camcorder models today, the charge-coupled device is a round chip about $\frac{2}{3}$- or $\frac{1}{2}$-inch in size.

The CCD image sensor in consumer camcorders consists of approximately 300,000 small microscopic light-sensitive elements. In camcorders used by professionals, the CCD sensors may have up to 500,000 elements.

The camera's lens projects an image that is to be recorded onto the charge-coupled device image sensor. The image that is projected on the CCD chip causes the cells to be electrically charged. The brighter light on any portion of the chip will cause a larger charge. The various light levels on the chip are then converted into video stgnal picture information.

There are many reasons that CCDs are now being used in all camcorders. First, they are lightweight and small so that the camcorders themselves can be made small and lightwaight. Second, the CCDs do not require very much power to operate. Also, CCDs provide excellent image quality; the pictures they generate are sharp and have very good color quality.

Another advantage of CCDs in consumer camcorders is that the are more shock resistant than the tube-type vidicons, so they can take the tough bumping around that may sometimes occur.

Another advantage of using CCD image sensors is that they have good sensitivity, but they do not cause any streaking, blurring, or burning of the screen coating as do vidicon tube-type cameras.

Development of the Video Signal

In a tube-type vidicon, an electron beam sweeps across the tube's faceplate, and the light image focused by the lens or this faceplate "screen" is converted into an electronic video signal. The beam scans the entire photo-conductive screen coating. The electron beam within the tube picks up enough electroms from each sell to neutralize any charge generated by the light emage. This action generates a signal that varies proportionally by the light emage appearing on the tube's photo-conductive coating. Thus, the video signal is produced.

The CCD image sensor process is developed with a second coating technique. Every $\frac{1}{50}$ of a second, the charge image of the chip sensor is instantaneously transferred to the second layer. Then, in the next $\frac{1}{50}$ of a second, as the next image is being built up, the cells of the second coating are sending out their charges one at a time. This transfer of charges results in a continuous electronic video signal, in which the direction and amplitude of current are proportional to the light charge, and thus portional to the light shining on the chip surface.

The video signal consists of both black-and-white and color information (Y and C signals). The black-and-white information (luminance Y signal) consists of three primary colors: 30 percent red, 59 percent green, and 11 percent blue.

Development of the Color Signal

In professional color TV video tube-type cameras, the light is devided into the various colors with a prism system, or a *dichronic filter*. A dichronic filter uses a thin film on s glass plate to separate the colors. Once the colors have been separated, three camera tubes are used to process the light, one each for the red, green, and blue primary colors.

In the newer-model color cameras and camcorders for consumers, only one pickup element, the CCD, is used. Stripe filters separate the image into the three primary colors. A complicated matrix circuit generates two color-difference signals from the three primary colors. This color information is then combined with the black-and-white monochroma information for the complete color video signal.

Repairing and Cleaning Your Camcorder

This section shows how to repair and clean your camcorder. As you know, the camcorder is a combination of camera and small VCR.

The most popular camcorder sold today uses the 8-mm tape format. The cassette for 8 mm is thinner than the VHS-C and is about the size of an audio cassette. For this reason, the 8-mm camcorder can be made much smaller in size and thus much easier to carry around. The units usually weigh less than two pounds. Figure 7-9 shows a Zenith model VM8300 8-mm camcorder.

FIGURE 7-9   A Zenith model VM8300 8-mm camcorder.

TAKING YOUR CAMCORDER APART

To clean, repair, and make minor adjustments, you will have to take your camcorder apart. When you do, be very careful because very small and delicate parts are contained inside. The screws are very small, so put them in a plastic cup or zip-top plastic bag, so as not to lose any of them. Also, remember where the screws come out of as they will be of different size and screw-thread types. You might want to draw a sketch as the case, screws, and parts are taken apart so that you will know how to put it back together after repairs and cleaning is completed.

How to take apart the cassette lid and deck   The following three drawings show all of the steps for taking apart a Zenith VHS-C camcorder. These same procedures can be used for most all models of camcorders.

1 Refer to Fig. 7-10. Take out the two screws (A) that hold on the cassette cover. Raise the cassette cover, as indicated by the arrow (B) to remove. With this cover removed, you can usually clean the video head cylinder and other rubber roller parts and even part of the tape path.
2 Take out the two screws (C) and remove the base assembly.
3 Take out the three screws marked (D) and one screw (E).
4 The front panel and side panel are engaged by a plastic rim. Carefully squeeze the portions of the side panel between your thumb and forefinger and raise the deck section slightly to disengage it.
5 Disconnect the connectors (F), (G), (I), and (J). The deck and operation sections can now be separated from the camera section.

Taking apart the lower case section

1 Refer to Fig. 7-11. Take out the screws marked *A* and *B*, and remove the insulator sheet.
2 Take out the screws (C), (D), and (E). Disengage the side panel from the lower case by shifting and raising it.

Taking apart the lower section of camcorder   Now, see how to disassemble the lower part of the camcorder that contains the lens and camera section.

1 Refer to Fig. 7-11. Take out screws labeled *A* and *B*, and remove the insulator sheet.
2 Take out the screws (C), (D), and (E). Disengage the side panel from the lower case by shifting and raising it, as shown by the arrow labeled *F*. Disconnect the connector indicated by G.
3 Remove screw H, screw I and wire clamp J. Then take out screws K and L.
4 Raise the camera section slightly and disconnect connectors M, N, and O, which are connected to the E-E and IND board, to remove the camera section from the lower portion of the case.

When performing these procedures, use care not to damage the wires and flexible cables.

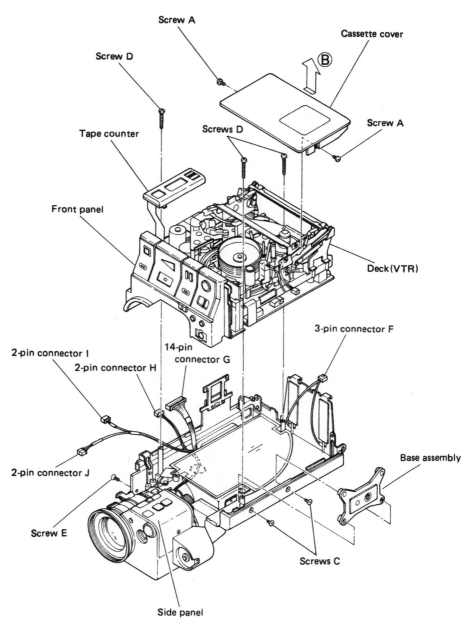

FIGURE 7-10   How the Zenith VHS-C camcorder can be taken apart. (Courtesy of Zenith.)

After removing several small screws, (Fig. 7-12) the case can be split in two parts for servicing.

Figure 7-13 shows the locations of the important components in a VHS-C camcorder deck that might require cleaning and replacement. If the tape transport does not wind, rewind, stop, or start properly the various sensors might need to be cleaned, adjusted, or replaced.

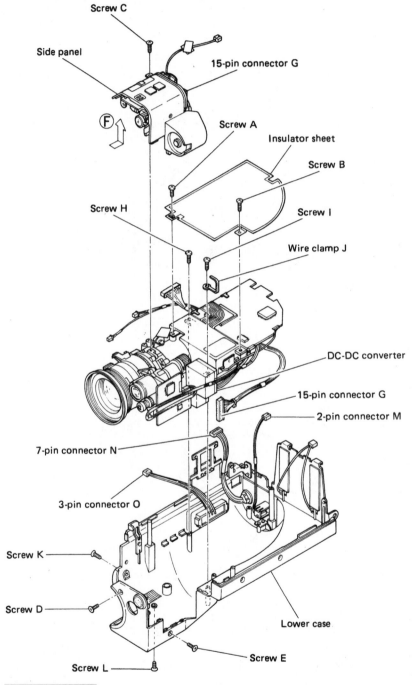

FIGURE 7-11   Taking apart the lower portion of a Zenith camcorder.

(Courtesy of Zenith.)

FIGURE 7-12 After several small screws are removed, this Zenith camcorder case can be split apart for repares. (Courtesy of Zenith.)

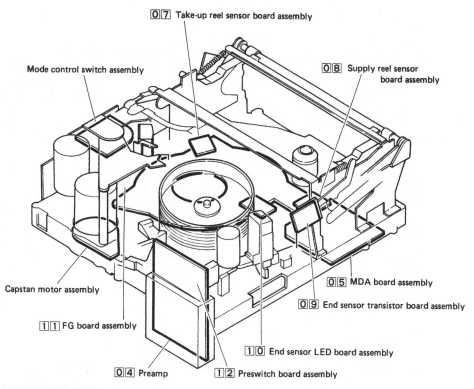

07 Take-up reel sensor board assembly

Mode control switch assembly

08 Supply reel sensor board assembly

Capstan motor assembly

05 MDA board assembly

09 End sensor transistor board assembly

11 FG board assembly

10 End sensor LED board assembly

04 Preamp

12 Preswitch board assembly

FIGURE 7-13 The location of various camcorder sensors found in a Zenith machine. (Courtesy of Zenith.)

Cleaning the camcorder heads You should clean the camcorder heads when the picture playback becomes snowy, noisy, fuzzy, or has streaks across the monitor screen. If you keep the heads clean, it will make them last longer and save on major repair cost later. The playback picture with streaks (Fig. 7-14) was caused by a very dirty video cylinder head. Oxide build-up on the tape head cylinder could cause the tape to be pulled, thus causing damage to other mechanical parts. The heads can be cleaned with a spray cleaner, cleaning fluid on swab, or a cleaning cassette. Head-cleaning spray is used in Fig. 7-15 and a cleaning cassette is shown in Fig. 7-16. However, a cleaning cassette might not clean the heads thoroughly. You can buy cleaning cassettes that will also clean the tape guide, spindles, and the rubber rollers. It's much better to use a swab soaked in a good head-cleaning fluid. On some camcorders, you can clean the cylinder head through the open cassette lid, but you can do a better job if you remove the door cover.

On most models, you can remove the cassette cover door and then get at the machine's mechanism for repairs and cleaning. You will usually just have to remove two small screws and the cover will come off. Some units have small plugs over the screws. Notice that the door has been removed for cleaning in Fig. 7-17.

With the cover removed, you will see a large shiny drum or cylinder that rotates. The tape heads are located on this drum assembly. A swab soaked in cleaning fluid is used in Fig. 7-18. The heads can be cleaned with a spray cleaner. These cans will usually have a small tube that you can use to control the area that you need to clean. However, use caution, do not get spray into other parts of the machine. The spray can technique will not do a very good or lasting cleaning job.

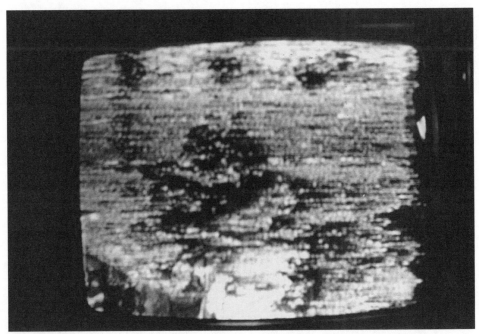

FIGURE 7-14   A dirty cylinder head will cause a picture that is noisy, has streaks, or looks very snowy.

FIGURE 7-15   Audio/video head cleaner spray is being used to clean a camcorders's cylinder heads.

FIGURE 7-16   A video head tape-cleaning cassette.

FIGURE 7-18   A sponge swab soaked in cleaning fluid is being used to clean the cylinder heads.

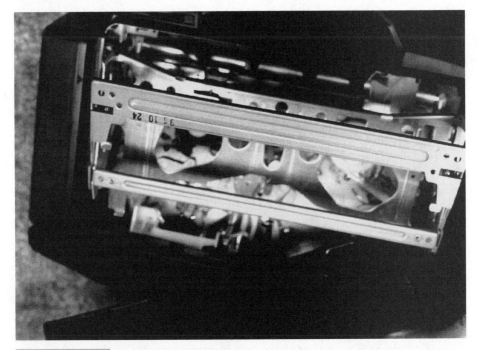

FIGURE 7-17   The camcorder tape door cover has been removed to make cleaning the heads and other parts easier.

If you use your camcorder a lot, you should clean the heads several times a year. As previously stated, the swab or chamois and cleaning fluid is the best technique. Do not use ordinary cotton swabs because the cotton material will pull and might damage the delicate heads. A lint-free cloth soaked with alcohol is good to use when cleaning other parts of the machine's rollers and tape guides. A cassette tape cleaner is used in Fig. 7-19, and the cleaning fluid is being applied to the cassette.

When using the swab or chamois, rotate the cylinder head from right to left several times. Always move the swab horizontally and not in a vertical motion so as not to damage the head's small tip assembly. And keep your fingers off of the drum because your body oil can cause damage. Always thoroughly clean all oxide dust from the cylinder heads and drum surfaces.

A dirty tape path, coated with oil and grease, could cause the tape to pull tight and break and/or wrap around moving parts. Some of these contaminations might even cause the camcorder to "eat" the tape. A defective cassette might have caused the tape to break, too. After the tape and cassette has been removed, thoroughly clean all of the rollers, capstan, drum, and tape path real good with an alcohol-soaked swab or a clean cloth. If the tape is so tangled up and you cannot remove the cassette, then remove the tape cover and the side or bottom covers of the camcorder. Remove any tape wrapped around the drum head and capstan. You might want to rotate the flywheel in a back-and-forth direction to remove the tangled tape. You might find many turns of tape around the capstan. I have had to cut the tape with a razor blade to remove the tape. Use care in doing this! Figure 7-20 shows the oxide dust and dirt being removed from the tape guides, rollers, and head drum after the cassette was removed.

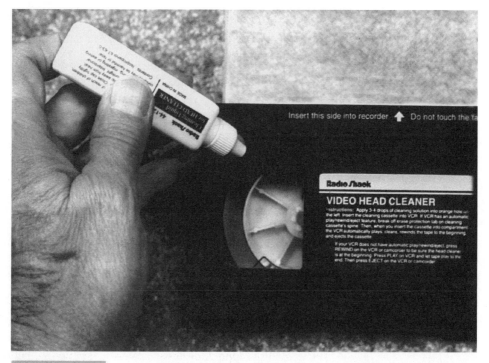

FIGURE 7-19   Cleaning fluid is being applied to a video head cleaner cassette.

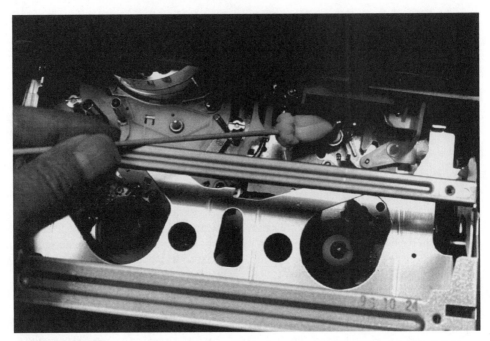

FIGURE 7-20   The various rubber rollers and tape path guides should also be cleaned when the tape head is being cleaned. You can use isopropyl alcohol for this process.

Tape will not move and no viewfinder picture   When you go to record and the tape will not move and all systems seem dead, you might have a defective power supply and/or a dead battery. First, check that the battery is charged and that it is installed properly. Figure 7-21 shows the battery being installed properly. Also, be sure that the battery contacts are clean. Now see if the low-battery indicator light comes on. Some camcorders might not have this indicator. If the indicator shows that the battery is low, plug in the ac adapter/charger, if you have one, and see if the camcorder now operates. If it does, check to see if the battery is bad or if the charger unit is working. A dc voltmeter can be used to see if the charger unit is supplying the correct voltage. If these items check out and there is still no operation, suspect that a fuse is blown or the power switch is dead. The motor could also be defective.

Camera auto-focus operation   The video camera uses infrared light rays to automatically keep the picture in focus. The infrared rays are generated on the front of the camera by an infrared LED and projected to the image that you are taping and then they are reflected back to a set of two photodiodes, also located near the camera lens. The photodiodes detect the infrared rays reflected back from the object and a time-lapse circuit calculates the current needed for a control circuit to "tell" the focus motor its running time for correct focus. You can check the infrared LEDs for emission by using an infrared detector card.

The auto-focus motor obtains it control voltage from an auto-focus IC processor, usually located near the camera lens. The focus/zoom circuit is shown in Fig. 7-22. For auto-focus problems, check for loose connections or dirty cable pin connections on this board. If it seems that the auto-focus circuit is working, check for voltage to the focus motor. If this

FIGURE 7-21 Be sure that the battery is installed properly and that the contacts are clean.

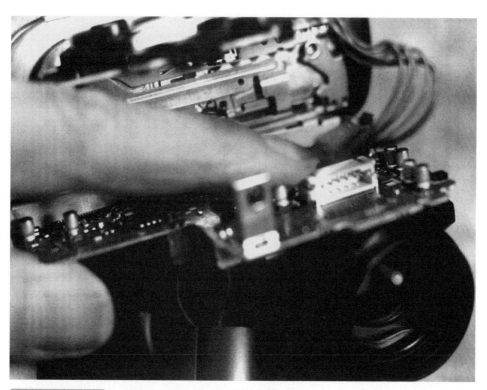

FIGURE 7-22 Inspect the zoom/focus circuit board for any poor connections and clean any flex cable connector contacts with isopropyl alcohol.

voltage is present, suspect the motor is faulty. Also, be sure that the focus sensor, located near the lens, has not been broken or damaged.

Some camcorders can be focused manually or by an auto-focus control. The lens can be rotated manually for good focus or you might want an out-of-focus picture for some special effects. When in the auto-focus mode, the lens rotates automatically. If the lens cannot be moved automatically or manually, suspect jammed gears, faulty focus motor, or even the control circuits. With the cover removed (Fig. 7-22), inspect the motor drive for stripped gears or dirt, grease, etc., jamming the gears.

Slide switches and control buttons Camcorders usually have several buttons for control and zoom and sometimes slide switches for power, etc. These buttons and switches might stick or not work properly. Some are mounted under a plastic membrane and can be taken apart for cleaning. The surface-mounted switches can be cleaned with a spray switch cleaner. When cleaning, slide the switch back and forth a few times to clean the contacts and work in the spray cleaner. This cleaning fluid will also solve sticky button problems

Cassette not loading properly Some camcorders load the cassette tape electronically when you place the cassette in the holder or press a button. In other models, you press a button to release the holder door and then close the door manually to load the tape. If the tape will not load properly, you might have a defective cassette or some object might have gotten inside the camcorder via the loading door. Check and see if the cassette holder door has not been bent, damaged, or broken. This could happen if the camcorder was dropped.

When the cassette is inserted, you should hear the loading motor working to pull in the cassette on such models. If you do not hear this, suspect that the loading switch is bent or out of adjustment. Also, check for a loose or broken drive belt or a broken or jammed loading gear assembly. The loading gears are being inspected in Fig. 7-23.

Intermittent or erratic operation If your camcorder has intermittent and erratic performance in various operational modes, such if it will not load or eject the cassette, it might not record or play back tapes at times, intermittent zoom lens operation, or at times, it will not work at all, then suspect a faulty flex cable with intermittent open wire runs or poor/dirty flex cable pin contact connectors. These flat flex cables are very delicate and the contact pins (Fig. 7-24) are very small. Clean the cable contacts and flex the cable while operating the camcorder modes and see if the problem comes and goes. If it does, you might have to replace the flex cable.

Camcorder motors In most camcorders, you will find three motors: one each for the capstan, cylinder drum, and zoom. Some of the older, more-sophisticated camcorders, have motors for loading, auto focus, and iris adjustment. However, some of the small 8-mm camcorders will only contain the loading motor, capstan, and drum motors.

If your camcorder has variable speed, the problem is usually a dirty or loose drive belt. Check the belt for being stretched, dirty, cracked, or worn. You can remove the belt and clean it with alcohol and a clean cloth. If this does not help your speed problem, then replace the belt with the correct one. If a belt has been slipping, it will usually appear very shiny on the pulley side. And be sure that you clean the idlers, pulleys, and capstan thoroughly with alcohol and a swab before replacing the new belt. A good cleaning and new belts will solve

FIGURE 7-23   Inspect for a jammed or broken loading gear assembly if the tape cassette will not load into the camcorder.

most erratic camcorder speed problems. Other erratic camcorder speed problems can be caused by a defective motor, defective or misadjusted brake-release system or, on some machines, a clutch pad release.

Sony Handycam servicing   Let's now look at the "tear-down" of the Sony Handycam recorder shown in Fig. 7-25. This Sony 8-mm camcorder has been placed on a service bench to undergo cleaning and/or repairs. Figure 7-26 shows how the Handycam case can be split apart for repairs and general cleaning or adjustments. A portion of the Sony Handycam's cover around the lens has been removed in Fig. 7-27 for repairing the zoom lens gearing system. After cleaning and repairs, the Handycam is put back together and will be ready for more videotaping.

Camcorder troubles and solutions   Many camcorder troubles are caused by some minor faults that you might be able to correct yourself with the following list of troubles and solutions.

Symptom/trouble   Will not record or playback.
Probable cause and correction   Check for a cassette in the unit. Try ejecting and reinserting the cassette. Cassette may be defective. Is dew indicator flashing? If it is, moisture is

FIGURE 7-24 Should you be having intermittent camcorder operation, check the flex cables and any plug/pin connections. Clean all plug connections.

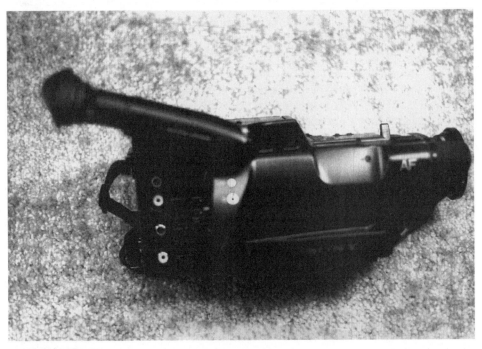

FIGURE 7-25 A Sony 8-mm camcorder, HandyCam model, on the service bench for repairs.

FIGURE 7-26 **The Sony HandyCam camcorder split apart for cleaning, adjustment, or repairs.**

in the camcorder. Keep it at room temperature for a few hours and retry. Also, the cassette tape might be at the end. Rewind the tape and try again.

Will not playback The VCR Play button must be in the VCR (Play) mode.

Symptom/trouble Will not record. The safety tab on the cassette is closed.
Probable cause and solution The VCR Play button must be in the Camera mode.

Symptom No picture is in the viewfinder.
Probable cause and solution The lens cap might still be over the lens.

Symptom/trouble Camcorder turns itself off.
Probable cause and solution Turn camcorder back on with the power switch. Some camcorders will turn themselves off if left in the Record or Play Pause mode for three minutes to prevent tape wear.

Symptom/trouble Camcorder will not work with remote control unit.

FIGURE 7-27   **A portion of the Sony HandyCam case has been removed for repairs of the zoom lens.**

Probable cause and solution Be sure that the remote control is aimed at camcorder's LED sensor. The lithium battery might be dead or not installed correctly in the remote-control unit. The remote-control sensor on the camcorder could be exposed to direct sunlight or strong artificial light.

Symptom/trouble Sound is very low or distorted on playback.
Probable cause or solution The person you are recording might be too far from the camcorder. Some camcorders have an external microphone that can be installed for greater range of audio pick up.

Symptom/trouble Very poor auto-focus operation.
Probable cause and solution The object that you are taping might not be in center of the viewfinder or two objects are at different distances. If your camcorder has a focus-lock feature, turn it to on.

Symptom/trouble The viewfinder displays are out of focus.
Probable cause and solution Be sure that the lens is clean and not smudged. The eyepiece focus control could be misadjusted. Some camcorders have three small control adjustments near the viewfinder. These are brightness, color, and focus. These might need to be readjusted. Use caution when adjusting these because they are miniature controls.

Symptom/trouble While recording, the camcorder will unload and then shut off.

Probable cause and solution   The tape has come to its end. Check for a defective tape-end switch.

Symptom/trouble   While recording, the color recorded is different than the actual color. The color is not true.
Probable cause and solution   Adjust the white balance.

Symptom/trouble   The picture is blurred when played back.
Probable cause and solution   The cylinder head is dirty, worn, or defective. See Fig. 7-28 photo.

Symptom/trouble   The external microphone is not working.
Probable cause and solution   Check and clean the microphone switch. Check for a broken microphone cable or plug connection.

Symptom/trouble   No picture is on viewfinder screen during tape playback.
Probable cause and solution   Be sure that the TV/video switch is in the Video position.

Symptom/trouble   The tape will not fast forward or rewind.
Probable cause and solution   The drive belt is very loose or broken, the drive belt is slipping, or the tape is at its end.

Symptom/trouble   The cassette starts to load, but then immediately shuts the camcorder down.
Probable cause and solution   The tape has not engaged properly. Try reinserting the cassette. The cassette tape has come to its end.

FIGURE 7-28   **Nose bars (streaks that are caused by a very worn tape cylinder head drum) are shown going across the monitor screen.**

Symptom/trouble   The ac adapter/charger has no ac operation.

Probable cause and solution   If you find, with an dc voltmeter, no or low voltage, check for an open fuse, faulty diodes, or a faulty transistor regulator. Check the ac line cord and plug. Check the power transformer winding for an open with an ohmmeter. Clean all switches and contacts.

Symptom/trouble   The ac adapter/charger will not charge the battery. The battery might be defective. Try a new one.

Probable cause and solution   Check the adapter's output voltage. If it's OK, the charging circuits might be bad. If the charging LED will not light, the charging circuits are the prime suspects, also. Check the transistors, LED charge indicators, and the zener diodes in the charging section of the adapter. If the charging circuits and voltage output is good and the battery does not hold its charge very long, the battery is defective.

Symptom/trouble   Intermittent video recording. When you are looking into the electronic view finder (EVF), the picture intermittently goes blank, streaks, or breaks up.

Probable cause and solution   With the camcorder case removed, try tapping the various sections of the PC board with a pencil eraser and see if the picture breaks up on the EVF. If it does, it might mean that a poor solder connection or a crack is on the printed circuit board. With a magnification light (Fig. 7-29), you might be able to locate the defect and repair it.

FIGURE 7-29   **A magnification lense and light are being used to locate poorly soldered joints, loose connections, or cracked PC boards, which can cause intermittent camcoder operations.**

Symptom/trouble   Picture is streaked across horizontally.
Probable cause and solution   May be caused by an improper adjustment in the "streaking" control located in the prevideo circuits.

Symptom/trouble   Tape is frilled on edges or warped.
Probable cause and solution   Tape guide or other tape path adjustments are needed.

Symptom/trouble   Recorded picture is too dark or light when taped under normal lighting conditions.
Probable cause and solution   Make sure the AIC level is properly adjusted.

Symptom/trouble   A dark border is present at side of the electronic viewfinder (EVF) screen.
Probable cause and solution   Adjust horizontal-vertical centering control.

Symptom/trouble   Image on EVF distorted at top and bottom.
Probable cause and solution   Make sure the vertical size is correct.

Symptom/trouble   Color balance is off, and images are distorted.
Probable cause and solution   Check for proper adjustment of the horizontal and vertical size and linearity controls. Adjust as required.

Symptom/trouble   Picture is smeared or washed out.
Probable cause and solution   For camcorders with a pickup tube, the beam current of the pickup tube is not adjusted properly or the pickup tube may be weak and needs replacement.

Symptom/trouble   Tape is broken, tape may be stretched out, will not rewind tightly, and/or is pulled out of cassette.
Probable cause and solution   This trouble is usually caused by dirty or worn brake pads. Clean and/or adjust the drum brake pads. If brakes are badly worn, replace the brake pads.

Symptom/trouble   Picture is out of focus in some portions of the zoom range.
Probable cause and solution   Check out the camcorder back focus adjustment. On some model camcorders, this is a mechanical adjustment.

Symptom/trouble   Lines across the picture and noise streaks. Otherwise, the picture is stable.
Probable cause and solution   Adjust the front panel or remote control customer tracking control. If this does not correct the problem, then make sure the tracking preset adjustment is correct.

Symptom/trouble   No picture when playing back a recording.
Probable cause and solution   Check and make sure the TV/video switch is in the VCR position.

Symptom/trouble   Camcorder starts to load cassette, then quickly shuts down.
Probable cause and solution   Cassette tape is at its end. Cassette is not installed properly, or tape is not engaged. Some type of interference infrared source is triggering the camcorder's remote mode.

Symptom/trouble   Camcorder will not fast forward or rewind.
Probable cause and solution   Check for a loose or slipping drive belt. The drive belt may be defective or broken and may need to be replaced. Tape may have come to its end.

Symptom/trouble   Camcorder cassette will not eject.

Probable cause and solution   Press eject switches to check if voltage is present at the loading motor terminals. If voltage is present, the motor is faulty. If voltage is incorrect or missing, check the power supply voltages. Recharge or replace battery. Check IC ejection system. Check for voltage at the loading motor control IC. If motor is operating properly, then check out the eject mechanism.

Symptom/trouble   Very noisy picture during playback.

Probable cause and solution   Check for dirty or defective drum heads. Clean if dirty. Check out control head recording circuits. Clean any plug connections going to drum head.

Symptom/trouble   Camcorder battery will not operate very long, even after being fully charged.

Probable cause and solution   Check and make sure the charger circuits are operating properly. Discharge battery completely and charge again. Check battery voltage to see if it has been fully charged. If the battery still does not hold a charge, it should be replaced.

Symptom/trouble   No color and random noise during playback.

Probable cause and solution   Clean heads on the drum and inspect the cylinder drum for any damage.

Camcorder Care Tips

- If possible, store your camcorder and tapes at room temperature.
- Always replace the lens cap when not using your camcorder.
- Before using your camcorder, be sure that your hands and face do not have any chemical residue, such as suntan lotion, because this could damage the unit's finish.
- Keep dust and dirt from getting inside the cassette door. Dust and grime are abrasive and will cause wear on the camcorder's head drum, gears, belts, and cassettes.
- The camcorder can be damaged by improper storage or handling. Do not subject the camcorder to swinging, shaking, or dropping.
- When the camcorder is not in use, always remove the cassette and ac adapter and/or battery.
- You should keep the original carton if you need to ship it for repair or to store it.
- Do not operate the camcorder for extended periods of time in temperatures below 40 degrees F or above 95 degrees F.
- Do not aim or point your camcorder at the sun or other bright objects because this could damage the CCD imager.
- Do not leave your camcorder in direct sunlight for extended periods of time. The resulting heat buildup could permanently damage the camcorder's internal components.
- Do not operate your camcorder in extremely humid environments.
- Do not operate your camcorder near the ocean because salt water or salt water spray can damage the internal parts of the camcorder.
- Do not use an adapter, adapter/charger, or batteries other than those specified for the camcorder. Using the wrong accessories could damage the camcorder.

■ Do not expose the camcorder or adapter to rain or moisture. If either component becomes wet, turn off the power and dry out or have it checked out by a service company.

■ Avoid operating your camcorder immediately after moving it from a cold location to a warm location. Give your camcorder about two hours to reach a stable temperature before inserting a cassette. When the camcorder is moved from a cold to a warm area, condensation could cause the tape to stick to the cylinder head and damage the head or tape. Some camcorders have a dew indicator and the machine will not operate until the moisture has been eliminated and the temperature has stabilized.

WIRED TELEPHONES, CORDLESS PHONES, ANSWERING MACHINES, AND CELLULAR PHONE SYSTEMS

Telephone System Overview

The telephone (telco) company lines from your home or office go to a central office or to a telco substation (also called *switching* or *call-transfer stations*). The calls over a pair of copper wires (twisted pair) are one of many pairs within a telephone cable. This cable could be located overhead on poles or buried underground. Most telephone calls are carried over fiberoptic cables from the substations to the central switching office. Fiberoptic cables are used between cities and other points across country. In a few selected areas, fiberoptic lines are used from some home and office phones.

From your local telephone central office your calls will go to routing centers to be transmitted across country to another area to complete your long-distance calls. These calls can go over AT&T long lines via copper wires, coaxial cables, microwave signals, satellite signals, or fiberoptic cables.

TIP AND RING CONNECTIONS

Regardless of what type of phone you use, the signal will start out over two copper wires. These two wires are called *ring and tip* on the phone plug and jack connections. Proper telephone wiring designates that the *tip* is the green wire and the *ring* is the red wire.

THE TELEPHONE RINGER (BELL)

The telephone "ringer" device is used to let you know when an incoming call is coming onto your phone line. When you have a call, the central switching office sends out a burst signal of approximately 85 to 110 peak-to-peak ac volts at a frequency of about 20 Hz. This burst lasts about two seconds and is off about four seconds. The old ringer bell uses two electromagnetic coils of wire that move a metal clapper back and forth from the magnetic force. The clapper hits two metal gongs, thus providing the ring. This is called an *electromechanical bell ringer*.

Most modern electronic phones use an IC to produce a more pleasing electronic ring. This electronic ringing sound is produced by a piezoelectric device or tone generator.

THE HOOK SWITCH

To clean and repair a conventional Western Electric telephone, you must remove two screws on the bottom (Fig. 8-1). Also note, in the upper left side of this photo is a thumb wheel that is turned to control the loudness of the phone bell ringer.

With the top cover removed (Fig. 8-2), you can now clean or repair most parts of this phone. The hook switch is being cleaned and burnished in Fig. 8-2. You can use very fine grit sandpaper or an emery board to clean these switch contacts. In this phone, the switch has several contacts that are pushed down by the weight of the handset that it is hung upon the cradle. The hook switch connects or disconnects the phone's voice circuit from the telephone line. Dirty or corroded switch contacts can cause noisy reception, intermittent phone operation, or complete loss of telephone operation. The old-style mechanical rotary dial on some phones also have contacts that might need to be cleaned.

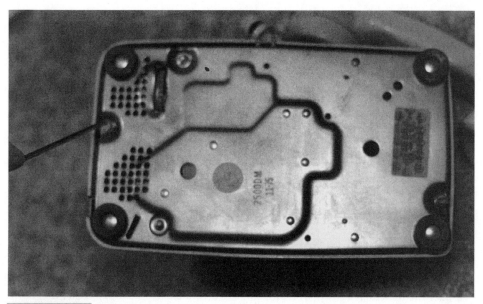

FIGURE 8-1   Remove two screws from bottom of a wired Western Electric phone to clean and repair this unit.

FIGURE 8-2   The "hang-up" switch contacts are being cleaned on this conventional wired phone.

TELEPHONE HANDSET AND TOUCH-TONE PAD

The phone handset usually houses the transmitter and receiver units that is connected to the main phone base unit with a coiled flexible cord. Because the cord is flexed a lot, it might fail and need to be replaced. The module plugs and the connectors that they plug into might become dirty and need to be cleaned. Some phones also have the Touch-Tone pad built into the handset. The Touch-Tone pad produces a unique *dual-tone multifrequency* tone *(DTMF)* that does all of the tone signaling in place of the old rotary dial pulse system. ICs are used in conjunction with the Touch-Tone pad to produce the DTMF tones.

Conventional Telephone Block Diagram

A conventional phone block diagram is shown in Fig. 8-3. The ringer (bell), hook switch, and dialer have been covered thus far. The dialer can be the old rotary dial or the modern touch-pad (DTMF) tone system. Now look at the speech circuit and see why it is needed. The speech circuit is used to couple the receiver and microphone transmitter in the handset, which has four wires into two wires for connection to your phone line. This is needed for full-duplex operation for you to talk and listen at the same time and use only one pair of wires. It also couples the dialer into the phone lines and produces a sidetone to the receiver that lets you control your speech level. The speech network consists of a hybrid transformer and a balancing network. A hybrid speech circuit is shown in Fig. 8-4.

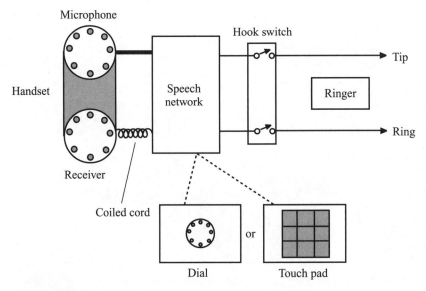

FIGURE 8-3   A block diagram of a conventional phone.

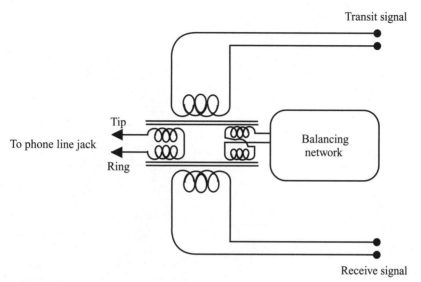

FIGURE 8-4   **Circuit for one type of hybrid speech system.**

Some Conventional Telephone Troubles and Solutions

If your phone is not working (dead), then check that phone jack with another working telephone. If that phone is also dead, you need to go outside your house and locate the phone box or telephone network interface housing. Figure 8-5 shows one type of telephone housing box.

USING THE TELEPHONE TEST NETWORK BOX

The telephone network interface test and connection box is provided so that you or a telephone technician can determine if the phone problem is in the home wiring, jacks, or the phone lines. This box has a convenient test jack that will help you to isolate telephone line troubles. You need to make this test before reporting a trouble to your phone company. This could save you an unnecessary dispatch and service charge.

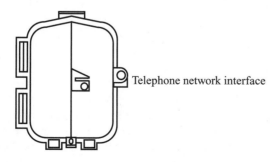

Telephone network interface

FIGURE 8-5   **A telephone network interface housing box. You can open this test box to help isolate your phone problems before calling for telephone repair service.**

FIGURE 8-6   Modular plug being removed from the test box's interface jack.

To make this test, remove the modular plug from the interface box jack and insert the plug from a working telephone set (Fig. 8-6). Now try your telephone. If a dial tone is heard, the problem is in your home phone equipment or house wiring.

After you have finished your phone test, unplug the telephone set (Fig. 8-7) and reconnect the modular plug back into the interface jack. Close the cover and screw the fastener down until the cover is snug and tight.

Now that you have confirmed that the phone line is OK, go back into your house to be sure that all telephones, answering machines, cordless phones, fax machines, DBS receivers, and computer modems have been unplugged. A problem in any one of these units could cause your complete phone system to fail. With all phone items unplugged, you can then plug an operating telephone back into a jack. If this phone now works, one or more of your other units you have disconnected is faulty. To find out which one, plug one item in at a time, then check your phone. If your test phone stops working, you will know which of your other phone equipment is faulty.

STATIC AND PHONE NOISE CHECKS

If your phone has static and noise, plug in another phone for a test. If the test phone is noise free, then your original phone is faulty. If you still have noise on the test phone, the wall jack might need cleaning or you have wiring problems. If the phone is faulty, then switch the line cord and clean any module connections. Next, check and/or switch the handset cord and clean any plug-in connections. If you still have noise, take the phone apart and

FIGURE 8-7   Unplug the test phone and reconnect the modular plug into the interface jack.

clean any switch contacts and look for loose connections. If the phone has a printed circuit or flexible circuit wiring, inspect it for poorly soldered connections and cracks.

LOW SOUND OR DISTORTION

For these symptoms, always check the cords, plugs, and any switch contacts. Another item to check is the transmitter and receiver diaphragm, located in the handset. The earpieces and mouthpieces can be removed, like you would take off a jar lid. They have threads on the caps that can be unscrewed. Clean the metal diaphragm and the electrical contacts. The microphone or receiver units might be defective and need to be replaced.

DTMF TOUCHPAD PROBLEMS

When any type of liquid (Fig. 8-8) is spilled into a telephone or other electronic equipment, various problems can develop. If any liquid gets into your telephone touch pad, you should immediately take it apart and flush it out with water or a good electrical contact cleaner. Then use a hair dryer to dry out the touch-pad circuit boards (Fig. 8-9). You can check the operation of the Touch-Tone pad by lifting the receiver and listening as you push

FIGURE 8-8   Liquid spilled into a phone can cause many problems.

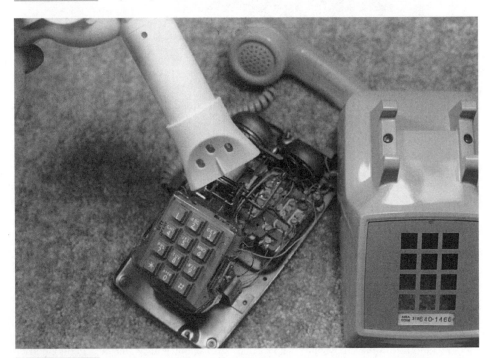

FIGURE 8-9   After flushing out the phone with clean water or an electrical contact cleaner, use a blow dryer to dry out the circuitry. This can be used for other types of electronic equipment also.

each button. You should hear a different tone for each button pushed. If you do not, clean the button contacts on the membrane pad and check all wiring from the pad unit to the main circuit board. The pad might have to be replaced or other components on the main board might be faulty if the tones are still not produced.

Electronic Telephone Operation

The electronic telephone contains diodes, capacitors, resistors, ICs, PC boards, and many other components. Refer to Chapter 1 for more details on these discrete components. Figure 8-10 shows the block diagram of the electronic phone, including a dialer IC, speech-network IC, ringer transducer, Touch-Tone keypad, and a voltage-regulating power supply. In most cases, the dc voltage is taken from the phone line to power the phone circuits. These phones will feature volume and voice level adjustments, multiple ringing, phone number memory bank, last-number-dialed memory, and many other features. Some advanced features are hands-free speaker phone systems, LCD display readout, and Caller ID.

The speech-network IC block (Fig. 8-10) is an IC that receives and transmits speech and the DTMF tone signals. The speaker/microphone is usually a electrodynamic type.

A zener diode protection device across the phone-line input is used to protect the phone circuit from voltage spikes and surges. The ringer IC is connected directly across the phone line and has a dc block to prevent loading down the telco line.

Most electronic phones have a dual-mode IC. This mode switch is labeled *T* and *P*. When in the T position, the IC sends out DTMF tones for dialing and in the P position, the old-type pulses are sent out for dialing.

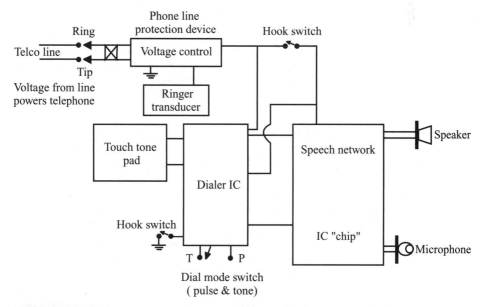

FIGURE 8-10   **A basic block diagram of an electronic telephone device.**

FIGURE 8-11   A Western Electric Princess phone taken apart for cleaning and repairs.

Most all components in an electronic phone can be mounted on one or two PC boards. Figure 8-11 shows a Western Electric Princess phone that has been taken apart for repair and cleaning. Notice one circuit board is in the base and another one is mounted in the handset.

Some phones use a microprocessor to enhance its functions and capabilities. With a microprocessor, many more features can be added, such as a visual display for clock time, Caller ID number display, call waiting, call transfer, call restrictions, answering-machine control, and many other features. These phones are very complicated and you should have them repaired at an electronic service company that specializes in telephone repairs.

ELECTRONIC TELEPHONE TROUBLES AND REPAIR TIPS

Electronic circuits, as well as parts layout, vary from one model of phone to another. As this section covers various troubles and solutions, you might want to refer back to Fig. 8-10.

Noisy phone operation   A typical electronic phone with the cover removed from its base is shown in Fig. 8-12. If you have any noise, popping sounds, or intermittent phone operation, you should check, clean, and tighten all of the screw terminals shown. Also, clean and tighten the module phone jack connection shown on the bottom in Fig. 8-13 or on the back or sides of other phones. Phones that have a Touch-Tone pad in the handset also have a pushbutton hook switch next to the pad. If these button slide switch contacts become dirty, it can cause noisy or intermittent phone operation. This switch and spring (Fig. 8-14)

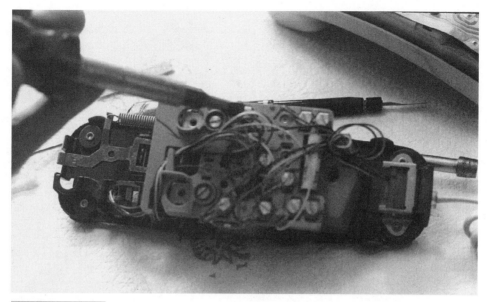

FIGURE 8-12   A circuit board located in the base of an electronic phone. Clean, check, and tighten any loose screw connections.

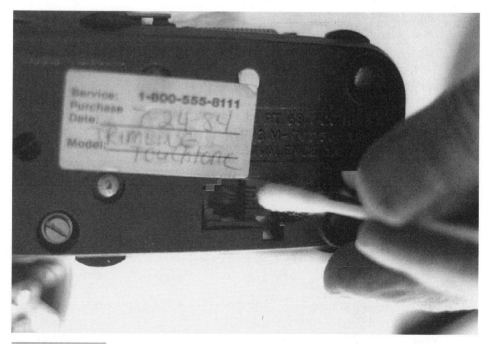

FIGURE 8-13   Clean the module jack with a good contact cleaner or with isopropyl alcohol.

FIGURE 8-14   Location of the hook switch slide contacts.

is located under the PC board. Use a spray contact cleaner for cleaning these slide switch contacts.

No phone operation (dead) Make these checks after you have determined that another phone works OK in this phone jack location. Some phones will have an external (power block) dc voltage supply that plugs in the ac wall outlet with a cable that then plugs into the back of your phone. Be sure that this power block in plugged in and then measure for 9 to 12 Vdc at its cable plug with an voltmeter. Also check the coiled cord that goes to the module phone jack at the phone's base and then connects to the handset. Phones with an internal built-in power supply might have a blown fuse or defective surge suppressor that will cause the phone to be dead. Make a visual inspection for loose connections or poorly soldered joints (Fig. 8-15). A small hairline crack on the PC board is often hard to locate, but can cause all types of telephone problems. A good bright light with some magnification can help you to locate these PC board defects. A small hair-line crack on the PC board is shown in Fig. 8-16. This is the circuit board you will find when the handset cover is removed. With the handset apart, check the wiring solder connections and tighten all screw terminals, resolder or repair the connections, as needed. Clean the receiver and transmitter elements and their electrical contacts.

Touch-Tone pad problems Most telephones with a touch pad have provisions for tone and pulse dialing modes. These modes are selectable by a slide switch labeled *pulse* or *tone*. Be sure that this switch is in the Tone mode if you do not hear tones as you dial. The switch might be in the center slide position and cause a no tone or a pulse dialing condition. The slide switch contacts might be dirty and need to be cleaned.

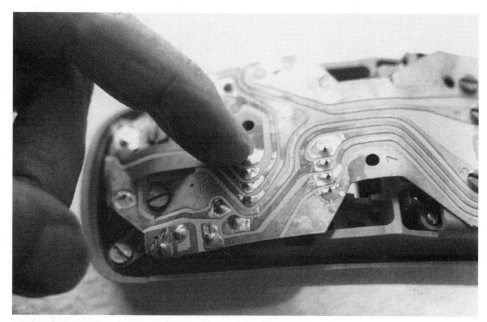

FIGURE 8-15   A poorly soldered connection might cause the phone to be dead or operate intermittently.

FIGURE 8-16   A PC board hairline crack is hard to locate, but can cause all sorts of phone problems.

If one or more tone buttons do not work, remove the phone cover and remove the keypad so that you can check all wiring and connections to this unit. Repair any broken wires or poorly soldered connections. Also look for any broken or damaged membrane switches. The keypad will often become dirty because liquids have been spilled into the pad. Carefully take the keypad membrane assembly apart and clean and dry out all components within the pad. You might have to replace the pad assembly if it is too badly damaged, cracked, or broken.

How a Phone Answering Machine Works

Telephone answering machines are found in many homes and offices. Of course, some people will not talk to a machine, but most find it an indispensable product. It's a time saver for receiving calls when away from home and screening those many nuisance calls when you are at home.

There are basically two types of answering machines. The old models that use one or two miniature cassette tapes. Figure 8-17 shows a cassette being installed in the tape compartment drawer. The unit can be a complete phone/machine combination (Fig. 8-18) or a

FIGURE 8-17   **A cassette being inserted into the tape compartment of an older model answering machine.**

FIGURE 8-18   **A typical stand-alone tape-type answering machine and telephone combination.**

machine that you can plug into your existing phone. The other type is the tapeless answering system that uses ICs to digitize and store the phone messages into IC memory and synthesized speech circuits.

A typical answering machine that uses one miniature cassette tape is shown in Fig. 8-19.

CONVENTIONAL TAPE MACHINE OPERATION

Your answering machine will have to detect a ring signal from the central office in order to tape a message. The ringer circuit detects and sends this incoming ring to a ring-detector circuit. This circuit converts the analog ring signal to digital logic for counting. This ring logic is counted by the microprocessor or CPU by the number of rings you select; this starts the machine tape with your prerecorded message.

CONTROLS AND FEATURES

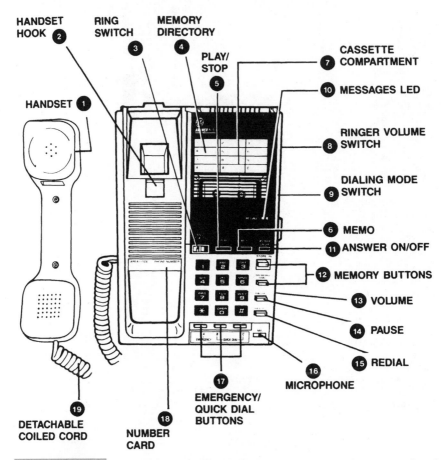

HANDSET HOOK 2
RING SWITCH 3
MEMORY DIRECTORY 4
PLAY/STOP 5
HANDSET 1
CASSETTE COMPARTMENT 7
MESSAGES LED 10
RINGER VOLUME SWITCH 8
DIALING MODE SWITCH 9
MEMO 6
ANSWER ON/OFF 11
MEMORY BUTTONS 12
VOLUME 13
PAUSE 14
REDIAL 15
MICROPHONE 16
EMERGENCY/QUICK DIAL BUTTONS 17
DETACHABLE COILED CORD 19
NUMBER CARD 18

FIGURE 8-19   Operating control locations and functions of a conventional answering machine that uses a cassette tape.

When the correct rings are detected, the CPU "tells" a relay to close, which seizes the phone line, starts the tape recorder, and connects the speech network. Figure 8-20 shows a simple block diagram of a single tape answering machine. The incoming call is amplified and you can hear the person calling, which enables you to screen the calls. Also, a microphone built in the machine allows you to record the outgoing messages (OGM).

Most late-model machines have a built-in DTMF decoder so that you can control your machine from any telephone by calling your home phone number. This DTMF decoder is connected to logic circuits and controls the CPU to give your machine the desired instructions. The CPU or microprocessor is the large-scale IC (LSI), with 36 or 42 pins (Fig. 8-21). By a preset code, you can retrieve your answering machine messages when away from

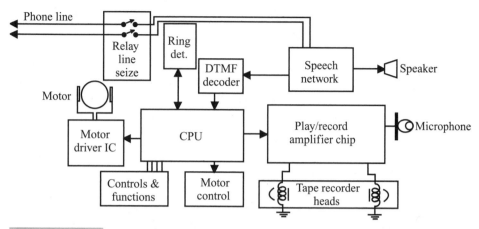

FIGURE 8-20   A simplified block diagram of a tape-type answering machine.

FIGURE 8-21   The LSI CPU or microprocessor found in an answering machine usually has 36 or 42 pins.

home. This is called a "beepless" remote-control system. With older machines, you carried a handheld beeper to control the machine over the phone system.

Play/record operation   Briefly, the component that actually handles the recording and replaying of tape messages is the play/record (P/R) amplifier (Fig. 8-20). The P/R amplifier controlled by the CPU is what "tells" the amplifier if it needs to handle outgoing or incoming messages and should the tape recorder be in the Play or Record mode. A beep tone controls the operating status with a logic signal to the CPU. The beep detect lets the machine find the beginning and end of each message.

Cassette tape operation overview   The job of a tape player in an answering machine is to run the magnetic tape back and forth at the proper speed and record and playback the correct portions of the tape as instructed by the CPU. In most tape machines, the drive motor (Fig. 8-22) is used to operate all of the mechanics for tape operations. A small rubber belt from the motor drives the cassette hub gears and the capstan. The hub gears are shown in Fig. 8-23, with the cassette tape removed.

Cleaning the tape mechanical system   For good record and playback operations, the tape must be kept at a constant tension and travel. To do this, a pinch roller is pressed against the capstan shaft with the tape passing between them. There can be one or more idler rollers. The pinch and idler wheels are made of rubber. The capstan and pinch roller is shown to the left of the pencil in Fig. 8-24 and the record/play head is to the right. All of

FIGURE 8-22   The main drive motor uses a small belt to operate all of the tape-recording mechanics.

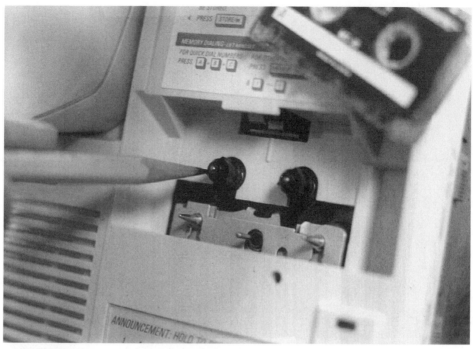

FIGURE 8-23   The hub gears turn the tape spools within the cassette unit.

FIGURE 8-24   The capstan and pinch roller is to the left of the pencil pointer and the record/play head is on the right.

FIGURE 8-25 Use a small brush and alcohol to clean all tape recorder mechanical parts.

the rubber rollers, capstan drive shaft, record/play head, and tape guides should be cleaned with denatured alcohol or a nonsolvent cleaner. After a period of time, all of these parts will have a build up of oxide from the tape. Dust and dirt should also be cleaned out with a small brush and alcohol from all of the tape mechanism (Fig. 8-25). As pointed out in Fig. 8-26, the record/play head can be cleaned with a cotton swab and alcohol without removing the top case cover from the answering machine.

For more information and service tips on the tape recorder section, refer to Chapter 2 on audio/stereo cassette player systems.

Digitized tapeless answering machines Many answering machines do not use a cassette tape, but utilize memory chips, analog-to-digital conversion (A/D), and speech digitizing processes. An answering machine that uses the digitizing system is shown in (Fig. 8-27) and is used with a conventional telephone plugged into the unit.

Refer to the simple block diagram in Fig. 8-28 to see how the tapeless machine works. These machines use a digitized speech network that records both outgoing and incoming voice messages. Some of these units can hold several outgoing messages that can be changed with a button touch, many incoming messages, a time/date stamp and you can rapidly select which message you want to hear in any sequence.

FIGURE 8-26 **The record/play head can be cleaned without removing the top case of the answering machine.**

A microphone is still used for your outgoing messages and these electrical signals are digitized as well as the incoming signal message on your phone line. These analog speech signals go to an analog-to-digital converter (ADC). The ADC samples the analog signal at a very high rate and converts it into digital words. To recover the incoming and outgoing messages, the digitized data from the memory chip must go to the digital-to-analog converter (DAC) for you to hear the messages. After filtering from the DAC, the reconstructed synthesized voice is very near that of the original. These tapeless machines can have more features and are almost trouble-free, compared to the cassette answering machines.

Some answering machine troubles and solutions To repair and clean the answering machine, take out the screws (Fig. 8-29) to remove the bottom of the case.

Problem or symptom Machine will not answer an incoming call.
Probable cause and correction Check the ring-detection circuit. Also check and clean the cord and module plugs from phone jack to the answering machine base. Figure 8-30 shows some zener diodes that are in series with the input phone line for protection. They might be defective because of lightning surges and could lower the ring voltage level. Use an ohmmeter to check these diodes. The problem could also be that too many phones are on one line, which can reduce the ring voltage level.

FIGURE 8-27   **A tapeless or IC memory/digitized voice mail answering machine.**

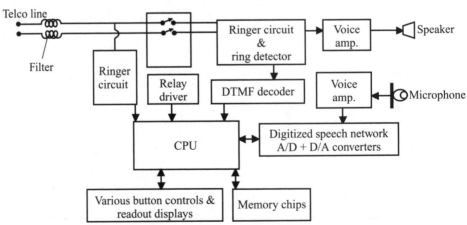

FIGURE 8-28   **A block diagram of an all-electronic digital answering machine.**

FIGURE 8-29   **For cleaning and repair, remove four screws from bottom of the machine.**

Problem or symptom   Cassette tape will not rewind.
Probable cause and correction   Check for a loose or broken belt from the motor to the hub spindle. Clean any dirt or grease from belts. Also, a broken or jammed gear could be the trouble. Check and clean any dirt or grease from the mechanical parts or rubber wheels.

Problem or symptom   New messages are being taped over all old messages. The message is not intelligible.
Probable cause and correction   The tape is not being erased between recording sessions. The erase head or its circuitry and wiring are usually at fault. Also, be sure that the erase and play/record heads are clean.

Problem or symptom   Tape will not move or moves erratically.
Probable cause and correction   You might have a broken tape or a damaged cassette. Remove cassette and replace with a new one. Check gears and spindles to see if they turn freely or are jammed. If tape is broken and tangled check the capstan shaft and pinch roller and see if any tape has been wrapped around them. Remove any tape and clean these components.

FIGURE 8-30 These zener diodes might be faulty
and cause the ring-detection circuit not to function, and
the machine will not answer or record calls.

Problem or symptom No dial tone.
Probable cause and correction Check all phone cords and module plugs. Check hook
(receiver hang-up) switch for proper movement.

Problem or symptom The message indicator flashes, but no message is recorded.
Probable cause and correction Replace the cassette tape with a new one and record another
message. Clean the record/play heads.

Problem or symptom Loss of memory modes.
Probable cause and correction The small battery (Fig. 8-31) could be worn out. This battery
usually plugs into a slot on the bottom of the machine and should be replaced. If the battery has become corroded, the connectors should be cleaned. Take the case off of the machine
and clean it with a brush and solvent (Fig. 8-32).

FIGURE 8-31   The small, thin battery used for chip memory is shown being replaced.

FIGURE 8-32   Battery terminals being cleaned with a soft brush and solvent.

Problem or symptom   The answering machine will not function. It beeps and the call-counter LED flashes.

Probable cause and correction   Machine has locked because of a loss of ac line voltage, surge, spikes, etc. Unplug the power block from the ac outlet for 20 seconds, then plug back in. This will reset or reboot the microprocessor (URT) within the answering machine.

Problem or symptom   The message sound level is too low.

Probable cause and correction   Check the setting of the volume control. Check and clean the play/record head and capstan.

Cordless Telephone Overview

The sales of cordless phones probably account for over half of all telephone sales. These phones set you free to roam around room-to-room, all over your home and even outside in the yard and workshop, etc.

SOME CORDLESS PHONE CONSIDERATIONS

Some portions of the cordless and conventional corded phones have the same operations. They both convert the sound of a voice into electrical signals and transmit them via telephone lines to another telephone receiving set. At the same time, the telephone converts these electrical signals of the person's voice "at the other phone" back into sound waves. Of course, the big physical difference with the cordless phone is that there is no cord between the handset and phone base.

With a conventional phone, the electrical impulses are carried by the cord between the handset and the phone base; then they are sent out over the telephone lines. However, with a cordless phone, the electrical signals travel between the handset and telephone base via radio waves.

The cord from handset to base has been replaced by a two-radio, which has duplex operation and allows two conversations simultaneously. A simple cordless phone drawing is shown in Fig. 8-33.

As with two-way radios, auto radios, and CB radios, the reception and interference can vary from location to location and from time to time. These same kinds of problems can be a factor with many cordless phones. This could be bothersome because we have all expected very clear reception over the fine telephone systems. Americans now expect phone privacy, excellent sound quality and high reliability.

Some cordless phone problems

Poor sound quality   Some phones might have poor audio response, receive interference from electrical devices and interference from other cordless phones close by.

Short range   The main appeal of a cordless phone is the ability to let you move around without pulling a cord. However, some phones have a very short range.

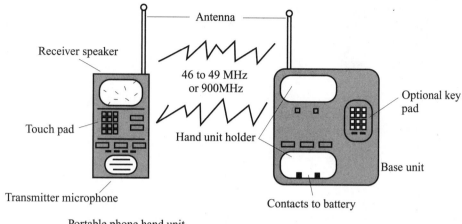

FIGURE 8-33   **Drawing of a basic cordless phone operation system.**

Your phone use time is limited   Because the cordless phone is powered by a battery in the handset, the phone might quit during a conversation because the battery needs to be charged. Also, most cordless phones will not work if the home power goes off.

Conversation privacy   Other people with a cordless phone on your frequency can listen to your conversation if they are within range of your phone. However, the newer cordless phones, some in the 900-MHz band, offer digital transmission with encoded speech information and also automatic channel switching if someone transmits on your frequency.

Cordless phone frequency bands   The early model and some even sold today work on the 46- to 49-MHz radio-frequency band. This is a small region between the CB band and TV Channel 2.

A new generation of cordless phones were developed in 1990 in the 900-MHz band. These phones operate at a much higher frequency (902 MHz to 928 MHz) and a greater transmitter output power. Phones operating at these higher frequencies have less interference and the band is not as crowded. Lucent Technologies (formerly AT&T) and Panasonic now have 900-MHz phones with a range of up to 4000 feet.

Two transmission modes   Not only do cordless phones operate in two frequency bands, but they have two different transmission modes to transmit and receive conversations. Early model phones used analog transmission, which is a continuous signal that varies in intensity like a radio broadcast station. An early model analog phone is shown in Fig. 8-34 photo. The latest 900-MHz phone technology uses digital transmission, which is a series of short, computer-coded signals that are decoded at the phone's receiver. Digital phone transmission reduces the noisy, buzzing, and crackling usually found in cordless analog phones; also, they are harder for someone to eavesdrop on.

FIGURE 8-34   **This cordless phone uses analog RF signals with frequency modulation on frequencies allotted from 46 MHz to 50 MHz.**

Digital phone modes The digital phone manufacturers use different ways to transmit their encoded narrow-band signal in the 900-MHz frequency range. Spread-spectrum cordless phones stretch (or spread) the narrow band signal over a multitude of different frequencies and are not as susceptible to interference. Spread-spectrum phones would be like many people talking identical messages all at the same time over the phone system. If one, or even quite a few, of these messages are blocked or interfered with, you would still receive the message from the others. Because the signals transmitted from these cordless phones are "spread out" over a wide bandwidth and with increased transmit power, these phones will have increased range and voice clarity. The 900-MHz phone (Fig. 8-35) has a range of more than one-half mile.

SOME DIFFERENT PHONE TECHNOLOGIES

Now look at some of the various cordless phone technologies that deal with clarity, privacy, and range. In regard to call security, the new digital technology now makes it possible to eliminate the eavesdropping problem.

Some security codes now being used To keep outsiders from using your cordless phone, almost all phones sold today use some type of code between the handset and the

FIGURE 8-35   **Young lady using a 900-MHz cordless phone. These phones can have a talk range of one-half mile or more.**

phone base to prevent unauthorized use of your phone line by someone using another handset on your frequency. However, many phones don't secure the call itself.

Basic scrambling This is a basic way of scrambling a conversation so that it is more difficult to decipher, except by its own receiving set. Just about any competent electronics technician could unscramble this code.

Digital encoding This one generates coded signals that are more difficult to decipher than a basic scrambling mode. Probably a competent electronics engineer could unscramble this coding and reconstruct the conversation.

Spread-spectrum encoding This is the most difficult code to crack, having been developed by the military for war applications. It would probably take a few years for a highly skilled communications engineer, with very specialized knowledge and very sophisticated equipment, to eventually bypass this security code. This spread-spectrum encoding goes by a trade name of *Surelink Technology*.

CORDLESS PHONE SOUND QUALITY

The characteristics of these two phone frequency bands has a lot to do with the quality of reception. Cordless telephones in the 46- to 49-MHz range are much more prone to interference from a much more cluttered frequency band. This band is much more susceptible to electrical interference and other radio services. Cordless phone receivers in the 900-MHz band must be a lot closer to an interference source to be affected.

Companding   This system is somewhat like the Dolby stereo audio system. The system essentially "loudens" the transmission to overcome the naturally occurring hiss, then brings it down to normal levels at the receiving end. This technique goes by trade names of *Sound Charger* and *Compander*.

Multichannel capability   Many cordless phones can operate over several channels. When you hear some interference, you can manually switch to another channel. However, some automatically move to another channel, looking for a channel with less interference. Because of the limited space between channels in the 46- to 49-MHz range, this technology is not very effective for these cordless phones. In the present FCC frequency band allotment, the number of channels is limited to 25 for low-band phones, 40 for high-band standard, and digital cordless phones have the equivalent of 100 high-band spread-spectrum channels.

Digital transmission   This transmission involves sending the message as a series of computer codes. Because each bit of code only has a designated value of "1" or "0," unlike analog radio transmissions that have infinite possibilities—the receiving set can more easily identify the incoming code in the presence of interference. However, if there is significant interference, the conversation might sound "choppy" because an entire code is lost.

Spread-spectrum transmission   The radio transmission technique also uses a series of computer codes. However, because the same signal is stretched out over a broad frequency band, the likelihood of "choppy" conversations is considerably eliminated. A receiver only needs to receive a part of the transmitted signal to reconstruct the original message. Spread-spectrum transmission will retain its quality—even if the 900-MHz frequency becomes more crowded. Digital spread-spectrum cordless phones often display the Surelink technology label.

Cordless telephone range   Range is a key characteristic of the cordless telephone—regardless if it operates in the low or high radio band and whether it is analog or digital. Different cordless phones transmit various levels of power, much the way radio stations use different levels of power output. Higher-powered cordless phones can transmit signals over greater distances. However, the phone also needs more battery power to do this. This will require larger batteries or a shorter use time. Spread-spectrum cordless phones get additional range at lower power levels because they use battery power more efficiently than nonspread-spectrum phones. The FCC also allows the spread-spectrum phones to operate at higher transmit power levels than conventional phones.

Analog phones   These phone systems have the shortest range and are the most likely to be affected by high buildings, hills, etc. These analog cordless phones, operating in the crowded low band, are restricted by the FCC to no more than 0.04 milliwatts of radiated power, and rarely exceed 500 feet of working range.

Standard digital phones   The inherent characteristics of this mode of transmission, plus the fact that they typically transmit in the 900-MHz band, increases their range up to 0.25 mile. The FCC limits the power output of these phones to no more than 0.75 mW.

Spread-spectrum digital The spread-spectrum system improves on the advantages of standard digital transmissions because of multiple signal transmissions. The FCC also allows far greater power output (up to 1 W), which have ranges of up to 0.5 mile.

Phone battery life The handset of a cordless phone has a battery pack for operating power, and it has to be recharged after being used for a period of time. The battery is recharged by placing the handset back on the base unit. The "talk time" for cordless phones is usually hours and the "standby" time, not being used or recharged, is in days. The time required to recharge a fully discharged cordless phone battery is approximately 8 to 12 hours. Of course, the phone cannot be used during this period. The base of the cordless phone is powered by ac power, with a plug-in power block. When you lose ac power, the cordless phone does not operate. However, if you use a *UPS (uninterruptable power supply)* you can plug your cordless phone power block in the UPS power supply and not have a loss of your cordless phone operation.

Quick charge capabilities Phones with a quick-charge system use more-expensive circuitry built into the phone. Most cordless phones with quick-charging features will recharge in one hour, but cost more than the standard, slower-charging systems.

Back-up battery units A phone-charging system for a backup battery, located in the base unit, provides a few advantages. First, the base of the cordless phone will have power if the ac power line fails. Second, the battery in the base of the phone and the battery in the handset can be "interchanged;" it will be kept charged, thus providing a "hot spare" battery for extended phone conversations.

DELUXE CORDLESS PHONE FEATURES

Now, with many cordless phones being sold, many manufacturers have opted to combine cordless phones with other telephone features, such as answering machines, speaker phones, or caller ID packages. These features can be useful, but they do not improve the sound quality of the basic cordless phone. You might want to consider buying a separate telephone with all of the deluxe features, and then buy a "stand-alone" cordless phone. One reason for this is that the more gadgets you build into one system, the more probability of a failure with a higher repair or replacement cost.

CORDLESS PHONE BUYING TIP

Your choice of cordless phone should be decided by what you need the most: Range, security, clarity, cost, or a combination of all four. Generally, the more limited the cordless phone capabilities, the less it will cost. If you reside in uncrowded rural areas and only require a short range, you might not want to consider digital spread spectrum. However, you will receive the best performance from the cordless telephone with the most sophisticated technology. For more information on Surelink spread-spectrum phone technology, call (800) 858-0663.

BASIC CORDLESS PHONE OPERATION

The cordless phone consists of two units that must work together somewhat like two two-way radio systems. The base unit transmits and receives RF signals and the portable battery-operated handset unit also receives and transmits RF signals.

The base unit has electronic circuits that connect into your local phone line. It also has a radio transmitter and receiver circuits. The base unit plugs into an ac outlet, usually a block power unit, to power the radio transmitter/receiver and has a built-in battery charger for charging the handset battery.

BASE UNIT CIRCUITRY

The cordless phone not only has to connect to the phone line, but also has to have a complete radio (RF) transmitter and receiver in the base (Fig. 8-36). The base also has a CPU, memory ICs, phone-line seize relay, ring-detector circuit, ringer, and some models will have a DTMF pad, tone-generator chip, and a back-up battery for power outages.

The base unit contains five blocks. These would be the power-supply/charger circuits, speech or interface network, microprocessor (CPU) controller, the radio receiver and transmitter sections. The RF carrier with modulation, which is transmitted back and forth between the base and handset is also modulated with speech (voice) and control signals. These control and speech signals are modulated in the transmitter and demodulated in the receiver. A duplex circuit is used so that the transmitter and receiver can use the same antenna. Interference and feedback is eliminated because different frequencies are used for transmitting and receiving. The receiver also has filters in its RF stages.

THE PORTABLE HANDSET UNIT

Figure 8-37 shows a typical cordless phone handset unit with all of the function control call-outs. As you refer to Fig. 8-38, you will notice that the handset contains most of the circuits found in the base unit. The handset has a transmitter, receiver, CPU control chip, ringer circuits, and DTMF circuits.

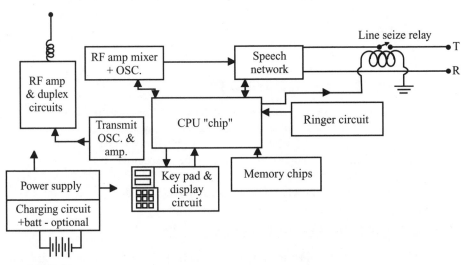

FIGURE 8-36   **A block diagram of a cordless phone's base unit.**

① **Antenna**

② **Volume Control**

③ **Channel Scan**

④ **Charge Contacts**

⑤ **Indicators**

RINGER – Lights when ringer is on and phone is in use, flashes when battery is low. (Indicator flashes even if ringer is off and battery is low.)

PHONE – Lights when phone is in use, flashes for hold and mute.

⑥ **Function Keys**

OFF   Hangs up phone when flip is open.

MUTE Mutes so other party cannot hear you.

PAGE INTCM Pages base.

HOLD Places calls on hold.

STO   Stores phone numbers.

RCL SECURE DEMO Recalls numbers from memory. Press and hold for Secure Clear™ demonstration.

RE-SND Redials last number dialed.

PHONE FLASH Accesses call waiting (if available). Press for dial tone if flip is open.

PULSE ✱ TONE With phone off, press and hold to switch to pulse dialing. Press and hold again to return to tone dialing. Changes temporarily from pulse to tone dialing during a call.

⑦ **Microphone**

⑧ **Flip with Telephone Number Card**

⑨ **Ringer Button** – Turns handset ringer on and off. If off, you will not hear your handset ring or beep when paged. Turning ringer off extends battery life up to 21 days.

⑩ **Numeric Keypad** – Enters dialing information.

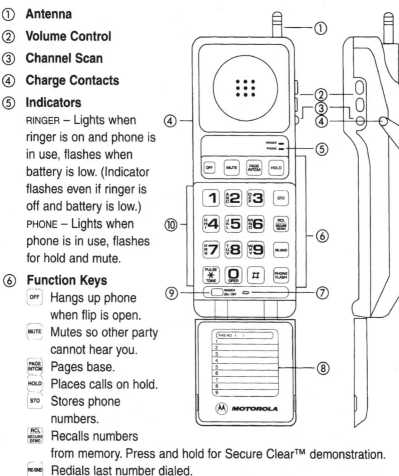

FIGURE 8-37   A drawing of a cordless phone hand set with all of the callout functions.

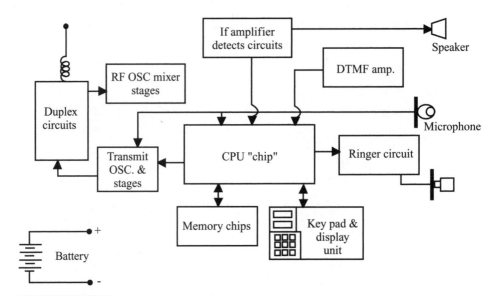

FIGURE 8-38   **A block diagram of the hand unit for the cordless phone.**

The CPU in the base and handset units controls all of the cordless phone operations. The CPU along with memory chips (ROMs or RAMs) keep track of all memory and program instructions for phone operations.

In the base unit, the CPU gives the instructions for the transmitter, the receiver, and sends control pulses to the portable hand unit, as well as interpreting control pulses sent back from the hand receiver/transmitter unit. The CPU also controls the phone-line seize relay, ring circuit detector, and DTMF dialing signals.

CORDLESS PHONE TROUBLES AND CORRECTION HINTS

Let's now look at some cordless phone problems. Some of these problems might be caused by electrical interference, other cordless phones, two-way radio interference, weak batteries, or no battery charging, not enough talk range, or other people listening in on your phone conversations. Other problems could be your phone not working at all (dead) or an intermittent operation problem.

Removing the phone case   To clean or repair the phone base unit, remove four or more screws (Fig. 8-39). This will give you access to the circuit board, power supply and charger section, and phone-line seize relay.

To take the portable hand unit apart remove the battery cover and take out the battery. Then remove the two screws (Fig. 8-40). Now lift up the battery end of the cover and swing the two covers apart (Fig. 8-41). This will then expose the circuit board, switches, and other components.

Now check the solder connections on the flat ribbon cable that connects between the two covers sections. If the phone has static noise or intermittent reception, check for circuit board cracks or poorly soldered connections (Fig. 8-42). Also clean any dust or dirt from any of the small switches (Fig. 8-43) and the membrane under the touch pad.

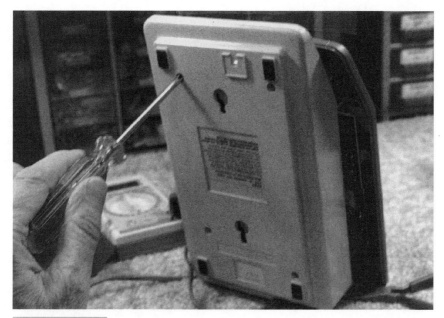

FIGURE 8-39   To repair or clean the phone, remove the four screws from the base unit.

FIGURE 8-40   Take off the battery cover and remove the battery. Then take out the screws to separate the two handset covers.

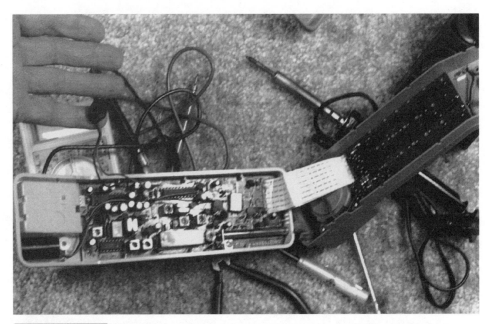

FIGURE 8-41   After the screws are removed, the covers will come apart.

FIGURE 8-42   Check for PC board cracks, broken components, or poorly soldered joints.

FIGURE 8-43   Clean dust and dirt from the switch contacts. Use a brush or a spray switch contact cleaner.

CORDLESS PHONE TROUBLE CHECKLIST

If you have phone problems, the following checklist should be helpful.

- Be sure that the power cord or power block is plugged in and the outlet has ac power.
- Be sure that the telephone line cord is plugged firmly into the base and the wall phone jack.
- Be sure that the base unit's antenna is fully extended.
- If the phone does not beep when the (phone) button is pressed, it could indicate that a battery needs to be recharged. Some phones have a (LO BATT) indicator light. Also, clean the contacts with a pencil eraser (Fig. 8-44) if the battery will not stay charged.
- Be sure that the battery pack is installed correctly.

Handset and base unit not communicating (two beeps)   Usually a "two-beep" signal indicates that the handset and base are not communicating properly. This could be caused by something as simple as being out of range when dialing a call. Try moving closer to the base unit and try the call again.

If moving closer to the base does not work, then it might be that the handset and base have different security codes. Try the following procedure:

Place the handset in the base, and check to be sure that the charging light is on. Wait 15 (or more) seconds, then pick up the handset and press the Phone button. The Phone lights on the handset and base should now go on, and the phone should now work normally.

FIGURE 8-44   Clean these contacts if the battery will not charge or if the phone does not work, but gives you a beeping tone that indicates a low battery.

Phone will not work (dead)

■ Verify that the modular jack is working by testing with a known-working phone.

■ Is the power cord or power block plugged in? When the base is plugged in and the handset is in the base, the charging light goes on to indicate that power is connected. If the charging light does not come on, you might need to clean the charging contacts with switch contact cleaner and a soft cloth.

■ Check the cordless phone line cord connected between the base and the modular wall jack outlet.

■ Are both antennas pulled all of the way out? Do this for the base unit and the portable handset.

■ Place the handset in the base unit to set the security code between the base and the portable handset.

■ Be sure that your phone is set correctly to either pulse or tone dialing to match your local phone service.

Noise or static problems   You are hearing noise or static when using your cordless phone. This is probably local electrical interference. Try pressing the channel button to change to a different channel.

Some phones have automatic channel-change circuits. If you still are receiving static or noise, try the following tips:

■ Move the handset unit closer to the base.

■ Be sure that both antennas are pulled completely out.

■ Try moving the base unit to another electrical outlet. Choose one that is not on the same circuit as other appliances.

Phone will not ring If the handset unit will not ring, try the following suggestions:

■ Be sure that the ringer button or switch is in the On position. Some cordless phones have a battery-saver switch; when it is on, the unit will not ring.
■ Check the cord and be sure that it is connected properly and that the power cord is plugged in.
■ Be sure that the antenna is pulled all the way out.
■ Move the handset closer or relocate the base.
■ Change the channels.
■ Unplug one of your other telephones. The strength of the ring signal is reduced if you have several phones on one line.

Phone will not work (dead)

■ Unplug and replug the power cord and telephone line plug. Then pick up the handset and place it back in the base holder. If the phone still does not work try some of the following tips: Place the handset in the base and be sure that the charging light comes on. Unplug the ac adapter from the wall ac outlet, wait 15 seconds, then plug it back in again. The charging light should go on again. Wait another 15 seconds, then pick up the handset and press the Phone button. The Phone lights on the handset and base should go on, and your phone should now operate properly. If not, then go to the next step.
■ Pick up the handset, open the battery compartment door and unplug the battery pack. The battery pack and small plug is shown in Fig. 8-45. Wait 15 seconds and then reinstall

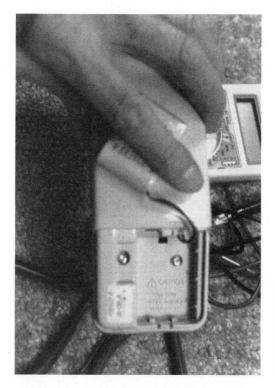

FIGURE 8-45   **If the phone will not work, unplug the battery (small red and black wires). Wait 15 seconds and plug the battery back in. The phone should now be operational again.**

the battery pack plug. Now, close the battery compartment door, place the handset in the base and check to be sure that the Charging light is on. Wait another 15 seconds, then pick up the handset and press the Phone button. The Phone lights on the handset and base should now go on, and your phone should now operate properly.

No dial tone   Recheck all of the previous suggestions. If you still do not hear a dial tone, disconnect the cordless phone and try a known-good phone in its place. If no dial tone is in the test phone, the problem is in your house wiring or with the local phone service. You can also plug your test phone into the outside phone junction box. If you do not receive a dial tone at this location, contact the local phone company repair department.

Phone interference review

- Be sure that the base and portable handset antenna is not broken and is fully extended.
- You might be out of range of the base.
- Press and release the Channel Change button to switch channels. This will not interrupt your call.
- Household appliances plugged into the same circuit as the base unit can sometimes cause interference. Try moving the appliance or base to another outlet.
- The layout of your home or office might be limiting the operation range of your portable phone. Try moving the base to another location. An upper story base location will increase range.

Cordless phone antenna replacement   You can easily replace a bent or broken antenna on your cordless phone. Most antennas are just screwed on or off (Fig. 8-46). Most electronics stores will probably have a replacement for your model phone. Also, if you cannot get your cordless phone operating, Radio Shack has a repair service for most all brands of cordless phones.

Phone surge protection   A phone surge- and spike-protection module that plugs into an ac outlet is shown in Fig. 8-47. These units can protect any type phone or answering machine from lightning spikes and surges. You might want to install one on each of your phones.

Mobile Radio Telephone Communications

For over 50 years, it has been possible to have 2-way communications via radio from a moving vehicle. This was first accomplished by a two-way radio system, then by radio telephone, and for the past decade or so with high-tech cellular devices. These cell phones can now be used just about anywhere in this country at an affordable price. And it is a great emergency device to have when you are traveling. You can think of the cell phone as a very sophisticated cordless phone system that was just explained previously in this chapter. The portable cell phone could be in your auto or coat pocket and the base unit would be the cell radio transmitter site that has a telephone interconnect to the local phone

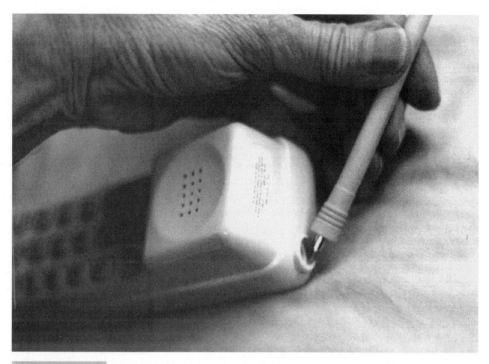

FIGURE 8-46   A new antenna is being replaced. Most older-model cordless phone antennas will screw in.

FIGURE 8-47   A telephone line-protection device can be easily plugged into an ac wall outlet. This provides good protection from spikes and lightning damage that can be coming into the phone line.

company land lines. These cell sites are spaced so that you usually have contact within a site's coverage area. As you drive, your cell phone signal is automatically passed along or "handed off" to the next cell site transmitter with the strongest signal strength. The cellular phone system operates in the 900-MHz frequency band.

TWO-WAY RADIO TRUNKING SYSTEM

About 1980, Motorola introduced the 800-MHz two-way radio trunking system, which also had provisions for using the telephone in a mobile vehicle or portable unit. This radio trunking system could be used as a two-way radio to talk to a base station or as mobile units to mobile units and also had telephone interconnect capabilities. These trunking systems utilized the 800-MHz to 890-MHz allocations (Fig. 8-48). This system was the forerunner of the now very popular nationwide cell phone networks.

800-MHz trunking system overview The 800-MHz trunking system consists of at least four voice transmitter/receive channels, one data control channel that sends and receives data at all times and a central system controller. A simple basic block diagram of a trunking system is shown in Fig. 8-49.

As with cell phones, the trunking system mobile and portable units have individual IDs that must be programmed into the trunking system's central controller before these units can use the radio system. Thus, to bring up the trunking system, the transmitting mobile's ID must be known by the central controller and data logger. The trunked central controller is shown in Fig. 8-50. All mobile transmitted trunked calls are initiated through the central controller channel.

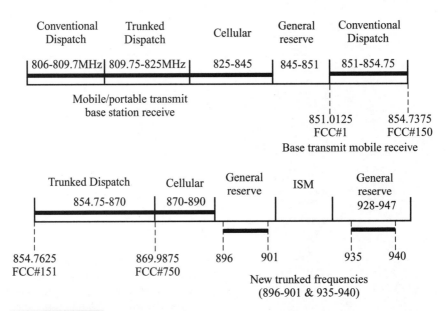

FIGURE 8-48 The radio trunking and cellular frequency band spectrum allocations.

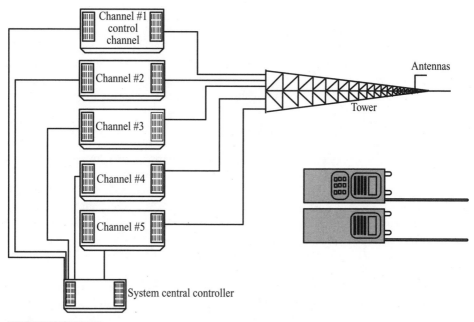

FIGURE 8-49   Basic five-channel radio trunking system.

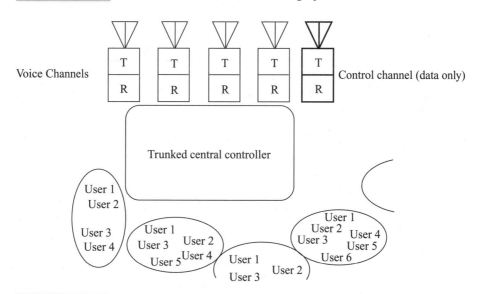

FIGURE 8-50   Block diagram of a trunked 800-MHz central controller.

When the mobile units microphone is keyed up and voice transmission starts, the mobile unit sends a subaudible connect tone at a low level, which the central controller recognizes and keeps the channel open or connected. When the mobile unit completes a call, a disconnect tone is transmitted and the central controller will issue a disconnect tone for the outgoing voice channel being used.

The 800-MHz trunking systems offer many features, such as private conversations (PCI, Privacy Plus) that lets a supervisor talk privately with an individual, System-wide call, Fleet-call, Status-message call, and call alert for selective paging of a specific person. These trunking systems also have a back-up called *failsafe* if the controller and data channel fail. When this occurs, the system reverts back to standard community radio repeater operation. All modes of operation are verified by data handshake signals.

Trunking telephone interconnect   The 800-MHz radio trunking system is also equipped so that a mobile telephone device can be used for making phone calls while on the move. These trunking systems have direct telephone interconnect with landline telephone networks. The Central Interconnect Terminal (CIT) extends the communications ability of a 800-MHz trunked radio system without changing the use of the system. Mobile or portable two-way radios that are equipped to generate Touch-Tone compatible signals (DTMF signals), can automatically access the local telephone landline system without going through a dispatcher. Also, regular phone customers who have been given an access code can call into the radio trunking system from any local or long-distance telephone system.

The cellular telephone radio system   The cellular radio telephone mobile operation is a highly sophisticated/complicated electronics system that has evolved over many years of development and huge investments of many communications companies. Now look at a basic cell phone system's unique operations.

The cell phone concept has very little resemblance to a conventional two-way radio communications or repeater system. Figure 8-51 illustrates how a regional cell network is laid out. The cell radio phone system breaks up the coverage area into small (cell site) divisions. Each cell site contains several low-power radio transmitters and receivers that are linked to a central cell computer controller equipment center location. When you start to use your cell phone, you will automatically be communicating directly to the closest cell site. As you travel along, your cell phone is "handed off" automatically from site to site. All of the phone calls from the cells in this group are then fed into the central cell computer-controlled switcher, which are now routed via telco lines (fiberoptics) to the local telephone exchange.

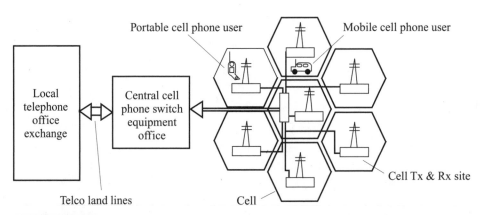

FIGURE 8-51   **Drawing of how regional cell system sites are laid out.**

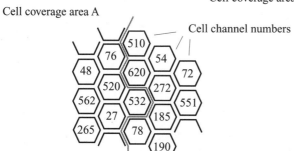

Cell coverage area B

Cell coverage area A

Cell channel numbers

FIGURE 8-52   **How cell phone channel numbers are allocated and laid out for a typical wireless phone system.**

The FCC has set aside more than 600 frequencies for cellular telephone operations in the 900-MHz band. A cell site can usually handle 40 or more full-duplex telephone calls from mobile cell phones simultaneously. Each of these calls need two different frequency channels for full-duplex (two-way) conversations. Figure 8-52 illustrates how area A and area B are divided up in cells of different channel numbers and frequencies to avoid radio interference. When a cell phone customer initiates a call, the closest cell site then automatically opens up two unused channels to complete the call.

HOW THE CELL PHONE OPERATES

The cellular phone packs a lot of sophisticated electronic components into a very small space. In fact, it is actually three devices in one unit. Not only is it a two-way radio, but a computer and telephone. And some cell devices have an answering machine and a pager. To repair them, you need to be a highly trained electronics technician and have very specialized and expensive test instruments. However, later in this chapter, some repair tips are listed that you can check on before calling on a professional service center.

Transmit/receive section   In Fig. 8-53, you will notice the radio RF transmitter and radio receiver sections that couple these sections with a duplexer to the antenna. The signal (voice/data) is received from the cell site and is filtered and processed to be heard in the speaker. The frequency synthesizer with instructions from the CPU tunes the cell phone to the proper receive and transmit channels. Of course, there is a touch pad and DTMF generator will enable you to make calls. Also, a read-out display indicates phone numbers that you dial and recall from memory.

CPU and memory logic   The heart of a cellular phone is the CPU (control/logic) and cell-control chip. The CPU receives program instructions for the ROM chip. The RAM chip is used for temporary data that is erased and updated in every day use. This could be phone numbers on a memory list, numbers to re-dial, etc. Every cell phone has an identification number and the EEPROM chip retains this and other permanent data. Not only does the cell phone process voice and DTMF tones, but it must receive and transmit a ream of data back and forth to the cell site. It also sends data to and from the cell control chip within the cell phone. The cell controller, after processing this data, sets up the correct transmit and receiver frequencies that the cell phone must operate on.

This should now give you a brief overview of what happens when you pick up and dial a cell phone or receive a cell phone call as the young lady is shown doing in the Fig. 8-54 photo.

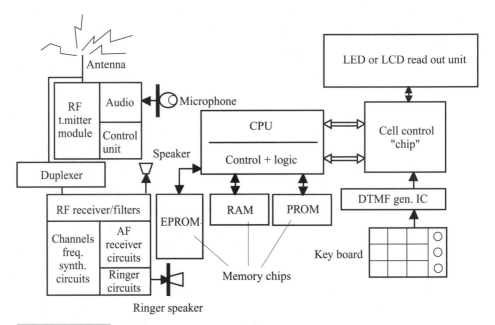

FIGURE 8-53   **A block diagram of a typical cellular phone unit.**

FIGURE 8-54   **A young lady in the process of making a cell phone call. Note: Auto is parked as call is being made.**

Some cell phone tips for poor, noisy or intermittent reception If you're using a portable cell phone, the problem could be a loose or broken antenna or your location. First, move to another location. With a mobile car phone, it could be poor coax cable, cable connections, or a broken or loose antenna. If the antenna has an on-glass mount, it might be defective or it could have been installed wrong. Do not overlook the possibility that the battery is weak or the battery contacts are dirty. Also, some computer chips in your car might cause interference to your cell phone—even if the engine is not running and the key is turned off.

Battery talk Most cellular portable phones use nickel-cadmium (NiCAD) battery packs. These packs are expensive and you need to take proper care of them. These battery packs are dated by the manufacturer, and will usually be replaced if found defective and failed at an early age. Keep the phone and battery in a cool place, if possible, and keep them properly charged. Read the instructions that come with the battery pack. Most are designed for a quick-charger system. The newer NiCADs do not have the memory problem like the older batteries. However, it is best if you can discharge them completely before recharging. You might want to keep a spare charged battery with you. These batteries do wear out after many charge and discharge cycles. Also, keep the pack dry and check/clean the contacts on a regular basis.

Drop-out and dead reception areas The ultra-high frequency of the trunking and cellular radio systems (800 MHz to 1000 MHz) are close to a "line-of-sight" RF signal transmission. For this reason, drop outs (loss of signals) occur when you are around hills, bridges, and large buildings. Your signal might fade in and out or flutter. In the worst case, your call might get disconnected. After you use your phone a while, you will get to know the various poor-reception areas. The dead zones will usually last longer when traveling in mountains, through hills and valleys, and in the larger cities with skyscrapers. These dead spots also depend on the proximity of the cell sites.

PERSONAL COMMUNICATIONS SERVICE (PCS)

An advanced cell phone wireless system called Personal Communications Service (PCS) is being rolled out in certain areas of the country. One of these PCS compact units is shown being used in Fig. 8-55. This PCS system is digital, of course, and a laptop computer can be used to upload and download on the net via a wireless connection. Also, the cellular carriers are quickly being converted over to digitizing their services. Generally, digital calls are clearer than the old-time analog system.

At the present time there is a lot of incompatibility of these PCS systems. These mobile formats are called CDMA, TDMA, and GSM, and which will be the one to survive would be a wild guess. Also, users of these three systems cannot tap into the cellular networks to make calls or receive calls on them.

The PCS incompatibility has to do with connections between your own phone and a mobile phone system. In many cases, if you can make the connection the first time, then you should be able to communicate with any other cell phone user. This should include any digital or analog cell phone system. At this time if you do not travel around and want to use the PCS service it can be cost-effective and convenient. However, for a wide area coverage and traveling then the cellular phone is the best bet.

FIGURE 8-55   Using the compact PCS unit.

BROWSING THE INTERNET

The AT&T PocketNet phone can be used to access the internet and pick up your e-mail while on the go. With the AT&T PocketNet phone you can browse the internet, send and receive e-mail, call up stock quotes, and check out the sports scores and news. This AT&T phone system uses Cellular Digital Packet Data (CDPD), the AT&T wireless data network, to transmit information without going through a server. Eliminating the server also does away with the conventional dial-up connections, which are usually busy. The multi-function "soft keys" will let the numeric keypad double as a keyboard for typing up e-mail messages; however, typing long messages is a little slow and tough with these "itty-bitty keys." The PocketNet Phone will let you track UPS and FedEx packages or send e-mail to a Fax machine. At this writing, the PocketNet Phone is now up and running in quite a few markets. I have found it does not always work in certain areas and does not always operate very well indoors. However, these problems can be corrected as the system is developed fully. All in all, the PocketNet Portable unit, which has blended voice and Internet wireless communication signals, is a big undertaking in the wireless technology.

EarthLink has a wireless service to provide Internet connections for hand-held computers that run the Palm and Pocket PC operating system. This EarthLink service uses the technology purchased from OmniSky wireless system.

EarthLink is the United States' number 3 Internet provider that offers monthly plans of $40 to $60 a month.

The company also offers wireless access to the Internet and e-mail via pagers from Motorola and Research in Motion.

SMART CELL PHONES

With chips and other electronic innards that keep getting smaller, the smart mobile phone has now become a reality. These smart phones, sometimes called personal digital assistants (PDAs), are designed to perform several different tasks. These devices have some features of the Palm hand-held computer, pager, cell phone, and lots of memory storage. The current smart phones are a little bulky and the keypads are not too easy to use. Thus, the smart phone has been "supersized" to include mobile phone, a web surfer, laptop computer, and lots of memory for phone numbers, etc. However, the small screens are tough to read. Also, with these smart ones you can send and receive e-mail while on the run.

Some smart ones   The three models I have used are the Kyocera model QCP 6035, the Samsung model SPH-1300, and the Handspring Treo model 180. Let's now take a look at the Treo-180, shown in Fig. 8-56. The Treo is a pretty small, rugged device and can easily be used with one hand. It is referred to as a communicator. The cost is about $400. When

FIGURE 8-56   **The Handspring Treo model 180 is a PDA smart phone with many features and a full keyboard.**

you open its clamshell case you see the BlackBerry thumb-typing keyboard. The Hand-spring "fast-lookup" software gives you access to many thousands of phone numbers that can be put away in the monster 16-megabyte memory bank with only a few thumb clicks. And you will find that other software will make sending short text messages very simple.

Cell Phones That Glow in the Dark

New on the market is a cell phone that glows in the dark, which means all movie and rock stars will need one. These are Motorola model i90c cell phones, shown in Fig. 8-57. One of the models has blue backlighting; the upscale exclusive i90c Limited Edition has a completely clear housing.

The Limited Edition is available only through select Bloomingdale's stores and sells for $399. The regular i90c lists for $199 at retail outlets nationwide and has a calculator, notepad, and the Sega game Borkov. Wow, what a small price to pay for a device of only 4 ounces and this much fun.

FIGURE 8-57 Motorola model i90c glow-in-the-dark cell phone.

Dual Cell Phones

What works better than a cell phone—well, of course, dual cell phones like those the busy DJ finds indispensable in Fig. 8-58 while taking on-the-air music requests for CD selections.

FIGURE 8-58 **A busy DJ using two cell phones to keep the music going.**

9

HOW REMOTE-CONTROL SYSTEMS WORK

How Remote-Control Systems Work

This chapter covers various remote-control systems, troubles that might develop, and what to do when they don't work.

The first TV remote "wireless" controls used ultrasonic frequencies in the 35- to 45-kHz range. Some of the Zenith remotes used hammers (clickers) to strike tuned metal rods (usually four) in the hand unit to produce (ring) one ultrasonic control frequency. With this set up, you could rattle a key chain and make the TV go off/on or change channels. These early model remotes produced only four to eight analog frequencies. Some later-model

Magnavox remotes generated ultrasonic control pulses and would have 10 or more remote control functions.

Some modern-day remote controls are shown in Figs. 9-1 and 9-2. Remote units now control TVs, VCRs, camcorders, stereo audio units, cable TV converter boxes, DBS satellite receivers, laser CDs, and much more. Of course, many multi-type remote controls will operate several different devices at the push of a button.

THE ULTRASONIC REMOTE TRANSMITTER

The ultrasonic remote control was used in the older TVs and very few of these systems are now in use. These remotes transmitted on an ultrasonic frequency range of 35 kHz to 45 kHz. A few models would generate 10 to 15 control pulses that could be decoded in the TV receiver for more logic control functions. A simplified circuit drawing of an ultrasonic remote-control unit is shown in Fig. 9-3.

THE INFRARED (IR) REMOTE-CONTROL TRANSMITTER

Modern remote controls use an infrared (IR) carrier frequency that is pulse-code modulated. The carrier frequency is approximately 35 kHz to 55 kHz. The pulses sent out are multiple cycles of usually 20 bits each that modulate the carrier. The logic coding is different for various devices, so only a particular remote will operate a device. However,

FIGURE 9-1   Some current-model remote-control hand units for controlling TV receivers.

FIGURE 9-2   A lineup of current remote-control devices used for DBS satellite receivers and cable TV control set-top boxes.

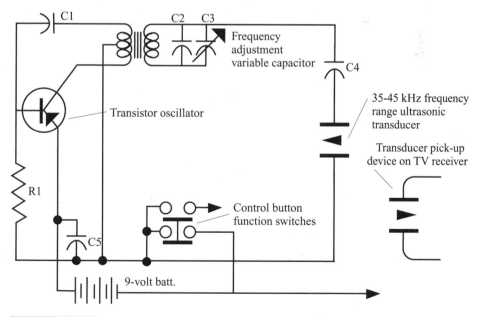

FIGURE 9-3   Simple circuit of an ultrasonic remote-control hand unit used to control older-model TV sets.

a universal remote can be reprogrammed for many different kinds of devices and multi-purpose remote control units are also supplied with many TVs and VCRs. A block diagram of an infrared digital remote-control transmitter is shown in Fig. 9-4.

Figure 9-5 shows the infrared remote receiver located within a TV, VCR, etc. The IR signal is picked up by an IR diode sensor on the front panel of a TV and is amplified and pulse decoded. The pulse codes are then sent to a remote-control microprocessor IC, which then sends control voltages to various parts of the TV circuits to control the set's operation, such as power on/off, volume control, etc. Figure 9-6 shows a typical color TV remote transmitter.

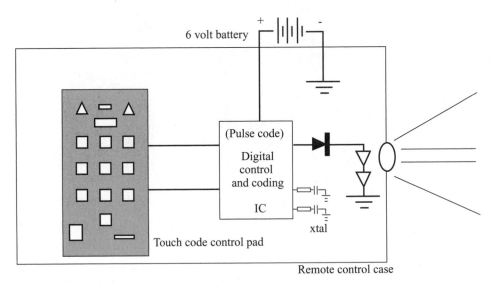

FIGURE 9-4   **A block diagram of an infrared (IR) digital remote-control transmitter.**

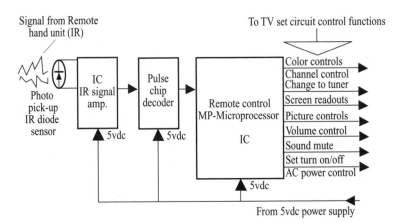

FIGURE 9-5   **A block diagram of the remote-control infrared (IR) circuits within the TV receiver.**

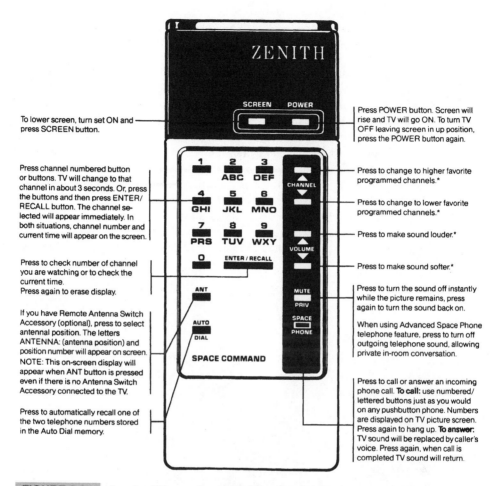

To lower screen, turn set ON and press SCREEN button.

Press POWER button. Screen will rise and TV will go ON. To turn TV OFF leaving screen in up position, press the POWER button again.

Press channel numbered button or buttons. TV will change to that channel in about 3 seconds. Or, press the buttons and then press ENTER/RECALL button. The channel selected will appear immediately. In both situations, channel number and current time will appear on the screen.

Press to change to higher favorite programmed channels.*

Press to change to lower favorite programmed channels.*

Press to make sound louder.*

Press to check number of channel you are watching or to check the current time.
Press again to erase display.

Press to make sound softer.*

Press to turn the sound off instantly while the picture remains, press again to turn the sound back on.

If you have Remote Antenna Switch Accessory (optional), press to select antennal position. The letters ANTENNA: (antenna position) and position number will appear on screen. NOTE: This on-screen display will appear when ANT button is pressed even if there is no Antenna Switch Accessory connected to the TV.

When using Advanced Space Phone telephone feature, press to turn off outgoing telephone sound, allowing private in-room conversation.

Press to automatically recall one of the two telephone numbers stored in the Auto Dial memory.

Press to call or answer an incoming phone call. **To call:** use numbered/lettered buttons just as you would on any pushbutton phone. Numbers are displayed on TV picture screen. Press again to hang up. **To answer:** TV sound will be replaced by caller's voice. Press again, when call is completed TV sound will return.

FIGURE 9-6   A color TV remote-control hand unit with the callouts for its various operational functions.

What to do when the remote control will not work   Remote-control units do not fail very often, unless they have been dropped, thrown around, or dunked in some kind of liquid. If your remote equipment does not work, make these quick checks:

1 If your TV (or other device) has a master on/off/manual/remote switch, be sure that it is in the Remote position.
2 If it has a multi-function control, be sure that it is not in the VCR function mode when you are trying to operate the TV.
3 If you have a universal remote unit, it might have become deprogrammed or programmed for the wrong model. Also, if the battery has been replaced, it will need to be reinitialized.
4 Speaking of batteries, you should now be sure that the battery is in good condition, with clean, corrosion-free contacts.
   Install a new known-good battery or use a dc voltmeter for a voltage check. The battery should be checked while under load in the remote unit. With the voltmeter connected to

battery terminals, press any button and see if the voltage drops more than 10%. If it does replace the battery with a new one. You can also use a battery tester meter because it puts the correct load on the battery for a valid test.

5 Be sure that you are using the correct remote unit because they look alike and some homes might have six or more various remote controls.

6 The problem could be in the TV or VCR that you are trying to operate. See if the TV or VCR will operate manually. If it does, try another remote, such as a universal one to determine if it is the remote control or TV/VCR.

7 To find out if your remote unit is transmitting an IR signal, you can use an IR detector card (Fig. 9-7). The card will show a red pulsing spot on the card if the control is transmitting. However, this does not indicate if it is sending out the correct pulse codes. These cards are available at electronic parts stores.

8 If the control unit has gotten wet, you might still be able to save it. As soon as possible, flush the unit in clean water and if you can take the case apart, use a hair dryer to completely dry out the case and circuit board. Then clean the battery contacts and install a new

FIGURE 9-7   **Testing a remote IR transmitter with an IR detection card.**

battery. It might keep on clicking. It's worth a try and you might not have to buy a new remote control.

UNIVERSAL REMOTE-CONTROL DEVICE

The Radio Shack universal remote control can replace many types of standard remote controls. These units are preprogrammed and do not have to "learn" commands from the original remote. Just "tell" this unit what remote control you want to replace (by entering the three-digit codes shown for many brands (Fig. 9-8) and the 3-in-1 universal remote control does the rest.

How to program the universal remote   Follow these steps to set up the 3-in-1 remote:

1 Check or install new batteries before programming the unit.

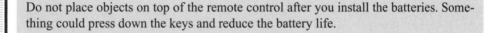

Do not place objects on top of the remote control after you install the batteries. Something could press down the keys and reduce the battery life.

FIGURE 9-8   The Radio Shack three-in-one universal remote-control unit. This can be programmed for many brands of TVs, VCRs, DBS receivers, cable boxes, DVDs, and other devices.

2 Refer to the device codes in Fig. 9-9 and write down this information.

3 Press the device key for the remote that you are replacing (TV, VCR, or CBL).

4 Press down and hold PROG until the red indicator blinks, and continue holding it down as you enter the 3-digit set brand code.

For example, to replace a Panasonic TV's remote control (code 051), you would press:

```
TV-PROG 0 5 1
```

5 When the LED indicator blinks twice, release PROG.

6 Point the remote 3-in-1 at your device you want to control and press Power. The device should turn on or off, if it was already on.

Repeat steps 2 through 6 for any additional devices.

The punch-through feature is automatically turned on for the TV's volume and mute controls. This means that when you select CBL and press one of the volume buttons or the Mute button, the remote actually sends the codes to the television and not to the cable converter box. If you want to use your cable converter's volume and mute controls, disable the punch-through feature for these buttons.

If the remote does not operate your device, try other codes listed in Fig. 9-9 for your brand of TV, VCR, or cable box.

The red indicator at the top of the 3-in-1 remote flashes twice when you enter a code that it recognizes. (This does not mean that it is the right code for your device, however.)

When the universal remote's range decreases or the remote operates erratically, replace with new batteries.

Be sure to have the fresh batteries ready to install before you remove the old batteries. The universal remote's memory only lasts about a minute without the batteries in place. If the memory is lost, simply re-enter the 3-digit code for your remote control.

If the universal remote stops working after you successfully test the control of each device (or if you are unable to get the unit to work at all), make the following checks:

■ Press the device key for the electronic device that you want to control.

■ Replace the batteries.

■ Confirm that your remote controls are working properly using manual controls or the original remotes.

Latest Code Lists for Programming the Remote Control

This sheet contains the latest brand and code information.
See your User's Manual for instructions on programming the remote to control other brands of equipment.

TV Codes

Akai	.002
Anam National	.038
AOC	.011,019,027,088
Candle	.011,027,093
Citizen	.011,027,053,064
Colortyme	.011,027,084
Concerto	.011,027
Contec/Cony	.056,057,040,042,064
Craig	.064
Curtis Mathes	.000,011,015,027,057
CXC	.064
Daewoo	.011,019,027
Dayton	.011,027
Electrohome	.006,011,014,
	.027,038,061,068
Emerson	.011,026,027,028,
	.029,030,031,032,057,042,053,
	.064,065,067,075,076,078,079
Envision	.011,027
Fisher	.017,021,059,041
Funai	.064
GE	.000,008,009,011,012,
	.027,058,068,086,089,091
Goldstar	.003,004,006,
	.011,019,027,037,050
Hallmark	.011,027
Hitachi	.009,011,027,036,
	.057,040,047,063,080
Infinity	.013
JBL	.013
Jensen	.011,027
JVC	.012,024,096,
	.057,040,048,051,074
Kawasho	.002,011,027
Kenwood	.006,011,014,027
Kloss Novabeam	.035,043
KTV	.078
Loewe	.015
Luxman	.011,027
LXI	.013,018,021,023,054
Magnavox	.006,007,
	.010,011,013,016,027,033,
	.035,043,049,066,087,089
Marantz	.013
Maranz	.011,013,027,069
MGA	.006,011,014,019,
	.022,027,041,056,061,068
Mitsubishi	.006,011,014,019,
	.022,027,041,059,056,061,068
MTC	.011,019,027
Multivision	.081
NAD	.018,023
NEC	.011,014,019,027,038,084
Panasonic	.012,013,038,086

Penney	.000,008,011,019,
	.027,040,068,077,086,088
Philco	.005,007,010,011,013,
	.016,019,027,033,035,
	.057,058,043,087,089
Philips	.002,006,007,010,
	.011,013,016,033,
	.055,057,058,043,066,073
Pioneer	.011,027,045,062,093
Portland	.011,019,027,037
ProScan	.000
Proton	.011,027,057,072
Quasar	.012,038,092
Radio Shack	.000,021,025,
	.056,057,059,064,078
RCA	.000,006,011,019,
	.027,034,038,044,046,088
Realistic	.021
Sampo	.011,027
Samsung	.006,011,014,
	.015,019,027,036,057,077
Sanyo	.017,021,059,
	.056,057,058
Scott	.028,057,064
Sears	.000,006,011,014,
	.017,018,021,023,027,
	.039,040,041,051,071,083
Sharp	.011,020,025,027,
	.057,052,053,059,060
Sony	.002
Soundesign	.011,027,053
Sylvania	.006,007,010,
	.011,013,016,027,053,
	.035,043,049,066,087,089
Symphonic	.064,076
Tatung	.058
Technics	.012
Techwood	.011,027
Teknika	.011,019,027,
	.053,056,057,040,066
Telecaption	.090
TMK	.011,027
Toshiba	.018,021,023,
	.040,071,077,085
Universal	.008,009
Victor	.051
Vidtech	.000,005,006,
Wards	.007,008,009,010,011,
	.013,019,025,027,028,
	.035,043,059,066,076,082,069
Yamaha	.006,014,019,027
Zenith	.001

Cable Box Codes

ABC	.022,046,053,054
Anvision	.007,008
Cablestar	.007,008
Diamond	.056
Eagle	.007,008
Eastern Int.	.002
General Instruments	.046
GI 400	.004,005,015,
	.023,024,025,050,056
Hamlin	.005,012,013,054,048
Hitachi	.037,043,046
Jerrold	.004,005,015,023,
	.024,025,050,056,045,046,047
Macom	.057,043
Magnavox	.007,008,019,021,
	.026,028,029,032,033,040,041
NSC	.009
Oak	.001,016,038
Oak Sigma	.016
Panasonic	.003,027,059
Philips	.007,008,019,021,
	.026,028,029,052,033,040,041
Pioneer	.018,020,044
RCA	.000,027
Randtek	.007,008
Regal	.005,012,013
Regency	.002,035
Samsung	.044
Sci. Atlanta	.003,022,035
Signature	.046
Sprucer	.027
Starcom	.046
Stargate 2000	.058
Sylvania	.011,059
Teknika	.005
Texscan	.010,011,059
Tocom	.017,021,049,050,055
Unika	.081,052,041
Universal	.051,052
Viewstar	.007,008,019,021,
	.026,028,029,052,033,040,041
Warner Amex	.044
Zenith	.014,042,057

VCR Codes

Aiwa	.015
Akai	.003,017,022,023,063,066
Audio Dynamics	.014,016
Broksonic	.010
Candle	.007,009,
	.013,044,045,046,052
Cannon	.008,053
Capehart	.001
Citizen	.007,009,013,
	.044,045,046,052
Colortyme	.014
Craig	.007,012
Curtis Mathes	.000,007,008,
	.014,015,044,046,053,064,067
Daewoo	.013,045,052
DBX	.014,016
Dynatech	.015
Electrohome	.027
Emerson	.008,009,010,
	.013,015,020,023,
	.027,034,041,042,047,049,
	.057,062,065,067,068,070
Fisher	.002,012,018,
	.019,045,048,058
Funai	.015
GE	.000,007,008,032,053
Goldstar	.009,014,046,060
Harman Kardon	.014
Hitachi	.005,015,035,036
Instant Replay	.008
JCL	.008
JC Penney	.002,005,007,008,
	.014,016,030,035,051,053
JVC	.002,014,016,030,046
Kenwood	.002,014,
	.016,030,046
Lloyd	.015
Logik	.031
Magnavox	.008,029,053,036
Marantz	.002,008,014,016,
	.029,030,044,046,061
Marta	.009
MEI	.008
Memorex	.008,009,012,015
MGA	.004,027
Midland	.032
Minolta	.005,035
Mitsubishi	.004,005,
	.027,035,040
Montgomery Ward	.006
MTC	.007,015
Multitech	.007,015,031,052
NEC	.002,014,016,030,
	.044,046,059,061,064

Panasonic	.008,053
Pentax	.005,035,044
Pentex Research +	.046
Philco	.008,029,053,056
Philips	.008,029
Pioneer	.005,016,035,050
Portland	.044,045,052
ProScan	.000
Quarta	.002
Quasar	.008,053
RCA	.000,005,007,008,
	.028,035,037,054,069
Radio Shack/Realistic	.002,
	.006,008,009,012,
	.015,019,027,043,053
Samsung	.007,013,022,032,042
Sansui	.016,071
Sanyo	.002,012
Scon	.004,015,041,049,068
Sears	.002,005,009,012,
	.018,019,035,045,048
Sharp	.006,024,027,059,045
Shintom	.017,026,031,055
Sony	.017,026,068
Sylvania	.008,015,029,053,056
Symphonic	.015
Tandy	.002,015
Tashiko	.009
Tatung	.050
Teac	.015,030,069
Technics	.008
Teknika	.008,009,015,021
Toshiba	.005,019,048,049
Totevision	.007,009
TMK	.067
Unitech	.007
Vector Research	.014,016,044
Victor	.016
Video Concepts	.014,016,044
Videosonic	.007
Wards	.005,006,007,008,009,
	.012,015,019,025,027,031,035
Yamaha	.002,014,016,030,046
Zenith	.011,017,026

Laserdisc Player Codes

RCA	.053
Pioneer	.053

1Q57 404-01A

FIGURE 9-9 A code listing for the Radio Shack three-in-one universal remote-control hand unit.

■ If some buttons do not function for your device, you might be able to scan to a better device code.

Remote-control care and maintenance Your remote controls are an example of electronic devices of good design and workmanship, and should be treated with care. The following tips will help you enjoy these electronic wonders for many years.

■ Keep the remote unit dry. If a liquid spills into the unit, dry it off as soon as possible and dry it with a hair dryer. Liquids contain products that can corrode the electronic circuits.
■ Handle the remote control gently and carefully. Dropping it can damage its circuit boards and case and cause the control to work improperly.
■ Use and store the remote control only in normal room-temperature environments. Temperature extremes can shorten the life of electronic devices and distort or melt plastic parts.
■ Keep the remote control away from dust and dirt, which can cause premature wear of parts.
■ Wipe the remote control with a damp cloth occasionally to keep it looking new. Do not use harsh chemicals, cleaning solvents, or strong detergents to clean the remote control.
■ To prevent any internal damage, do not twist the remote unit.

Remote-control extenders The remote-control extenders are radio-frequency (RF) devices that let you control TVs, lights, radios, and DBS satellite receivers from other rooms or even outside your home. Most remotes use infrared light signals and cannot go through walls of a building. Basically, they are a "line-of-sight" control device.

Now look at a remote-control extender made exclusively for the RCA DSS satellite dish receiver. This unit is made by Windmaster (904-892-7815) and is shown in Fig. 9-10. This remote extender will let you control the DSS receiver from any room in your home.

Transmitter and receiver extender installation The transmitter attaches to the remote control and senses the infrared signals from the remote control. The transmitter converts the IR control signals into RF waves. Figure 9-11 shows the extender being installed.

The base receives the RF signal from the transmitter hand unit and converts these waves back to infrared signals.

Place the base receiver in front of your RCA DSS satellite receiver (as shown in Fig. 9-12). The base receiver must be located so that no obstructions will be between it and the receiver. Figure 9-13 shows the extender installed on the remote and ready for use.

What to do if you have trouble with the extender

1 Be sure that the base receiver is plugged into a working electrical ac outlet.

2 If the red LED light inside the transmitter does not light when the remote control is operated, check the following:

FIGURE 9-10   **The Windmaster remote control DBS extender control unit on the front of the DBS control unit and the receiver unit.**

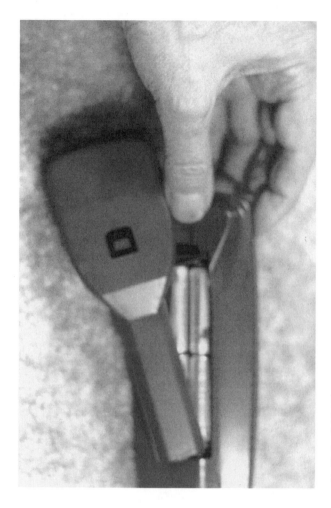

FIGURE 9-11   The battery cover is removed from the DBS remote unit. Just snap the remote extender onto the RCA hand remote-control unit.

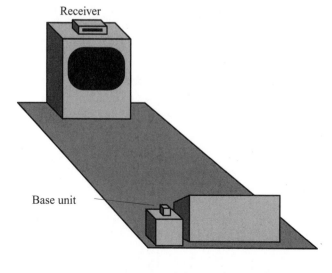

Receiver

Base unit

FIGURE 9-12

Placement and location of the remote extender base unit.

FIGURE 9-13 The remote DBS Windmaster extender installed and being used for RF control.

■ Be sure that the battery inside the transmitter unit is good and installed correctly.

3 If your receiver does not respond, but the red light inside the base receiver lights when the remote control is used, be sure that nothing is blocking its light path.

4 If the red light on the transmitter lights when you operate the remote, but does not light in the base receiver, then check the following:

■ Move the antenna to a different position for better reception.
■ Lower the antenna rod.
■ Move the base receiver to another location and try to operate it again.

To order a DBS remote extender, call 1-800-624-4112.

INTELLIGENT REMOTE-CONTROL SYSTEM

The Philips Pronto Intelligent remote system has been developed to control several electronic devices with only one control unit. This Pronto Remote is shown in Fig. 9-14. The remote is lightweight but has more heft than a standard remote control unit. When turned

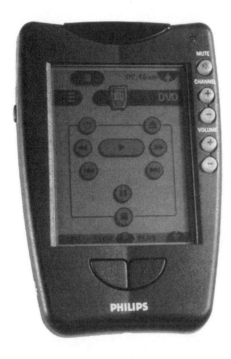

FIGURE 9-14   **The Philips Pronto Intelligent control hand unit.**

on, the Pronto has a high-resolution (320 × 240) LCD touch-screen with a back light. All of the icons are present for ease of operation. Every imaginable button you would want is featured along with access buttons along the outside of the LCD screen that works even when the remote is in the off mode (volume, channel change, and mute). All that you have to do is give it a slight tap and you are ready to control your equipment.

The Philips Pronto operates on four AA batteries, and a recharger unit is available for them. It would probably be wise to have the recharger, as the Pronto will draw more battery power than a standard TV remote control unit. When it is in actual use, I noted the battery level meter was taking up battery power at a pretty good clip, which would be expected with all of the operational modes that the Pronto can perform.

The Pronto is easy to program. The first step is to set the time, day, screen contrast, and how long you want the remote to stay on after being used. I think you will find that programming the Pronto is quite easy when using the 36-page user's operational booklet.

SONY'S RM-AV2100 UNIVERSAL LEARNING REMOTE

The best universal remotes, I have found, come with preprogrammed codes and the ability to learn or record the commands that are not already programmed into it. As an example, you can put in the rewind button of your VCR, even if the built-in code does not have it listed.

Note that "macros" are buttons you can program to carry out a vast sequence of commands with just one touch of a button of the control. By using the macros, as an example, you only have to tap one or two buttons to go to a cable TV mode.

With the Sony RM-AV2100 touch-screen remote, shown in Fig. 9-15, every component has its own button layout. And each button can be named. And it's easy to find the right button,

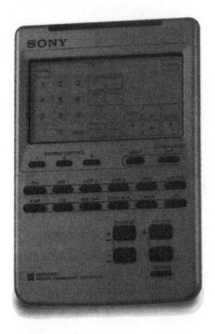

FIGURE 9-15   Sony's RM-AV2100
universal learning remote control that makes
remote control easier and eliminates using
a multitude of remotes.

because of the bright blue display, which automatically goes dark after a few seconds after no
more commands are entered. You also have the option of tapping the "Simple" button if you
would like to hide the obscure, set-and-forget commands.

One of the nice advantages of the RM-AV2100 is its ability to make macro operations
faster. Generally, macros execute one or two commands a second. This means a 15-step
macro will last 7 or 8 seconds. You will find that the Sony RM-AV2100 can complete a
15-step macro in less than 3 seconds.

PROGRAMMING THE LEARNING REMOTE

To program the RM-AV2100, round up all of your remotes from various units that you
want to consolidate into one. You should do this where there are no bright lights. Also,
have fresh batteries installed in your single-function remotes.

Now, line up the "eye" of the learning remote with the emitter of your "teaching" remote
that you are using for a TV set, VCR, dish receiver, etc. Check the instructions that come
with the learning remote, then push a few buttons, and the infrared code emitted by the teaching remote is received by the learning remote.

Always test the first command you teach a remote before trying any others. If it does not
work properly, the most likely causes are as follows:

1 Batteries may be weak. Replace with fresh ones.
2 Change the distance between the teaching remote and the sensor of the learning remote.
   Try it at different distances from ½ inch to 4 feet.
3 Try to teach the command by briefly tapping it, instead of pressing and holding it.
4 If you are programming on a shiny surface, try moving the remote so that reflections
   will not affect it.

5 Check and make sure the room lighting is dim.

6 If you have a plasma TV or monitor operating nearby, turn it off.

7 If you are teaching from a two-way remote, move to another room.

8 Sometimes a remote will not learn a command because you have already filled the memory space that is available. Go back and try to save space by reteaching commands by using the "tap" techniques as in tip 3 above. If the code still works, this will save a lot of memory. The tap technique will not work with volume up and down.

Tips on Macro Programming

Macro programming is a process of recording a sequence of button pushes. Every button pushed is counted as a step. It's tempting to cram as many steps into each macro as possible, but that may not always be a good way to go.

Let's now see what can go wrong when you overdo this macro technique. A VCR stop command, as an example, might cancel a recording that you may want to use at another time. Another problem would occur if you should program all of the power-on commands into a single macro. This is what might occur.

The VCR is turned on automatically when you put the tape in. Now when you tap the macro, it will turn everything else ON, but will turn the VCR OFF. Now, if you tap the macro again, the VCR turns ON, but all other devices turn OFF.

It would work better if you would turn on the TV and the receiver first. Then, leave those steps out of your "look at cable TV" and "watch DVD" macros.

DESIGNING USER-FRIENDLY MACROS

You will note that some buttons will turn a component ON, but will not turn it OFF. In one case, if you tap the play button on most DVD players, the player will turn ON. Tap the play button again, and the DVD player stays ON.

Sometimes a remote's input buttons (called TV, VCR, or DVD) do not just select the inputs, they also turn on, but will not turn off, the corresponding components.

These "on-only" buttons help you make macros much easier to use. Using them to turn on a component, instead of the on/off power button, makes the macro much more reliable.

Programming the Sony Universal Learning Remote

First, you turn on the TV and receiver.

1 Pick up the receiver's remote control unit.

2 Aim the remote control at the receiver and press the power button.

3 Pick up the TV remote control.

4 Aim at the TV and press the power button. Now perform the following procedures.

TO VIEW CABLE TV PROGRAMS

1 Pick up the receiver remote control.
2 Aim at the receiver and press the TV button.
3 Pick up the TV remote.
4 Aim at the TV set and keep tapping the input button until the TV displays ANT A.
5 Pick up the cable box remote unit.
6 Aim at the cable box and press the power button.

Use the cable box remote for changing channels and the receiver remote for volume.

FOR VIEWING DVD PROGRAMS

1 Pick up the receiver remote control unit.
2 Aim at the receiver and press the DVD button.
3 Pick up the TV remote.
4 Aim at the TV and keep tapping the input button until the TV displays EXT 1.
5 Pick up the DVD remote control.
6 Aim at the DVD remote player and press the power button, then the play button.

Use the DVD player remote control for play, pause, fast-forward and the receiver remote for volume control adjustments.

FOR VIEWING VCR TAPES

1 Pick up the receiver remote control unit.
2 Point it at the receiver and press the VCR button.
3 Now, go to the TV remote unit.
4 Point it at the TV set and keep tapping the input button until the TV displays EXT 2.
5 Next, pick up the VCR remote control unit.
6 Aim this unit at the VCR and press the power button, then the play button.

Use the VCR remote for play, pause, fast forward, etc., and the receiver remote for volume level control.

SCROLLING COMMANDS FOR THE SONY RM-AV2100

When you have to continually tap the remote to change the TV set's input modes, this is referred to as a "scrolling" command. Because you do not know what input your TV set was left on, you cannot program a macro to perform with a certain number of taps to obtain the required input.

To eliminate this problem, a preprogrammed universal remote can have some codes that will specifically select an input of your TV set, even if the supplied remote did not have such a command. You can try all of the codes listed for your brand of TV receiver. Then, with each code, try all of the buttons on the control. You may find a button that will directly select the input you want. Also, I have found on some TV receivers, the channel up control will reset the input to ANT A.

Should you locate a button that will directly select a certain TV input, you may have a key program "niche" that can be used. This niche code may let you program steps that will let you build up your own secret command. Let's now review how this can be accomplished:

1 Designate the sequence with the TV channel up command. Thus, regardless of what the TV input was on, it is now ANT A.
2 Now scroll with one tap of the input button, and the TV set is on ANT B.
3 Scroll with one tap of the input button, and the TV receiver is on EXT 1.
4 Scroll with one more tap of the input button, and the TV set is now programmed for EXT 2.

If the Up Channel Commands will not do as stated above you can try the following program steps:

1 Try entering a channel number and the enter command when the TV receiver is set to an external or auxiliary input. If it goes to the antenna position, you have your niche code. This seems to work on RCA and other TV brands, but not all of them.
2 Try entering 0, 2, enter, and then channel down. If the TV goes to external or aux input, then you have another niche code.
3 Another technique you can try is to enter the highest TV channel number and then channel up. If the TV goes to external or aux input, you now have another niche code.
4 I have tried entering a niche code on my RCA ProScan TV receiver and it does work. These ProScan TV receivers use channel 90 for input, 91 for another input, and 92 for the last input code.

Remote Control Selection

You probably can improve the odds that a preprogrammed remote unit will have all of the codes you can use if you obtain one that has lots of buttons and a massive built-in code memory capacity.

Of course, what you want to do is trim down the amount of remote control units you need to operate all of your entertainment equipment. When you program a multibrand remote unit that we just explained, you may not cut down on the number of steps it takes to switch from cable TV viewing to operating your DVD, but you have been able to eliminate the number of remote controls you have to operate.

Radio Shack VCR Programmer

With the Radio Shack VCR programmer shown in Fig. 9-16, you just follow the six steps printed on the unit to program most VCRs. Simple on-screen prompts help ensure supereasy setup. With the large buttons you can't miss. Big LCD displays show day, date, time, channel, and recording lengths—so you can always double-check to see if all is programmed correctly. This VCR programmer will now automatically turn on your VCR, cable box, or satellite receiver box and then switch to the desired channel when it is time to record. This VCR remote control unit requires three AAA batteries for operation.

Backlight display.
Read simple step by step instructions for programming your VCR.

Step 1.
Start programming sequence to record your favorite show.

Step 2.
Select the day of the week you want to record.

Step 3.
Choose the time you wish to start recording.

Step 4.
Scroll through recording time lengths.

Step 5.
Select channel.

Step 6.
Lock in and save your settings.

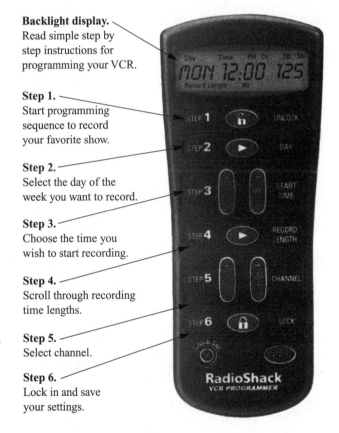

FIGURE 9-16 The Radio Shack VCR programmer remote-control hand unit.

10

PRINTERS, COPIERS, AND FAX MACHINE OPERATIONS

(Continued)

Daisywheel Printer Operation

The daisywheel printer system is used in *personal word processors (PWPs)* and electric typewriters. A drawing of the daisywheel layout is shown in Fig. 10-1. A photo with two daisywheels is shown in Fig. 10-2. Figure 10-3 shows the actual placement of the daisywheel in a PWP printer. A typical PWP is illustrated in Fig. 10-4.

The daisywheel printer is an all-electronic machine that prints faster and is quieter than a typewriter. These printers are faster because they print, bi-directionally, to both left and right on alternating lines at 20 characters per second. However, these daisywheel printers work at a much slower speed than the other printers covered later in this chapter.

The heavy-duty printwheel (Fig. 10-1) consists of spokes or petals. Each spoke contains a unique character. The printwheel will actually rotate very fast to the left or right, depending upon the character that has been typed from the keyboard, or with a printer, the wheel is controlled by the PC. The printwheel and print hammer is shown in Fig. 10-5. An electronic digital IC interface that is controlled by the keyboard tells the printwheel when and where

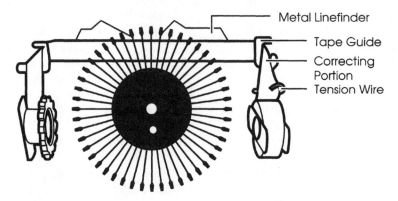

FIGURE 10-1   **The daisywheel found on some printers and typewriters.**

FIGURE 10-2   A close-up view of two daisywheels.

FIGURE 10-3   A daisywheel being installed on an Olivetti printer.

to spin and then activates the hammer solenoid to strike the printwheel character and make its mark. These wheels come in various styles and sizes that you can interchange to produce different looking documents. These wheels come in "pitches" of 10, 12, or 15 characters per inch.

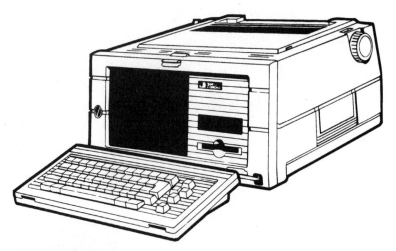

FIGURE 10-4 A typical older-model personal word processor (PWP).

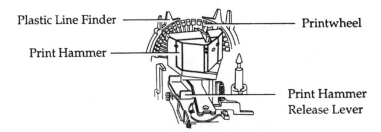

Plastic Line Finder — Printwheel

Print Hammer —

Print Hammer Release Lever

FIGURE 10-5 The printwheel and print hammer that transfers images on the daisywheel to the paper via a ribbon cartridge.

These printwheel printers use ribbons and can be made of fabric, carbon, or a correctable ribbon. Some machines even have a correcting tape when they are used as a typewriter. Most of these machines will print documents with either a ragged right margin (uneven) or a justified margin (perfect straight right margin).

Daisywheel Printer Tips

DATADISKS

Although the DataDisks are not fragile, certain precautions should be followed.

- Do not place the DataDisk near any magnetic object.
- Do not expose the DataDisk to temperature extremes.
- Do not bend the DataDisk.
- Do not store in any power cord storage compartment because they are usually close to an electromagnetic device, such as a power transformer.

KEYBOARDS

To clean covers or keyboards, sponge it off with a mild ammonia or soap solution. Do not use household cleaners containing chlorinated components.

PRINTWHEEL

To remove any residue from the printwheel, dip the characters wheel edge into a small container of ethyl or isopropyl alcohol (rubbing alcohol) and wipe with a clean dry cloth. Do not soak the printwheel.

PLATEN CLEANING

Wipe the platen surface off with a mild soapy solution. Do not use household cleaners containing chlorinated compounds.

MONITOR SCREEN

The monitor screen should be cleaned with the power turned off. Dust with a dry, soft cloth or use a good-quality CRT screencleaning kit that will neutralize static and will not streak or scratch the monitor screen.

Checks for PWP Machines

If your PWP does not function properly, then perform the following checks:

- Check for proper position of the correction tape spool.
- Does the ribbon cassette cartridge need to be replaced?
- Is the top lid closed tightly?
- Does the correcting tape need to be replaced?
- Has the print carrier been released?
- Has the printwheel been installed correctly?
- Has the printwheel been installed?
- Has an object fallen into the carriage and jammed its operation?
- Has the print hammer device been positioned correctly?
- Is the monitor screen dim or blank? Try adjusting the contrast and brightness controls.

How the Ink-Jet (Bubble) Printer Works

Now, find out how the ink-jet printer works. The ink-jet printer is also referred to as a *bubble-jet printer*, which will become obvious as the system is explained. The Cannon Model 1000 is used for the ink-jet system operation (Fig. 10-6). When connected to your PC, the Model 1000 does not only make print documents, but can make copies,

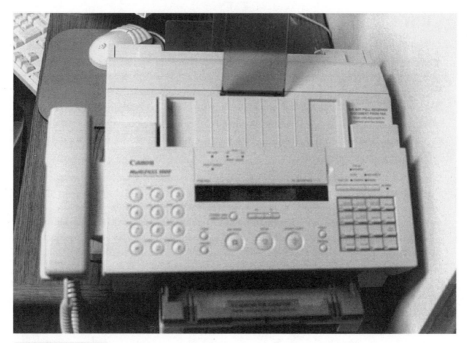

FIGURE 10-6   The Canon model 1000 printer, copier, fax, and scanner multimachine.

faxes, and scans. The ink-jet printers are very compact, lightweight, and have about the same print resolution as a laser printer. Compared to the standard typewriter and daisy-wheel machines, the ink-jet is very quiet during operation. The ink-jet machine prints the paper by squirting small droplets of ink out of the print head nozzles. Figure 10-7 shows the print head and ink cartridge being replaced on the Cannon Model 1000 multi-pass machine. Figure 10-8 gives you details for replacing the ink-cartridge/print-head assembly.

An ink-jet printer contains four major blocks (Fig. 10-9). These blocks include the print head, paper handler, carriage transport, and electronics logic control board. A motor starts the print head moving along a track and IC printer circuits send a voltage pulse to each head nozzle, which then leaves the proper mark on the paper.

PRINT CARTRIDGE AND NOZZLES OPERATION

The printer cartridge contains many ink-filled chambers that feed into each ink-jet nozzle. Figure 10-10 shows the print head, which contains many fine nozzles, whose diameter is smaller than a human hair. Each ink chamber and hole has a thin resistor (or heating element) that is fed a controlled electrical pulse from the logic control board at the proper time to heat the ink to more than 900 degrees for a few millionths of a second.

Thus, the ink is heated to form a bubble vapor. When the heated bubble expands, it is pushed through the hole. The pressure of the vapor bubble shoots the ink droplet onto the paper. The

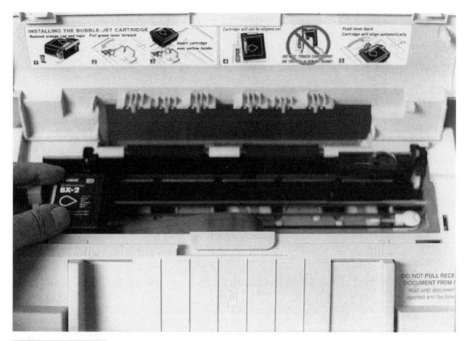

FIGURE 10-7   The print head and ink cartridge being replaced on the Canon model 1000 machine.

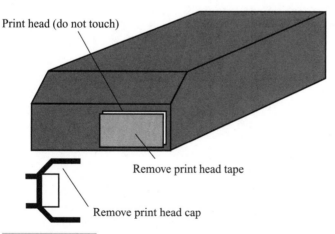

FIGURE 10-8   This illustration shows the print head and ink cartridge and ink cartridge replacement details.

print character is then formed by an array of these droplets pushed through the micro holes. The more and finer nozzle holes in the print head, the better the printer resolution or sharpness will be.

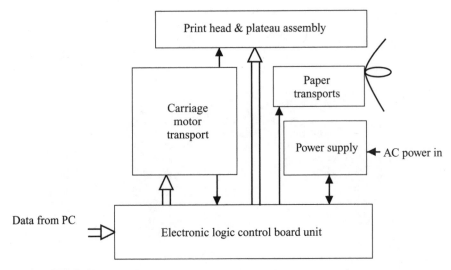

FIGURE 10-9   **A simplified block diagram for the ink-jet printer operations.**

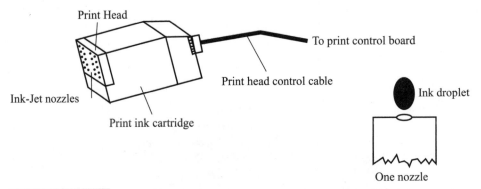

FIGURE 10-10   **A simplified close-up view of the print head and ink-jet nozzles.**

INK-JET HEAD PROBLEMS

These ink-jet printers will print on many different types of paper surfaces. They have a higher printer speed than daisy-wheel printers and are very quiet. If your machine is not printing properly, be sure that you are using quality paper for ink-jet printers. Porous paper will absorb the ink and make your printed documents look faded and dull.

INK-JET PRINTER PROBLEMS

Now take a look at a few common printer problems. Many print problems are caused by a defective print head or it has run out of ink. The ink-jet machines use a print head/cartridge combination module that is very easy to snap in and out. You might see some of the print characters partially missing and this is probably caused by some of the nozzles being plugged up in the print head. In this case, you must replace the head cartridge. Other printing problems can be caused by a defective IC in the logic control unit or head-driver electronics. Try cleaning all cable plug pins and push-on connectors. If the machine will not print at all, check the status lights and see if it is on line. Also check the interface cables and connector plugs from your computer. If the print head moves back-and-forth, but does not print, suspect a defective print head, out-of-ink cartridge, defective print head cable, or dirty or broken connectors to the print head. Then recheck and be sure that the print head is installed correctly.

PAPER-HANDLING PROBLEMS AND CHECKS

Printers use two different types of paper-feed systems. These are friction feed, like a typewriter uses, and the tractor feed, which requires special paper with notches on each edge of the paper (sometimes referred to as *computer paper* or a *continuous-feed paper*). Paper-feed problems can be caused by paper not installed properly, wrong type of paper, or mechanical problems. Be sure that the correct paper is being used and that it is installed properly. Then check and clean the platen and pressure rollers. Check and clean any drive gears and chain or belt drives. Be sure that the gears mesh and move freely. The drive feed motor might be defective or a fault might have developed in the logic IC control board.

PRINT-HEAD CARRIAGE ASSEMBLY PROBLEMS

The ink-jet printers move the print-head from left to right and back again on a rail to print across the page surface. Figure 10-11 shows the BX-2 print head cartridge on the left side; as it prints, it moves to the right side on a rail guide and is pulled by a belt or chain drive powered by the carriage motor.

If the print head does not move across the carriage rod or moves in jerks and does not position itself at the left side when the printer is turned on, it probably has a mechanical problem. This problem could be a loose or broken carriage belt, chain, drive gear, or pulley, and possibly a faulty carriage motor. If the belt is loose or broken, you will need to replace it. If your machine uses a carriage chain or gears, they might only need to be adjusted or cleaned. Also, check, clean, and tighten all cable plug-in pin connections to the motor and control PC boards.

If the belt, cable connections, and motor checks out OK, then the problem will be in the electronics logic control PC board's drive circuits, the optical encoder, or a power supply problem. Always check the power supply for correct dc voltages when you have a print head, carriage transport, control or drive circuit, and paper-feed problems. To prevent ac line surge damage, plug your printer into a protection device (Fig. 10-12). On fax machines, always check and clean the phone module plugs that are shown in Fig. 10-13.

SOME MULTI-PASS TROUBLESHOOTING TIPS

This section for the Cannon MultiPass machine looks at software problems.

Printout is wrong

- Check and see that the cable connections are clean and tightly secure.
- If you are printing in DOS, check that the printer control mode matches the driver that you selected.

FIGURE 10-11 The BX-2 print head is shown on the left side. It moves to the right side on a carriage rod to print lines across the paper.

FIGURE 10-12 A surge-protection device for your printer that plugs into an ac outlet. It also has connections to plug in phone and fax equipment for phone-line protection.

FIGURE 10-13   For intermittent fax problems, be sure all module phone plugs are clean and tightly connected.

Print job vanishes   If you have a print job that vanishes or are printing garbage, the problem might be because another Windows application on your PC is trying to communicate with the printer port of the MultiPass Server it is using. This conflict will generally result in strange printed garbage.

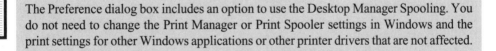

The Preference dialog box includes an option to use the Desktop Manager Spooling. You do not need to change the Print Manager or Print Spooler settings in Windows and the print settings for other Windows applications or other printer drivers that are not affected.

BUBBLE-JET PRINT JOBS DISAPPEAR UNDER WINDOWS

The MultiPass software includes a driver named MPCOMM DLL. This driver adds to the standard Windows communications driver named COMM.DRV. Under Windows 3.1, MPCOMM.DLL serves to improve bubble-jet printing performance and ensures that the MultiPass software does not interfere with any serial communications that is set properly.

CHARACTERS ON SCREEN DO NOT MATCH PRINTED CHARACTERS

Many graphics characters and special symbols are produced by different ASCII codes on each make of computer and printer. You will need to reset the correct character table and printer control codes if you encounter this problem.

PRINTOUT DOES NOT MATCH PAPER SIZE

■ Be sure that the paper size you selected in the software matches the paper loaded into the unit.

■ Be sure that the width of the paper on which you are printing matches the width indicated by your software so there is always paper between the print head and the platen. The print head might be damaged if it prints directly on the platen. If the print head prints on the platen, feed a few sheets of paper through the printer to clean the ink off the platen.

THE MACHINE WILL NOT PRINT ANYTHING

■ Be sure that the printer is plugged in and that no error light conditions are displayed.

■ Be sure that the MultiPass server is loaded in.

■ Check print manager and delete all pending jobs, then retry the print operation.

■ If you are printing from a non-Windows program, be sure that the unit is online and in the Printer mode.

YOU CANNOT PRINT FROM THE FILE MENU IN A WINDOWS APPLICATION

Check to be sure that the unit is correctly connected to the computer and that it is turned on. If it is still not printing, perform the following:

■ Open the control panel/printers in Windows.

■ Set the Canon MultiPass printer as the default printer.

■ Click the connect button to be sure that the correct port has been selected, then click OK.

■ Doubleclick the Canon MultiPass name. This sets the Canon MultiPass printer as the default printer.

THE PRINTOUT IS TOO LIGHT

If you are printing in HS mode, the print quality might be too light or the print settings in Windows might be set to draft. Choose the HQ mode.

DISCONNECTING THE PRINTER PORT

You might want to use the printer port on your computer for other equipment. If you want to do this, you must disable the MultiPass Server software before disconnecting the Multi-Pass printer.

To disconnect the MultiPass printer perform the following items:

1 Turn off your computer.
2 Unplug your computer from all electrical sources.
3 Unplug the MultiPass printer from all electrical sources.
4 On the back of your computer, remove the cable connector from the parallel printer port.
5 On the MultiPass 1000, release the wire clips and remove the cable connectors from the port.
6 Now plug your computer back into the electrical outlet ac socket.

7 Then unplug the phone cable from the MultiPass 1000. Your Canon printer is now disconnected from your computer.

UNINSTALLING THE MULTIPASS DESKTOP MANAGER

You must use the Uninstaller program when you want to remove the Desktop Manager and related scanner, printer, and fax drivers, and install a new program version or want to use another type printer with your computer.

UNINSTALL PROGRAM FOR WINDOWS 95

Perform the following steps for uninstalling the MultiPass server for Windows 95.

1 Close the MultiPass Server.
2 Click the Start button and point to Programs. Click MultiPass Utilities. Then click MultiPass Uninstaller.
3 Then you follow the instructions that come up on your computer screen. When the program is completed, you can return to the desktop program. The files will all be deleted. If you want to install a new version, you can do so at this time.
4 To completely delete the MultiPass desktop manager, use the Windows Explorer.

DIAGNOSING SOFTWARE AND HARDWARE PROBLEMS

If you are having some type of printer problem, you can use the MultiPass diagnostics to identify software configuration problems as well as hardware installation problems.

Using MultiPass diagnostics for Windows 95

1 Click the Start button and point to Programs. Click MultiPass Utilities. Click MultiPass diagnostics. The program will then start and the diagnostics will begin. When the diagnostics are finished, a message appears, stating that all tests were performed successfully. If there were any problems, messages appear suggesting solutions.
2 A dialog box appears asking if you want to view a log file. The log file contains important trouble information and solutions.
3 Click Yes to view the log file. The MultiPass Diagnostics window appears.
4 To save this file, choose Save from the File menu. Select a drive and directory if you do not want the file saved in the MPASS directory. In the File Name box, type a name.
5 Click OK. The file is saved as a plain ASCII test file.
6 To exit this window, choose MultiPass Diagnostics from the Exit menu. The window closes and you return to the desktop.

Plain-Paper Fax-Machine Operation

The plain-paper fax machine is used to transmit and receive images, printed or graphics, over regular telephone lines at a speed of approximately 10 seconds per page. A typical fax machine is shown in Fig. 10-14.

FIGURE 10-14   Photo of a typical fax machine.

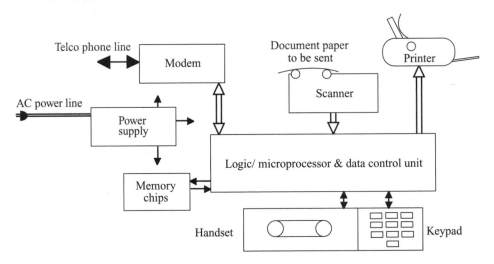

FIGURE 10-15   A block diagram of a plain-paper fax machine.

As you are given a simplified explanation of fax-machine operation, refer to Fig. 10-15. These blocks represent the scanner, printer, telephone modem, logic/microprocessor control unit, memory chips, and power supply.

The scanner is used to scan your document and send this digital data to the logic/microprocessor unit, where it is processed and sent to the modem. The modem is used to translate the digital signal into an audible sound or tones so that the fax information can be sent over a phone line. The reverse occurs when your fax machine receives a document. The modem receives the telephone tone signals and converts them into digital information that the logic/microprocessor unit can understand. The processed digital/logic information is then sent onto the printer, where it is then printed onto plain paper.

The memory chips are used as a buffer to store or hold data if it comes in over the phone line faster than it can be printed or if a document is scanned faster than the data can be sent out over the phone line. Also, it is used to store fax data if the printer runs out of paper, then it will print the fax when the paper cartridge holder is reloaded. The power section block is used to supply a regulated dc voltage to all of the other fax operational blocks.

FAX MODEM OPERATION

Because digital information cannot be sent over your telephone line, a modem is used to convert the digital pulses into audible or tone signals. This continuous processing of modulation and demodulation between your fax machine and phone line is performed by a *modem (mo*dulation/*dem*odulation). The block diagram for a fax modem is shown in (Fig. 10-16).

SOME FAX MODEM PROBLEMS

Look at some common fax problems that are actually modem- or phone-line related. You might have a problem where you cannot send or receive fax messages. This could be a very simple problem of the modem being "locked-up." The modem will actually stop the transfer of data. This lock-up will usually occur when you have had some power-line or phone-line glitches. This could be lightning-induced spikes or power surges. The modem can usually be unlocked by just removing the ac power to the fax machine for 1 or 2 minutes and then plugging it back in again. This will reset the modem operation again.

Noise on the phone line can cause modem lock-up, a brief interruption of the fax message, or garbled printed text. This same problem might occur if you have too many phone devices connected onto one phone line and are loading it down.

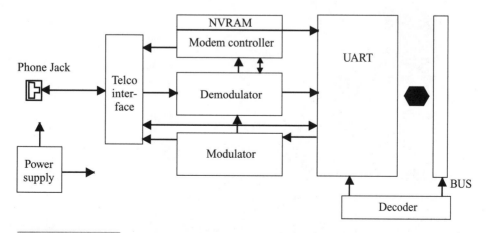

FIGURE 10-16   **A block diagram for a fax machine modem.**

Some fax machines can be programmed to automatically switch over to the fax machine when any fax tones are received, then switch back to your phone when the fax is completed. This allows you to only need one phone line. You can also have a special "ring" programmed from your local phone company to let you know if you have a phone call or a fax coming to your phone line. This special ring will also cause the fax machine to come on line for a fax message. If your fax machine does not have these features and you only have one phone line, you can install an automatic fax/phone switch. The fax switch (Fig. 10-17) can be purchased at Radio Shack and will automatically switch from the phone to the fax machine.

FAX MACHINE OPERATIONAL PANEL

A typical fax-machine control panel is shown in Fig. 10-18. Refer to the numbered key points and see what these various controls are used for.

FIGURE 10-17   With this Radio Shack automatic fax machine switchover unit, you need only one phone line for voice phone and fax machine operation.

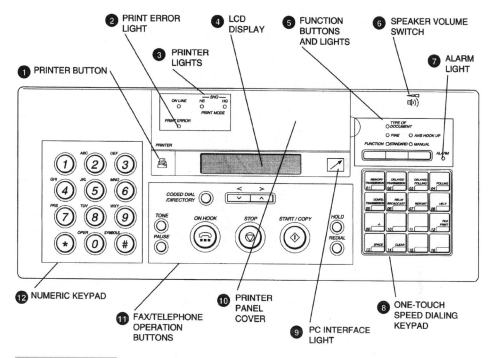

FIGURE 10-18   **Key control buttons and their locations for a typical fax machine operations panel.**

You will use the Start/Copy button to send a fax after dialing up the fax number. The Stop button is used to stop the machine for any reason, such as paper being jammed, empty ink cartridge, etc.

- *Printer button*   Use this button when you need to perform print head cleaning or when you want to print from a non-Windows application.
- *Print error light*   It lights when a paper jam occurs, or when sending or receiving a fax.
- *Printer lights*   These lights will indicate the status of the fax printer.
- *LCD display*   Displays messages, print errors, and other fax machine settings.
- *Function buttons and lights*   Use these buttons for fax and telephone operations. The lights indicate the status of the fax machine.
- *Speaker volume switch*   Use this switch to adjust the speaker's volume. This switch works in conjunction with the On-hook button.
- *Alarm light*   Flashes when an error occurs, when the printer is out of paper, out of ink, or when a received fax document is stored in memory.
- *One-touch speed dialing keypad*   Use these buttons for one-touch speed dialing and to perform special operations.
- *Printer panel cover*   Lift to access the printer panel, which you use to control fax printer operations.
- *Fax/telephone operation buttons*   Use these buttons for fax and telephone operations. On-hook or off-hook operations.

■ *Numeric keypad*   Use these buttons to enter numbers and names when loading information and to dial fax/telephone numbers that have not been entered for automatic dialing.

SOME FAX PROBLEMS AND SOLUTIONS

The following list contains some common fax problems and solutions:

You cannot send documents

■ Be sure that you are feeding the paper properly into the automatic document-feeder device.
■ Check and see if the receiving fax machine has paper installed, machine is turned on and on line, and the fax machine is in the Receive mode.
■ Check to hear a dial tone when you lift the handset.
■ Be sure that the dialing method, Touch-Tone or pulse, is set correctly.

The images you have sent are dirty or spotted

■ Be sure that the document scanning glass is clean.
■ Properly clean the scanning glass if it is dirty.

You cannot receive documents automatically

■ Be sure that all fax machine connections are tight and clean.
■ Be sure that the fax machine is set to receive documents automatically.
■ Be sure that you have printed out any documents that have been received and stored in memory.
■ Be sure that paper is installed in the paper cassette holder.
■ Check any of the read-out displays for any error messages, then clear them.

You cannot receive documents manually

■ Be sure that you have not fed a document into the automatic document feeder.
■ Be sure that you press Start/Copy before hanging up the phone receiver.
■ Be sure that you have printed out any documents in memory before sending or receiving manually.
■ Check any of the read-out displays for any error messages and clear them out.

Nothing appears on the printed page

■ Clean the print head several times.
■ Check and be sure that the ink cartridge is properly installed.
■ Be sure that the Ink detector option is set to On.
■ Still not printing? Then install a new ink cartridge.

You cannot make copies

■ Be sure that the handset is on the hook.

- Be sure that the document is set into the automatic document feeder.
- If your fax machine has a self-diagnosis feature, then print out an activity report and see if any faults have occurred that need to be corrected.

Fax machine will not work (dead)

- The fax machine might have overheated and has shut itself down. Let the machine cool down and then try again.
- Unplug the fax machine. Wait 20 to 30 seconds and then plug it back into the ac socket. Now try to operate it again.
- At times, the problem might be with the party's fax machine you are sending to. If yours works with other machines, have the other party check out their fax machine. Also, be sure that you both have compatible fax machines.

Fax machine paper jammed   For the following jammed paper problems, refer to Fig. 10-19.

Problem   The paper document is jammed.
Solution   Remove the paper document you are trying to send and start again.

Problem   The fax machine tried to receive instead of send because you did not feed the paper in properly.
Solution   Feed the paper document into the machine and start the operation again.

Problem   Your fax machine tried to poll another unit, but the other fax machine did not have a document to send.
Solution   Contact the other party and have them set their fax machine document for polling.

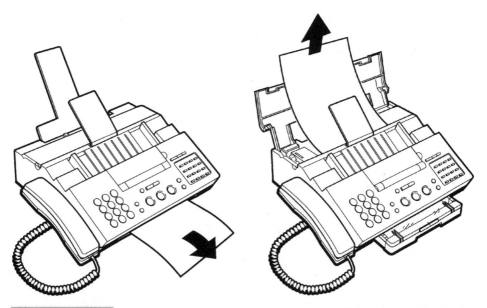

FIGURE 10-19   You should pull any jammed paper out of the fax machine in the direction shown by the arrows.

Problem The paper has become jammed.
Solution Clear the paper jam (Fig. 10-19).

Problem The paper cassette has not been completely inserted into the fax machine.
Solution Insert the paper cassette all the way into the machine, then press Stop.

Paper jammed in printer area

Problem With the paper jammed in the printer area you should open the top cover. The printer cartridge will move over to the left side.
Solution Now gently pull the jammed paper out the top. Be sure the printer light is off. If the printer cartridge is not all the way to the left side, gently move the cartridge to the left and then remove the paper.

Paper jammed in bottom of fax printer If you have looked in both the paper cassette area and the printer area and have not been able to locate the jammed paper, perform the following steps:

- Remove the paper cassette tray and any document supports.
- Tilt up the front of the fax machine.
- Now look into the opening at bottom of unit where the paper cassette was removed.
- If the paper is jammed back inside this opening, carefully remove it at this time.
- Now lower the Printer back on the table.
- Replace the paper cassette holder and any documents support items.
- Press the Stop button.

Dot-Matrix Printer Operation

The dot-matrix printer has been in use for many years for making hard copies from computers and home PCs. They are slower printing and quite noisy as the print head pins are fired, as compared to ink-jet and laser printers. The dot-matrix printer is lower in cost, uses an ink ribbon and can make multiplE paper copies. The dot-matrix printer produces characters by firing a bundle of plunger pins, between 9 and 24, with magnetic coils onto inked ribbon. These dots are all generated within the print head by electrical pulses from the printer's microprocessor.

DOT-MATRIX PRINTER BLOCK DIAGRAM

The dot-matrix printer contains six main blocks (Fig. 10-20). These block's operations are as follows:

1 Microprocessor printer control unit.
2 The paper-transport device.
3 Printer ribbon transport.
4 The dot-matrix print-head assembly.
5 Carriage transport system.
6 The power supply.

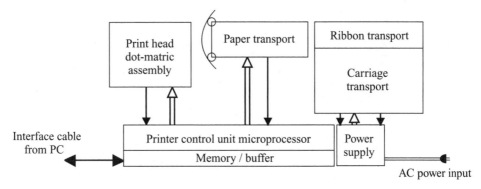

FIGURE 10-20   **A block diagram of a dot-matrix printer.**

PRINT-HEAD OPERATION

Each dot of the dot-matrix print head is formed by a print-pin that is magnetically pushed down by a coil solenoid (Fig. 10-21). An electronic current pulse from the control-unit ICs "shoots" the pin out by the coil's magnetic force. When no pulse is present, this will cause a loss of the magnetic field and the return spring pulls the pin back into place for the next pulse to occur. The coils and print pins are very small and are offset from each other within a bundle. The moving pin strikes an ink ribbon and thus marks the paper. The pin coils are, at all times, receiving different timing pulses for the characters to be printed as the print head moves across the page. The results that proper characters are formed and printed on the paper. A partial drawing of a 24-pin dot-matrix print head assembly is illustrated in Fig. 10-22. The impact printing of the dot-matrix system is very noisy and the current required to drive all of the pin solenoid coils generates considerable heat.

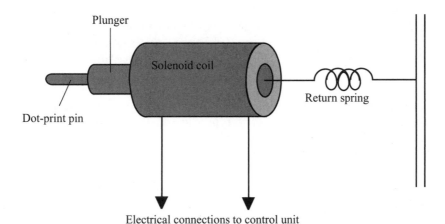

FIGURE 10-21   **One print-head configuration of a 24-pin cluster mounted used on a dot-matrix print-head assembly.**

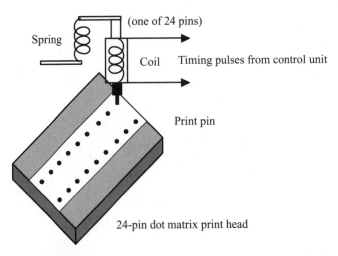

FIGURE 10-22   **A partial drawing of a 24-pin dot-matrix print-head assembly.**

OVERALL SYSTEM OVERVIEW

Now take a brief look at the overall operation of your computer and dot-matrix printer operation. Refer to Fig. 10-23 to follow along on this simplified explanation.

When you type a page, the computer is putting digital codes into memory. When you "tell" your PC to print this information, it sends ASCII digital codes over the computer interface cable to the printer's buffer ICs. These ASCII codes gives the printer the proper characters to print, carriage returns, tabs, and other control information. Because your PC will "dump" the digital codes much faster than the printer can print, the buffer (which consists of RAM chips) is used to store this data into memory until it is called upon and can be used by the printer.

The printer's microprocessor takes the ASCII codes and properly processes them to activate the print-head pins, make carriage returns, control movement of the platen, and print head position. The current pulses generated from the processor actually activates (drives) each electromagnet pin within the print head to produce a readable text on the paper. To make a "Bold type format print," a second set of dots are offset slightly from the first ones.

To simplify, data is sent from your PC and is interpreted by the printer processor control unit and converted to a series of vertical dot patterns that is imprinted on the paper.

The operation of the friction paper-feed drive, the tractor paper-feed drive transport, the carriage transport, and ribbon transport systems are all very similar to the daisywheel and ink-jet printers that have been explained in the earlier part of this chapter. Please refer back to them for their operation and repairs.

Printers/print head troubles and tips

Problem   Intermittent printing.
Solution   This could be caused by poor or intermittent print head cable connections. Unplug the printer and check and clean all cable connections and wires. Also, check any ground

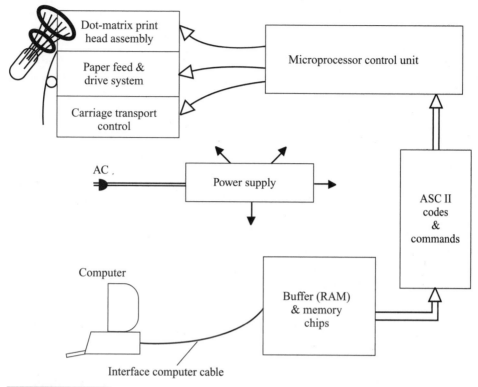

FIGURE 10-23 The overall block diagram of a dot-matrix printer system.

wire connections. Use an ohmmeter to check the cable wires for any open lead wires. Also, check the voltage from the power supply that goes to the print-head drivers.

Problem   Print head moves back and forth, but will not print.
Solution   Check the ribbon to be sure it is not out of ribbon and the cartridge is seated properly. Also, note if the ribbon is advancing properly or may have become jammed with the print-head pins. Clean, check, or install a new ribbon cartridge.

Problem   Printer will not print at all.
Solution   Be sure that the printer is plugged in, is receiving ac power, and check for any blown fuses in the power supply. If these items check out, then be sure that the printer is on line. An on-line light should be on the printer's control panel. The printer will not print if it is off line. The printer might also be out of paper. Next, recheck your computer printing programs. Then check the interface cable between your PC and the printer. Clean and tighten the plug-in and pin connections. Try another cable if you suspect it to be defective.

Problem   Print dots appear missing or faded.
Solution   The most common cause for poor or faded print quality is a ribbon problem. If in doubt, install a new ribbon cartridge. Also, the spacing on the print head might need to be adjusted. This adjustment can give your letters more intensity. Check and clean all parts of the pins and print head and be sure that all of the pins can be moved freely and do not stick.

Problem Dots are missing. The missing dot syndrome might be an intermittent problem. *Solution* If the printer is working OK in all other respects, this problem is probably caused by one pin not firing, a stuck pin, current not getting to the firing coil, or a bent or broken pin. Inspect and clean the print head. Check for a stuck or bent pin. Check all wiring for any shorts or breaks. For a worst-case situation, you will have to replace the print-head assembly.

How Laser Printers Work

The laser printer, also referred to as an *EP (ElectroPhotographic) printer* is not at all like the ink-jet or dot-matrix printers with a moving print-head carriage that were previously explained in this chapter. The laser printer uses a photosensitive drum, static electricity, pressure, heat and chemistry to produce sharp-printing house-quality detail copy. And the printing is fast and quiet.

To do this quality printing, the laser machine must perform these following operations, all at the same time:

- Digital data language from your PC must be properly interpreted.
- The PC data language and instructions are then processed to control the laser-beam modulation and movement.
- The paper movement must be precisely controlled.
- Drum position and rotation speed has to be controlled.
- Paper must be sensitized so that it will accept the toner that produces the printed images.
- The last step is to fuse the toner image to the paper with heat.
- The microprocessor dc controller must control all of these procedures flawlessly.

LASER PRINTER BLOCK DIAGRAM
OPERATION EXPLANATION

At the start of printing, the controller sends a command to load in a sheet of paper from the tray holder. As shown in Fig. 10-24, the paper is drawn through the paper pickup and feeder rollers. Then, as the paper goes around the drum, the laser beam starts to form the static image on the drum surface. (More details on the photo-sensitive drum later.)

The scanning mirror is used to deflect the laser beam back and forth across the surface of the cylinder drum. Also, during scanning, the beam is turned off and on from the modulation, which is controlled by whatever digital information is sent from your PC to the printer's microprocessor controller. All of this laser-beam action is what produces the proper dot pattern on the drum.

During this operation, toner (black powder) is applied to the drum. The toner has a negative electrical charge and is attracted to the positive-charged dots that the laser beam created on the drum's surface. The toner sticks because of an electrostatic charge on the drum in small dots to produce the printed pattern.

The paper, now with toner on it, receives the image from the drum and is then fed into the fuser system.

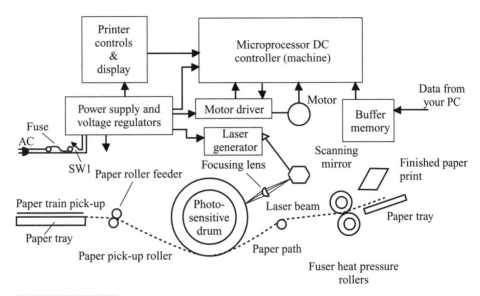

FIGURE 10-24   Simplified block diagram of a laser printer.

The motor drive and motor is used to turn the photosensitive drum. Also, a series of gears or belts from the motor turns the paper-pickup loader, feeder rollers, and the heated fuser rollers.

The information that you want printed from your PC goes to a buffer and then into the laser printer microprocessor or dc controller. The dc controller is the heart of the laser printer. The microprocessor controls all printer operations, such as paper travel, paper-tray amount, temperature of the fuser heating system, drum and paper speed, and the laser generator. The microprocessor controls the sweep of the laser beam and also quickly turns the laser light beam off and on to paint the image to be printed on the drum's surface. If problems develop the microprocessor will shut the printer down and, in some cases, give you an error on the control display panel.

The power supply develops all of the required voltages, some regulated dc voltages, to perform all of the printer operations. Power is required for drum rotation, paper-feeder rollers, pressure rollers, laser generator, scanning mirror, paper movement, and to heat the fuser roller.

PHOTOSENSITIVE DRUM OPERATION AND CARE

Figure 10-25 illustrates in more detail of how the photosensitive drum produces images on paper inside the laser printer. Each time that a new sheet of paper is printed the drum must be electrically erased and cleaned of any toner particles. A rubber cleaning blade is used to remove any toner from previous images. An erase light is also used to neutralize the drum before the laser beam can produce a new image. In order for a new image to be written, the drum must be charged or conditioned. A high-voltage static charge is used to condition the drum.

To rewrite this clean drum, a laser beam is swept back-and-forth across the drum's surface while the beam is turned on-and-off to create an image with a series of dots. This dot process is somewhat like the way a dot-matrix printer makes an image.

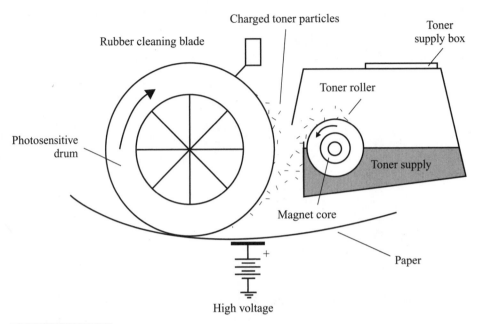

FIGURE 10-25   **How the toner (powder) is transferred to the photosensitive drum and the paper image is developed in a laser printer.**

The rotating drum then passes very close to the toner roller. The toner roller, which has a magnetic core, has toner particles attracted to it, which, in turn, deposits the toner onto the drum's surface. The toner powder then sticks to the charged areas of the drum, which then creates the image to be printed.

Now the toner image on the drum surface must be developed. To do this, the toner is transferred from the drum surface onto the paper, which is fed under the drum. The drum rotation is held to the same travel speed as the paper. The toner is transferred to the paper with electrostatic ionized air, which acts as a magnet to the plastic resin and iron toner particles. The drum is then cleaned for the next image with the rubber cleaning blade.

Next, the toner-covered paper goes onto the fuser assembly to be permanently bonded to the paper. Heat and pressure is used for the fusing process. The paper passes between two rollers, the top roller is heated and melts the toner, while pressure from the bottom roller presses the toner into the paper fibers and the printed page is finished. The paper will feel warm after it is printed.

LOOKING INSIDE THE LASER PRINTER

Let's now take a peek inside a typical laser printer and review the type of parts that make up this printer. In most models of laser printers, they obtain operating power from switch-mode power supplies. The high-voltage power supply of a laser printer or copier has a switch-mode power supply, a primary corona, and primary corona grid. The primary corona deposits negative charges on the surface drum and the grid makes sure that the negative charges are distributed evenly over the drum surface.

THE PRINTER CONTROL CIRCUITS

For operational control of the printer or copier, the dc power supply voltages are used to coordinate all of the electronic and mechanical operations that takes place during the printing process.

In these printers the dc control circuits drive the laser beam, monitor drum sensitivity, check on laser beam motion data, and match dot pattern data with the proper paper size. Plus, the control circuitry controls and monitors paper motion, the fuser unit temperature, erase lamps, motor operations, and high-voltage performance perimeters.

CONTROLLING THE PRINTER WITH THE MICROPROCESSOR

Because of the many things that go on inside a laser printer and the requirement to keep operation simple, the microprocessor is used to keep every operation in order. The electronics consist of a microprocessor, ROM, static RAM, dynamic RAM, and peripheral circuits like timing controllers, in/out controllers, and various interfaces. Let's see what the microprocessor and associated circuits are called on to perform.

- Displaying information on the control panel
- Storing font usage information
- Page type formatting
- Storing configuration information
- Monitoring control panel key operation

HOW IMAGES ARE TRANSFERRED TO PAPER

To copy an image on paper with the laser printer, the process requires the interaction of electronic circuits, optics, and electrophoto graphics. Generally, the actual process of copying or printing a document may be less than 30 seconds. The procedure consists of six stages of operation that pertain to the photosensitive drum. These six stages are as follows for printing one copy:

- Conditioning
- Cleaning
- Writing
- Developing
- Transferring
- Fusing

All of the components used in the image development process are subject to wear and tear and will be degraded as a result of the printing process.

NOTES ON CARTRIDGE USAGE

Laser printers have a replaceable cartridge that contains almost all of the components that will have the most wear. The replaceable cartridge contains a photosensitive drum, primary corona, developing station, toner cavity, and cleaning station. The cartridge is made of extruded aluminum and is coated with a layer of organic photoconductive material (OPC). The photosensitive drum has properties that allow an image to form on the drum surface and can then be transferred to paper. The aluminum base of the drum connects to ground. Microswitches within the laser printer control laser power so that the power matches the sensitivity of the drum.

COLOR PRINTER OVERVIEW

The low cost of quality color printers in the last few years, coupled with fast home computers, has brought quick, high-quality color printing to the masses.

Of course, color printers can be complicated, thus to simplify them some tradeoffs are made for cost reasons. You will find that a low-cost unit will be an ink-jet color printer. It operates very much like the dot-matrix printer, covered earlier in this chapter, but without the impact and with four times the amount of color. Ink-jet color printers are now in the price range of the black-and-white ink-jet printers. Also, the sharpness and detail of the color ink-jet printers are almost as good as the laser printers. However, the ink-jet printer is slow and the ink-filled print head has to be changed and cleaned. I use a color ink-jet printer and it does a good job for the printing I require. Thus, the color ink-jet printer is ideal for the home office for quality printing at a low cost.

For better printing quality and faster work in an office the laser printer or the color thermal printer would be a better choice. The office color thermal printer uses heat to transfer colored waxes from a wide ribbon to the copy paper. You will find that this process provides vivid colors because the inks used do not bleed into each other or soak into the coated paper that must be used. However, the thermal's four-pass technique is slow and takes a lot of ink. The color laser printer produces a very fine detail copy but is high cost because it requires four print engines that are timed to apply color toner to the print page for only one color at a time.

All color printing is produced by using different combinations of light. Color printing uses four pigments, as follows:

1 Black, which reflects no color
2 Magenta (purple-red)
3 Cyan (blue-green)
4 Yellow

You will find some low-cost color ink-jet printers that do not use a black print head, but use equal parts of magenta, yellow, and cyan to produce black. However, the resulting black is not always that good, thus for a personal printer you should purchase one that has a black ink print head. We will now cover the operation of a color laser printer.

COLOR LASER PRINTER OPERATION

At this time you may want to go back and review the black-and-white laser printer operation that was explained earlier in this chapter. As with the black-and-white laser printer, the color laser printer, shown in Fig. 10-26, starts by making the image of the page to be printed on a revolving photosensitive drum by quickly turning a laser light beam on and

FIGURE 10-26   A photo of a Canon color laser printer.

off. Thus when the laser light strikes the drums, it will create an electrostatic charge. The drum makes several revolutions, and at these times the laser fires a dot pattern for the four colors used in printing. These colors are magenta (red), yellow, cyan (blue), and black.

Now refer to the color laser printer drawing in Fig. 10-27 and its numbered callouts. Its operation is as follows, by callout:

1 The corona wires will set up the drum for the next pattern of electrostatic charges.
2 The photosensitive drum assembly.
3 When the drum turns, it will come into contact with the toner cartridge, which contains a powder. This toner cartridge contains four sections. Each contains one color of toner. These colors are red, yellow, blue, and black. Tandem laser printers use a different format and will be shown later.
4 With every rotation, the laser-charged drum will pick up a different color and this will be transferred to the transfer belt.
5 This is the laser light generator and focus lens assembly. The laser light beam scans the photosensitive drum.
6 The fuser roller devices will make the toner stick to the paper.
7 When all of the print colors are on the belt the copy paper is taken from the storage bin and is moved along beneath the feed belt. The rollers then press the paper against the belt, transferring all of the color toners to make a completed color copy.
8 The tandem color laser has separate lasers, as shown in Fig. 10-28, and photostatic drums for each color. The colors are transferred to the feed belt in one turn of the belt instead of the four that is required by a single-drum color laser printer.

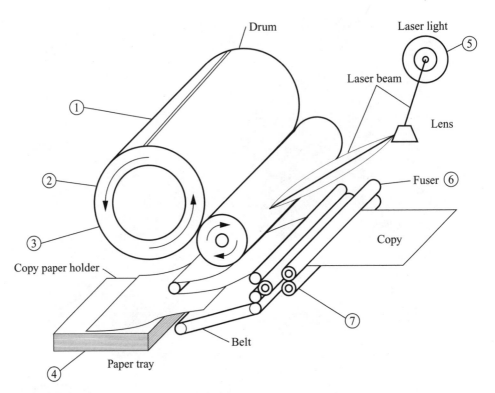

FIGURE 10-27 A drawing of a laser printer with callout numbers, keyed to explanations in the text.

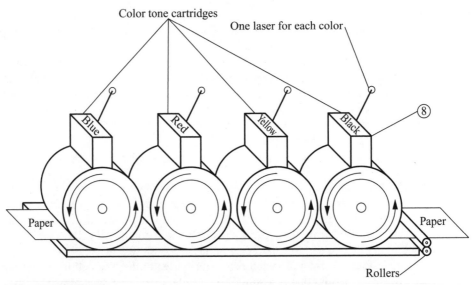

FIGURE 10-28 A drawing of tandem color laser printer.

LASER PRINTER PROBLEMS AND TIPS

The laser printer is a very sophisticated machine and all sections must perform correctly for you to have a crisp, sharply detailed print copy.

Printer will not turn on (dead)   First, check for the presence of ac power at printer socket. Is the laser printer plugged in? If this is OK, check for any blown fuses. If you have power and the cooling fans are working, check for any error readout messages for the self-test check. These error messages give you some clues of the problem. Some printer shutdown problems can be caused by a power-line "glitch," which can lock up the microprocessor and can be corrected by unplugging the printer power for 30 seconds and plugging it back in. This will reset the printer microprocessor controller.

The error message might indicate a communications problem between your PC and the laser printer. This might be caused by an incorrect software program or a fault in the data cable between the PC and printer. Check the cable, clean and tighten the plug pins and contacts. You might have to replace the cable with a new one.

Paper is jammed or has tears   Paper jammed or feeder problems could be caused by paper not placed in the tray properly or the wrong kind of paper being used. Other paper jam problems can be caused by loose belts, broken or worn gears, and misadjusted feed and pressure rollers. And do not overlook a defective toner cartridge for not only poor prints, but also paper jams.

Prints have splashes and specks   Turn the printer off. Check and clean the fusing rollers. Also, be sure that the fusing roller is being heated. The laser printer should be cleaned often because of the toner dust that is always present. Check the adjustment of the rubber cleaning blade and see if the cylinder drum is being cleaned properly because this will cause the copies not to be clean and sharp.

A laser copier is shown in Fig. 10-29 with the cover tilted up to clean and service the machine. The laser printer in Fig. 10-30 is used in conjunction with a PC to make hardcopy printouts.

As you can see, the laser printer is a very sophisticated electronic and mechanical printing machine. Except for minor repairs and cleaning, you might want to consult a qualified laser printer service technician or company. In many cases, some very expensive diagnostic equipment is needed to solve the laser printer problem. Use caution and work with care on any of the printers covered in this chapter.

Scanners

Scanners are used to convert printed images from paper, photos, etc., into a digital format that can then be fed to a computer for memory storage or reproduced on the PC monitor screen. A quality scanner and a fast PC can produce an image that is difficult to tell from the original copy. A scanner can also be used to scan a page of text and bring it up on the PC word processor program, where the text can be edited.

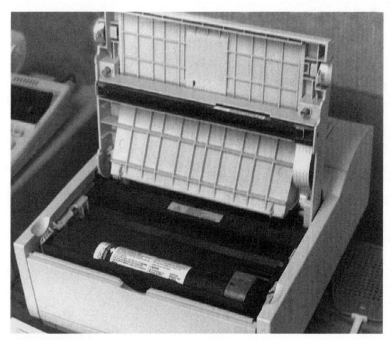

FIGURE 10-29   A laser copier with the cover tilted up so as to be cleaned or serviced.

FIGURE 10-30   A laser printer that is used with a PC for printouts. The cover has been lifted up to clean and/or service the unit.

THE THREE TYPES OF SCANNERS

The three types of scanners are the hand-held unit that you pull across the page of print, the sheet-fed device, and, the most popular, is the flat-bed scanner. With the flat-bed scanner you place the page to be copied on top of a glass plate, and the scan head will move back and forth beneath the glass. With the sheet-fed scanner, the paper is fed into rollers and the paper goes past the scan head.

THE FLATBED SCANNER

We will now take a detailed look at the flat-bed scanner operation. The flat-bed scanner uses a stepper motor to move the scan head across the document to be scanned. Light from the lamp shines on the image, is reflected to mirrors, and then strikes the sensor elements. How much resolution the scanner can obtain is determined by the amount of sensors across the scan head width and the action of the stepping motor. In a low-cost scanner, all the programming is done by the computer, as determined by the software supplied with the scanner.

Now refer to the drawing of the flat-bed scanner shown in Fig. 10-31. By callout number, the scanner works as follows:

1 The light source shines on the paper sheet that is placed face down on the glass plate, which is located above the scan head. Black or white images will reflect more light than color prints/photos, etc.
2 A motor is used to drive the scan head under the glass plate. As the scan head moves back and forth, the scan head picks up light reflected off the page to be scanned.

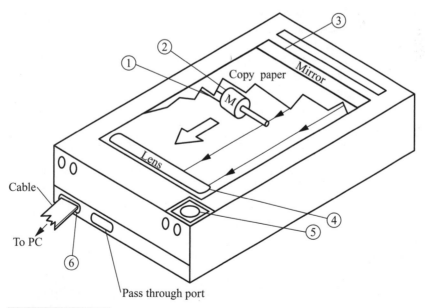

FIGURE 10-31   A drawing of a flatbed scanner with callout numbers that are referenced to the text operating explanations.

3 The light from the paper sheet to be scanned is reflected from mirrors that must move to always keep the light beams in alignment with the lens.

4 A lens is used to focus the beam of light onto sensor diodes that convert the light level into an electrical current. The higher the light, the more voltage and/or current that is generated. For scanning color pages the reflected light goes through red, green, and blue filters that are mounted in front of separate pickup diodes.

5 This is the location of the analog-to-digital converter (ADC) that stores all of the analog voltages to digital pixel light levels that are reflected from the document that is being scanned.

6 Digital information is sent via this multiconductor flat cable to software in your PC. The data will be stored in a format as an optical character recognition or graphics program.

TOP-OF-THE-LINE SCANNERS

The more expensive scanners are easier to use and scan much faster. These scanners have much more computer power and a lot more electronics inside them. This type of scanners is usually a stand-alone unit found in very busy offices. Most of these scanners are in the off mode until a photo or paper to be scanned is slipped into the loading bed and the start button is pressed. These scanners contain all of the processor power required to operate the scan head, retrieve data from the sensors, and store all of the digital images into their memory chips. Some will have a document feeder, and a whole load of pages can be put in the hopper; the machine does the rest.

CONNECTING TO THE COMPUTER PORT

One of the most common ways to connect a scanner is through your PC's parallel port, the port the printer is usually plugged into. Now, let's see how you can make both work off of the same port. Note that the scanner has a pass-through port. Thus, you connect the scanner with a cable to your PC port and then connect the printer to the scanner's parallel output port. This is like a "daisy chain" hook-up technique.

SCANNER REVIEW

Think of the scanner as a device made to convert an image, print, drawing, or photo into a digital file that can be stored in your PC memory and manipulated if you so desire. Try to obtain a scanner with the best resolution for the price you can afford to invest. Note that the more and smaller the dots, the sharper the image scanned will be reproduced. However, the downside of a sharper image will be that you need a larger digital memory file for each image.

11

DIGITAL VIDEO DISC (DVD) SYSTEM OPERATION PLUS SERVICE PROBLEMS AND SOLUTIONS

Note:

You may also wish to refer to Chap. 3, "Audio CD Player Operation and Service Maintenance," as a lot of the mechanical operation is very similar to that of the DVD player.

The DVD Video Player

The concept behind the DVD player was to develop a way to condense signals that have the same pattern as satellite video signals. It was also a way to put the complete video and multitrack sound of a movie on a disc the same size as an audio CD. These signals are then concentrated and put into a digital recorded format in order to have clearer, sharper video that can be used for HDTV-quality viewing.

The engineers have always run into problems when trying to compress video signals. DVDs use the same compression technology as satellite digital transmissions. This technique results in much more space to store video and audio data, which, of course, will produce much sharper video images in the HDTV format and will make it possible to add many more features on the DVDs. The DVD system is able to produce over 1000 horizontal lines of video resolution.

The DVD disc condenses video by looking at any repetitive image signals, such as the background of a static camera shot in a movie, and then only using it one time. Eight hours of digital video can be placed onto a double-sided dual-layered disc.

Panasonic was the first major electronics company to bring out a DVD for the consumer electronics viewer. Sony was the first company to have movie films converted on to the DVD format. These revamped motion pictures are packaged and sold along with Sony's own line of DVD player machines and discs.

DVD Machine and Disc Technology

As we look at how the DVD system works, follow along with the callouts shown in Fig. 11-1.

1 Let's look at a few of the DVD technology features.
   Sound channel: The DVD player and disc will handle six sound channels for outstanding Dolby surround sound.
   Subtitles: Options will include multilingual dialog and various subtitles.
   Parental Lock: This feature will allow the operator to skip scenes according to desirable ratings.
2 As stated above, the DVD discs are of a multilayered construction. A dual-focus lens enables the laser beam to read two different layer levels on the disc at the same time. Later in this chapter we will take an in-depth look at the laser system.
3 The audio CD has much larger pits located farther apart than the DVD.
4 The smaller pits on the DVD store the digital video information. One layer on the DVD will contain 7 times more information than an audio CD. Later in this chapter we will review the operation and construction of the DVD disc player system.

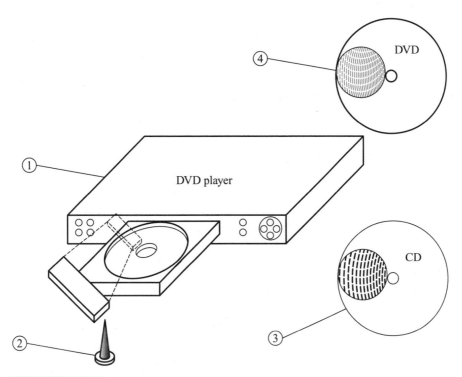

FIGURE 11-1   Callouts for various features and functions of a DVD player.

DVD Player Operation

A DVD player system has a wide range of circuits that include audio encoders and decoders, servos, optics, amplifiers, microprocessors, digital signal processors, and power supplies. Use the DVD player block diagram shown in Fig. 11-2 as we delve into some simplified circuit operation discussions.

SIGNAL PROCESSING

The video signal begins with data taken by the laser beam pickup assembly from the disc. From this point, the signal passes through the RF signal processor to the digital signal processor, referred to as the *DSP block*. The output signals coming out of the DVD DSP block travel to the system microcontroller (microprocessor chip) and the MPEG decoder.

SERVO AND OPTICAL PICKUP ELECTRONICS

Optical electronics contain the optical unit, the servo motor assembly, and a motor drive IC chip. The motor drive chip uses a transformerless BTL driver to drive the tracking actuator, sled motor, focus actuator, tray motor, and spindle motor. The motor drive unit has built-in thermal shut-down, voltage lockout, mute circuits, and overvoltage control protection.

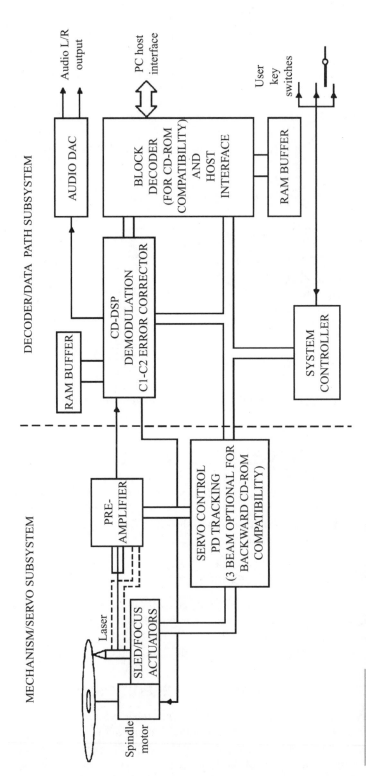

FIGURE 11-2 Block diagram of a DVD player.

It also has a fully integrated digital servo that controls the spindle motor speed and keeps track of the disc rotating speed.

THE RF SIGNAL PROCESSOR

The video and audio information in RF form is transferred from the OPU to the main circuit board via a flat ribbon cable to the RF signal IC processor. The RF signal processor IC amplifies and equalizes the RF signal before it exits the DVD digital signal processor (DSP) IC. In addition, the circuit includes internal RF automatic gain control (AGC) circuits, an internal APC circuit, an internal disc defect detector, and an internal focus protection circuit to guard against a disc defect problem.

DVD DIGITAL SIGNAL PROCESSOR

The IC301, shown in Fig. 11-3, provides a number of the DVD video player functions. The chip's analog front end converts the high-frequency input signal to the digital domain by using an 8-bit analog-to-digital converter (ADC). An AGC circuit working before the ADC circuit sets the gain control required for having optimum performance from the converter. An ADC CLC circuit provides the clocking sync for this operation.

Now when the amplified and equalized RF signal is fed into the DVD DSP circuit, a part of this DSP signal serves as a data slicer, which functions as a 16- to 8-bit decoder, and sets up the error correction. With the playback information still carried within the RF signal, the data slicer pulls out the embedded data and temporarily stores the data in the memory chip. From this point, the data goes to the 16- to 8-bit demodulator chip.

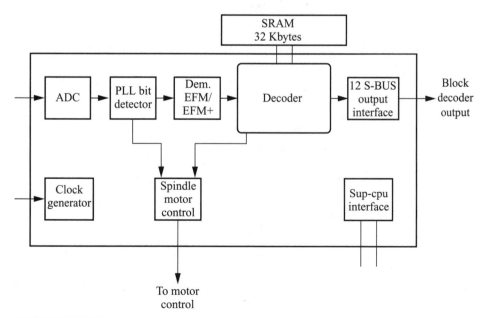

FIGURE 11-3   Block diagram of the DVD DSP.

A phase-locked loop (PLL) and bit detector block form a subsystem that recovers data from the channel stream. This block also corrects asymmetry, performs noise filtering and equalization, and then recovers the bit clock and data from the channel via the PLL. This advanced bit detector offers improved data recovery for multilayer discs and contains two extra detection circuits to increase detection of recovery errors.

DVD and MPEG-2 Technology

DVD recording technologies rely on MPEG-2 encoding and decoding to ensure the high-quality reproduction of movies and other video programming. Each disc contains one track of MPEG-2 compressed digital video. This may be in a constant-rate or variable-bit-rate format. The encoding process uses 24 frames per second of the progressive material from the film to be reproduced. Thus, the MPEG-2 encoder places flags within the video stream to ensure compatibility with either 60-Hz or 50-Hz video TV standards.

When building a DVD machine, the engineers have the option of including additional video and audio so the disc will operate in either an NTSC or PAL standard player. This additional video and audio information will decrease the amount of space for program playback. Usually, MPEG-2 video will be stored in either the NTSC or PAL format.

DVD units using the PAL/SECAM standard can play NTSC-formatted as well as PAL-formatted discs. To do this, the DVD converts the NTSC signal to a 60-Hz PAL signal. With some of these conversions, an NTSC formatted disc will play in a PAL standard player while a PAL formatted disc will not play in an NTSC standard player.

ENCODING AND DECODING

Briefly, the video encoding and decoding section of the DVD player contains the MPEG A/V decoder IC, a Philips NTSC/PAL encoder, a couple of amplifier chips, a clock generator, and two SDRAM ICs. As you refer to the block diagram in Fig. 11-4, note that the data travels to the circuits from the microcontroller section, which is not shown. Outputs for the circuits shown in the block diagram are the audio IF signals, the audio/video IF connectors, and the servo IF signals.

The audio decoding portion, not shown, includes two audio digital-to-analog converters and four operational amplifiers. Input signals travel from the system's microcontroller and the MPEG A/V decoder. Dolby Digital AC-3, Linear PCM, or MPEG-2 audio signals travel to the audio/video connectors and can be used to drive your home video theater system.

Laser Injection Diodes

At this time we will review the operation of the laser diodes that are used in DVD players to obtain data from the discs. Laser, which stands for light amplification by stimulated emission of radiation, was developed by the old Bell Labs that was part of Western Electric Company (AT&T). The makeup of this laser diode begins with top bottom sections composed of a metallic substance, as shown in Fig. 11-5. Between these two metallic parts

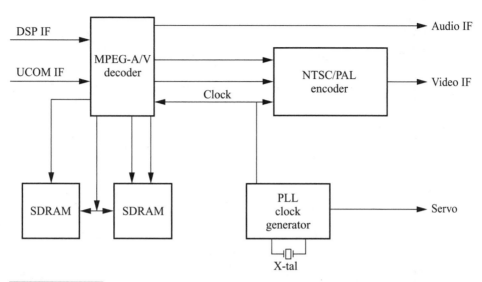

FIGURE 11-4   **Simplified block diagram of the video encoding/decoding portion of a DVD player.**

is a p-type material that is forward biased when in operation. This forward bias causes holes and electrons to be injected into the active layer. Thus the energy is kept in the active layer by the adjacent layer barrier. When this mixing of energy takes place, photons stimulate other photons until the current reaches a reaction point at which light is emitted and it "lases." Optic lenses are placed at each end of the diode in order to reinforce the beam to produce a laser beam.

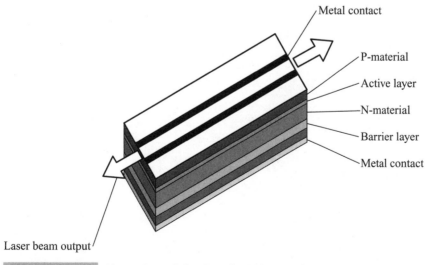

FIGURE 11-5   **How a laser injection diode is constructed.**

Diode lasers are generally manufactured in this way and are ready for installation. Thus, the laser and all of the optics are hermetically sealed in one module. As to the module, it may appear to be a simple product; however, it is quite a complex and accurately made device. For most laser modules, it will have a collimator (this is to focus the diverging light beam), then a prism pair, to change the beam from elliptical to circular, and then an expander and a focus lens. The drawing in Fig. 11-6 will illustrate this laser module for you. And we should note that the laser diode is actually the smallest part of this module. Of course, the various modules will have different optical specifications that the DVD player manufacturers will require for their units. An example of these differences is the adjustment for the beam's focal point with regard to the final optical component. This accuracy is essential in order for the CD player to read the disc properly.

Always use caution when working around a device that is using a laser. Even low-power laser could possibly cause eye damage under the right conditions. Never look directly at a laser beam or point one at another person.

CLEANING THE LASER UNIT

Cleaning or servicing a laser device should be done with care. A laser diode can easily be destroyed or damaged even when handled carefully. Laser diodes are static and voltage sensitive and the module assembly must be treated with great care. When you first look at the laser module assembly you will see a small window. In most DVD units this small window is the final focusing, or objective, lens.

At times the little window may become dirty from dust or dirt and other particles that are found on the DVD. The best way to clean this window is by blowing it off with a compressed air blower. Another way is to use a pressurized can of "clean" air. This is available at electronics parts stores and can also be used to clean circuit boards and other electronic devices that become dusty and dirty. You can also use alcohol on a cotton-tipped swab; however, you must use caution because any stress on the floating tracking module may

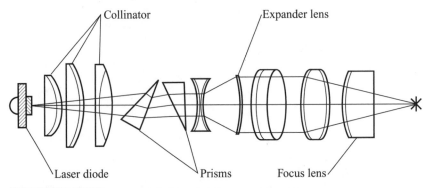

FIGURE 11-6   **A drawing of a typical laser diode used in a DVD player.**

damage it. You must be certain the window is very clean and has no streaks or it will distort the beam and the DVD will have reading errors.

Construction and Operation of the DVD Disc

The data on a DVD optical disc is encoded as indentations called *pits* and spaces called *lands,* which are accomplished by a stamping process. Aligned into spiral tracks, each transition from a pit to a land is a binary one, while each constant land or constant pit is a binary zero. Refer to the cutaway DVD disc in Fig. 11-7.

The double-sided DVD disc is made up of seven layers. The center one is a polycarbonate plastic, then each side has an opaque layer, a transparent film, and then an outer surface of a protective plastic. The data on these discs include video, audio, text, or other program material. Because of the small pit sizes, the DVD discs will hold up to 8.5 MB of data.

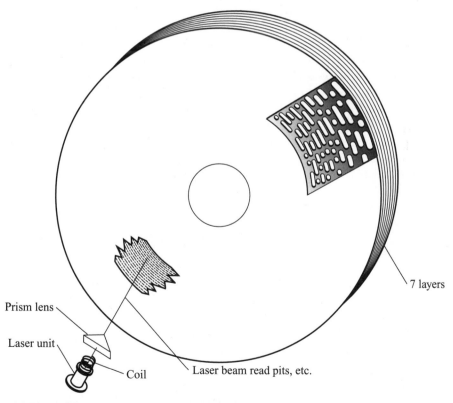

FIGURE 11-7   **Cutaway view of how a laser DVD disc is constructed and operates.**

The DVD drive uses a laser beam to read the lands and pits. The DVD laser has a much shorter wavelength than an audio CD, which makes the beam very narrow and accurate enough to read the much smaller pits and lands. Coils around the laser beam enable the head to focus the beam only on the transparent film.

When the beam hits a pit, the light is scattered in all directions. However, when the beam strikes a flat area, it will be reflected back to the reading head. The prism then deflects the beam to a device that converts the bursts of laser energy. These bursts are interpreted by the computer as code and data information.

Note that the capacity of a single-sided DVD is doubled when the same layers of materials are applied to the other side of the disc. However, to obtain this capacity the DVD player must have two laser units and reading heads, one for each side. Of course, you could flip the disc over and read the other side.

The two spiral tracks, not shown in Fig. 11-7, of data recorded on two layers of the DVD disc turn in different directions. The read head will follow the first track until it reaches the center, then with the disc still turning at the same rotation, the head will read the second layer spiral track back to the edge of the DVD disc. This dual-track technique removes any delay in the data flow and is very important when the DVD contains multimedia material.

DVD POINTS OF CONSTRUCTION REVIEW

As shown in Fig. 11-7, the DVD disc has two thin discs bonded together into a single unit. These two discs are only 0.6 mm thick. The data density that can be recorded on a DVD is much greater than what an audio CD can hold. So much more density is obtained by making the pits smaller, which means many more can be put on the disc. The smaller pits require a thinner laser beam spot from the prism that is focused on the pits. The thinner disc construction of the bonded-disc technique lets the smaller pits be read more accurately by this thinner laser beam spot size.

Another advantage for bonding two discs of the same material is that one disc may tend to warp because of temperature change and moisture, but two discs together will offset this, as they will warp in opposite directions. Thus the disc will keep its flat shape and data obtained will be more accurately retrieved.

DVD Troubleshooting Information

Player will not work   Reset the DVD player by unplugging the ac power cord for a few minutes and then plug it back in the socket. This will reboot the unit.

Disc will not play

1 Insert a disc with the label side facing up.
2 Check the type of disc that you are using. Some units will only play DVD video discs, video CDs, or audio CDs.
3 If the disc is a DTS music CD, it will require a DTS decoder. If you have a DTS receiver hooked up to the DVD player and you still get no sound, make sure the Trusurround option is turned OFF.

No power to unit   Make sure both ends of the power cord are plugged in. Also, make sure that you have ac power at the wall socket.

The DVD unit will start to play and then stops

1 The disc may be dirty. Clean the disc or try another one.
2 Condensation has formed inside the player. Allow some time for it to dry out.

No picture

1 Turn the TV set to the correct channel.
2 Make sure the TV receiver is turned on.
3 Check and make sure all equipment is connected properly.

No sound or distorted sound

1 Make sure the DVD player is connected properly. Check that all cables are inserted properly into the correct jacks.
2 You may need to readjust the digital output setting. Check the service manual for this adjustment.
3 Make sure your TV set is tuned to the correct input channel.
4 Remember, sound is muted during still, frame advance, or slow motion play modes.
5 If you have connected an audio receiver to your DVD player, make sure you chose the correct input setting on the receiver.
6 If the disc is a DTS music CD, it requires a DTS decoder. If you have a DTS receiver hooked up to your DVD player and you are still not receiving sound, check and see if the Trusurround option is turned OFF.
7 If you are *not* using a stereo TV receiver for your sound source, turn OFF the Trusurround option.

Remote control will not operate

1 If you are using a universal remote, you may be in the wrong mode. To operate the DVD player, press DVD on the remote before you press any other buttons. If you want to operate the TV set, press the TV button first, etc.
2 Batteries may be weak. Replace with new ones.
3 Operate the remote control no more than 20 feet from the device to be controlled.
4 Remove any obstacles between the remote and the DVD player or other components. If your DVD player is in an entertainment cabinet, and it has glass doors, they may be keeping the LED from the remote reaching the TV set or DVD. Open the doors.
5 The remote control may need to be reset. To do this, remove the batteries, and hold down one or more buttons for about 1 minute to drain voltage from the microprocessor inside the remote unit to reset it. Reinstall the batteries and try the remote operation again.

Cannot advance through first part of a movie   You cannot advance through the opening movie credits and warning information because the disc is programmed to prohibit such advancing action.

The Ø icon appears on screen   The feature or action cannot be completed at this time because:

1 The disc's software restricts it.
2 The disc's software does not support the features, angles, etc.
3 The feature is not available at this time.
4 You have requested a title or chapter number that is out of range.

The DVD picture is distorted

1 Has the DVD been correctly connected to your TV set?
2 Have you connected the VCR to your DVD player? If so, disconnect it.
3 The DVD disc may be damaged. Install a known good disc.

Picture is distorted during forward and reverse scan   This is a normal occurrence during scanning.

A screen saver icon appears on the TV screen   Most DVD players are equipped with a screen saver that appears on the TV screen after your player has been idle for several minutes. There are a few ways to make the screen saver disappear from the screen and return to the player's main menu. For example, press stop or go back buttons on the remote unit.

Subtitle and/or audio language selection   The subtitle or audio language is not the one selected from the initial setting. If the subtitle and/or audio language does not exist on the disc, the initial settings will not be seen or heard. The disc's priority language is selected

FIGURE 11-8   DVD player, front view, with the slide drawer open ready to receive a disc.

instead. Set the subtitle and/or audio language manually through the info display on the DVD player menu.

The disc will not begin playing   The rating of the title on the disc exceeds the rating limit set in the ratings limits when you press play. Unlock the player or change the rating limit in those menus.

No forward or reverse scan

1 Some discs have sections that prohibit rapid scanning or title and chapter skip.
2 This part of most movie discs is programmed to prohibit skipping through them.

Desired angle cannot be changed   Some discs do not have the multicamera angle system, and some discs have it only in certain parts of the movie.

Picture is too tall and thin   Change the aspect ratio using the TV image settings in the display menu.

Picture is too short and wide   Change the aspect ratio using the TV image settings in the display menu.

Forgotten password bypass   Some DVD players are equipped with a "backdoor" unlock sequence. Press and hold the stop button on the front of the DVD player and the stop button on the remote control unit at the same time. Now, hold both buttons down at the same time for at least 3 seconds.

Cannot copy discs to videotape (VCR)   You cannot record DVD discs onto videocassettes because the discs are encoded with anticopy protection.

Disc tray (Fig. 11-9) will not open   Disengage the retail lock feature. Press and hold a combination of buttons on the front panel at the same time. Press and hold skip, eject, and TS surround for at least 3 seconds. The tray should now open unless it's jammed or has motor problems.

Disc will not eject   When you put a disc into the player, the player may take up to 15 seconds to read the disc. You will not be able to eject the disc during this time. Wait 15 or more seconds and try to eject again.

No video or audio

1 Check for a dirty turning mirror. Clean the mirror.
2 Dirty or scratched lens. Clean the lens or replace.
3 For intermittent audio or video, check and clean the flat cable (see Fig. 11-10) and connectors or replace the cables if defective.

Laser beam will not track properly   May have a defective focus actuator. Replace the actuator if cleaning or the adjustment will not correct the problem.

FIGURE 11-9   The tray has been removed from the DVD player to repair a problem with the tray slide not operating properly.

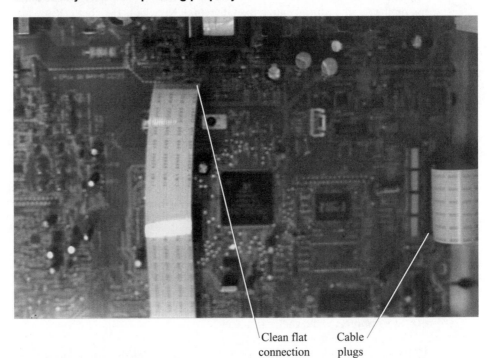

Clean flat
connection

Cable
plugs

FIGURE 11-10   Loss of audio and video may be intermittent because of loose or dirty flat cable connectors. Clean cable connectors and plugs.

Door will not open and/or disc will not load

1 Dirty mechanism or broken or worn gears. Clean the slide assembly or replace the drive assembly.
2 Dirty drawer switch. Clean the switch assembly.
3 Shorted motor assembly. Replace the loading motor.

Disc has erratic speed   Dirty or dry spindle. Clean and lubricate spindle.

Laser beam will not track properly

1 Dirty or dry spindle. Clean the sled assembly.
2 Motor may be defective. Replace the motor.

DVD Player Precautions

- Before connecting any other components to your player, be sure all other components are turned off.
- Do not move the player while a disc is being played. The disc may get scratched or broken, and internal parts may become broken or misadjusted.
- Do not place any container with liquid or any small metal objects on the unit.
- Be careful to not place your hand into the disc tray.
- Do not place anything other than a disc into the disc sliding tray compartment.
- Outside influences such as lightning, power line glitches, and static electricity can affect normal operation of a DVD player. If this occurs, turn the unit off and then on again with the ON/OFF buttons, or disconnect and reconnect the ac power cord to the power outlet. This will reboot the player and it should operate normally.
- After using the DVD player you should remove the disc and turn off the unit.

DISC HANDLING PRECAUTIONS

- Do not touch the disc's signal surfaces. Hold them by the edges or by one edge and the hole in the center.
- Do not place labels or adhesive tape to the signal surface of the discs.
- Do not scratch or damage any portion of the disc.
- Do not use a damage (cracked or warped) disc.

CLEANING DVD DISCS

- Dirty discs can cause reduced video and audio performance.
- Always keep discs clean by wiping them gently with a soft cloth from the inner edge toward the outer perimeter.
- Should a disc become very dirty, wet a soft cloth in water, and wring it out well. Wipe the dirt away gently, and remove any water drops with a dry cloth.
- Do not use record-cleaning sprays or antistatic agents on DVD discs.

DVD DISC CLEANING CAUTION:

Do not clean the DVD discs with benzene, thinner, or other volatile solvents that may cause damage to the disc surfaces.

DVD Player Front Panel Control Locations

Refer to the DVD player control callouts shown in Fig. 11-11 as we review their operation:

Disc tray   Press open-close to open and close the disc tray.

Skip back   Allows you to move to the beginning of the preceding title, chapter, or track on a disc, thus skipping that particular title, etc.

Skip forward   Allows you to move to the beginning of the preceding title, chapter, or track.

Play/pause   Begins disc play (and closes disc tray if open). When pressed during playback, pauses disc play.

Stop   Stops the disc from playing.

Front panel display   Reads out information for all functions of the player and disc.

Random   Changes play mode to random (plays the disc tracks or chapters in a random order).

TS surround   Use the TS surround button to simulate surround sound. Each press of the button toggles the setting between ON and OFF.

ON/OFF button and ON/OFF indicator light   This button turns the player on/off manually. The on/off indicator lights up when the DVD player is ON.

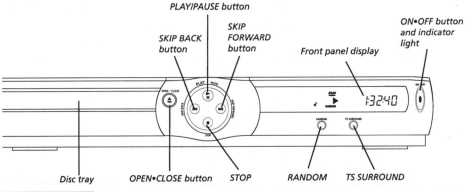

FIGURE 11-11   **Callouts of controls found on a typical DVD video player.**

Personal Video Recorders (PVRs)–TiVos

The personal video recorder (PVR) is a machine that uses a hard disc like that found in a personal computer (PC) to let you record TV programs, at the time they are on, automatically, without having to make complex programming as with a VCR unit. This cutting edge technology is certainly not a glorified VCR-type taping machine TV program recorder.

As an example, the PVR machine lets you record any upcoming program, just by selecting the title from an on-screen program guide and instructing the TiVo to record the show each time it is aired. These localized TV program guides are automatically downloaded to the PVR via your phone line or in some cases from the dish for satellite service.

After they have been recorded they can be played back from an on-screen menu. The PVR machine can also search out and record selected types of movies or programs, such as "mystery types," even on less popular channels at any time of the day or night.

Another plus is that the PVR unit can automatically record whatever live program is currently being transmitted and put it into its data buffer for 30 minutes or more. This feature lets you go back and do your own sports replays or pick up viewing the program after you may pause it to answer the phone or leave the TV set for more important reasons. And another feature, one that advertisers won't like, is the ability to fast forward right through commercials while watching recorded programs or live shows that you have paused for a short time.

You will also find combination PVR units, such as Motorola's and Atlanta Inc.'s, that have cable set-top boxes with the recording hard drive features. There is also Microsoft Corp.'s UltimateTV, a DirecTV receiver with a personal video recorder as well as Web-surfing capabilities. Sonicblue has plans to integrate a PVR system in its ReplayTV machine. At this point this technology can be used to manage digital video for all kinds of information sources into the customer's home.

REPLAYTV 4000 SERIES PVR SYSTEM

The ReplayTV 4000 PVR machines have some expanded capabilities, such as enough memory for 320 recorded hours. The ReplayTV 4000 consists of four models: the 4040, 4080, 4160, and 4320. The numbers after the "4" indicate the amount of digital memory storage in hours; the numbers also indicate the sizes of the internal hard drives: 4040 has a 40-gigabyte hard drive, the 4080 has an 80-GB hard drive, 4160 has a 160-GB hard drive, and the 4320 has two 160-GB hard drives. All of the other features are the same on the 4000 series machines.

As noted previously, the PVR's ability to receive a program guide via a phone line or satellite download is the big advantage over a conventional VCR, which is tough to program, not to mention getting rid of the flashing "12:00." Using this guide feature, you can set the PVR's built-in microprocessors to automatically record programs you want to view with a one-time setting, which instructs the machine to store the program on the hard drive. In other words, you set it and forget it. In this way TV viewing is not limited, but you can record and play back programs at your convenience.

Another feature of the ReplayTV 4000, as with other machines, is its ability to skip seamlessly through commercials. This is called commercial advance (CA), which allows a recorded program to be viewed commercial-free. It has been available on VCR units, but there is quite a difference in the way a PVR machine works. With a VCR when a commercial is detected, the VCR will go into a scan mode, which causes a short or long pause. With the ReplayTV in commercial advance, the commercials just disappear, like with magic.

However, we must note that the CA feature may not always work. The commercial advance feature, when operating in the real world of video TV, is effective between 70 and 90% of the time. It seems that it will not work during the first or last 2 minutes of a program. It does do a remarkable job of taking out commercials, but it seems to vary with the types of programs that have been recorded. At times the ReplayTV PVR will show the first 2 or 3 seconds of a commercial break, then come back into the last 2 seconds of the commercial. At other times, the commercials will not be shown at all. It's not 100% perfect, but it sure beats seeing all of those endless commercials. All in all the ReplayTV 4000 series consists of very good recording PVR machines.

12

BASIC CONSUMER ELECTRONICS SERVICE AND MAINTENANCE

Introduction

This chapter is devoted to tips for locating, repairing, and adjusting common problems that "crop-up" in consumer electronic products that are usually found in the home or office. Also, there are notes on some maintenance procedures that will help you keep these products working trouble free and longer.

ADJUSTMENT AND SERVICING CAUTION:

Most electronic devices sold today do not have a power or isolation transformer, which means the chassis ground is connected directly to one side of the ac line voltage. This equipment is referred to as having a "hot chassis," and touching these chassis points could cause a deadly shock. Always unplug the device before checking out a problem, such as replacing a fuse or component. The device you are working on can also be plugged into an isolation transformer; however, this is not always foolproof either.

Service Notes or Manuals

A service manual is a very helpful item to have when you are checking out or adjusting any electronic device. Save any of the printed information that comes with the equipment, or better yet, purchase a service manual. These can be quite helpful and may quickly solve any problem, plus give you all of the correct adjustment procedures. Some will have a section on the equipment test procedures, check outs, and any faults that may have occurred for this device. And there may be included a list of common troubles and hints on solutions to these problems.

Points to Consider before Starting

Let's take a few minutes and go over a few points before you start any repairs on your electronic equipment.

- Have a clean and well-lighted work area, with a rubber pad, and several small containers to keep any screws or small parts that you have to remove.
- Take all of the safety precautions for working with your electronic equipment.
- Take your time and think through what you are going to do and how.
- Make sure you have all of the proper tools, etc.
- *Do not* go any further with the repairs than what you are capable of doing. If you take equipment too far apart or make adjustments you do not understand, you may do more damage and undergo more repair cost than if you had taken it to a professional service center. A simple problem could turn into a very costly one.

- If you are going into the circuit boards with a volt-ohm meter probe, use extreme caution, as just one slip with solid-state devices can be costly or render the equipment not repairable.
- Be very careful when using a hot soldering iron.

Circuit Boards and Solder Connections

A good many electronic devices develop problems, sometimes intermittent, because of poor solder connections and these can be affected by temperature changes (the problem develops after a warm-up period) or by some type of vibration that causes the device to malfunction.

THERMAL PROBLEM

The heating up and cooling down of the circuit board and components may cause a solder joint to fail and an intermittent condition to appear. You can try locating the problem area by flexing the board, moving various parts, or heating with a hair dryer different sections until the intermittent develops. Also, try using a cooling spray as it will serve the same purpose. The area you are heating and cooling when the problem occurs is where the defect is located. Also, look for any cracks in the printed circuit (PC) board. You can now try resoldering the connections in this area that has been pinpointed.

LARGE OR HEAVY COMPONENTS

If your electronic equipment has some large components mounted on the PC boards, carefully inspect or resolder all of their connections. This is a quite common problem in some electronic equipment, especially if it has been subject to lots of vibration and has been carried around a lot.

CIRCUIT VIBRATION

As stated above be on the lookout for equipment that is portable and has been subject to lots of vibration, and carefully inspect the PC boards for cracks and poor solder joints.

Intermittent Problems

Let's take a look, in more detail, at some other reasons your electronic equipment may have developed an intermittent problem. An intermittent problem could be caused by an outside interference problem if the device is a TV receiver, cell phone, cordless telephone, stereo amplifier system, or AM/FM receiver. This interference could be coming in as an RF carrier over the airwaves, or transmitted over the power lines and even the telephone line coming into your home. You can try filters on the power line or telephone lines to see if they do the trick. Also, try moving the equipment to another location; if the intermittent problem goes away you know it's an outside RF-type interference problem.

The other intermittent problem would be within the electronic device's internal circuits. The first test is to gently tap on various parts of the case and see if the intermittent condition can be duplicated. If this makes the intermittent condition show up, then you may want to remove the device's case. Then use a small wooden dowel to press around on various components and circuit boards. As noted previously, try some heat or cooling spray to make the problem appear. Always be on the lookout for "cold" defective solder joints. If there are any cables or cable connections present, flex and wiggle them and/or clean any plug-in connection. The cables themselves may be defective and need to be replaced. Transistors and ICs will also fail internally and the heat and cold treatment will usually make these components start acting up.

Using Electronic Equipment Flowcharts

At times you may find flowcharts along with circuit diagrams and other service information packed with your new electronic equipment. If you purchase service manuals for your devices, they will sometimes include flowcharts. When trying to determine what is causing a problem in your equipment, try thinking logically how the circuits work and what are the trouble possibilities. That's when the flowcharts can be of value because they will give you a simpler understanding of how the circuit flow throughout the device is accomplished. It is also a good idea to make the simple checks first and think of the most probable faults that will occur.

USING THE SIMPLE FLOWCHART

The electronic equipment flowchart is actually a simple block diagram of the much more complicated, detailed circuitry schematic. How these blocks function, their main purpose, and how the circuits are interconnected is usually shown on these flowcharts. The blocks will indicate their subcircuit functions. After studying these blocks and their subcircuits this should help you to note various equipment failures and determine which section is likely to be at fault. As you refer to the drawing in Fig. 12-1, you will see a simple flowchart of a color TV receiver.

For any type of electronic equipment, especially if the unit is dead, the power supply block is a good place to start. Check the fuses or circuit breaker and any power plugs/cords. If you have a voltmeter, then some voltage checks can pinpoint the trouble to the power supply or to another circuit block. A faulty power supply or its filter and regulator circuits can cause many different symptoms. Some flowcharts can be very complicated looking, but you can redraw them in a more simple way that you can understand.

Another tip is to break down the complete device, such as the color TV flowchart in Fig. 12-1, to the one section that you are having a problem with, after the power supply is performing properly. As an example, should you have a sound problem in a TV set, you would zero in on the flowchart or block diagram of the audio circuit, shown in Fig. 12-2. After a preliminary check of the audio flowchart you should then go to the actual circuit drawing or to a more detailed subflowchart for more testing. In the detailed subflowchart you can look for the key components such as transformers, capacitors, transistors, and IC chips. The detailed flowchart in Fig. 12-3 has some of the key components identified for

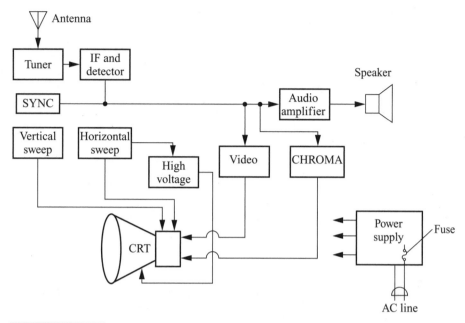

FIGURE 12-1   A simplified flowchart for a color TV receiver that should help pinpoint circuit problems.

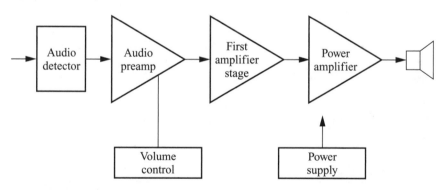

FIGURE 12-2   Flowchart of the audio section of a TV receiver that is used to track down audio problems.

further testing. If your TV set has only sound trouble, and the power supply is OK, then you do not need to start checking out other flowchart blocks. Just stay with the ones that pertain to the audio circuitry, speaker wiring, and speakers. In a remote-controlled TV receiver, do not overlook a fault in the sound mute circuit, or that the TV set has actually been muted with the remote hand unit.

When drawing your own flowchart, each individual capacitor or resistor will not have to be noted. The active or key components, such as transistors, ICs, and transformers, are the

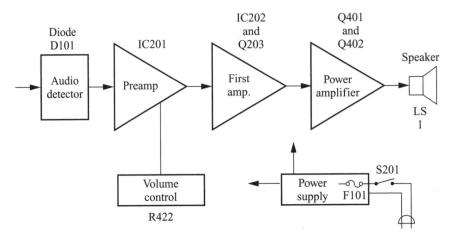

FIGURE 12-3   You can write in the key components on the flowchart of the stages in question, to help narrow down the location of the faulty component in the troubleshooting procedures.

main concerns. Any of the other "passive" components next to the active ones will be in the same flowcircuit block.

Many blocks in the flowchart can be eliminated by your process-of-elimination thinking. As an example, in a color TV receiver, if you have good sound and a perfect black-and-white picture, you start looking for a problem in the color or chroma circuits. You would not consider checking out the tuner or sound circuits. So, use your old standby, the trick of the process of elimination and logical thinking. A good flowchart, some logical thinking, and trouble symptom considerations will let you track down the defect that is causing the problem.

Fuses and Circuit Breakers

When a fuse or circuit breaker fails, there can be several reasons for this to occur. It may well be an ac line voltage surge, a momentary overload in the electronic device, a spike or glitch on the power line, a fuse or breaker that's actually defective (weak), or a defect in the equipment itself. It's OK to replace the fuse with one of the correct or an exact value as called for in the service data. Professional service technicians do this as a standard procedure. A replacement fuse of a lower value will keep blowing and a higher-value fuse will not give proper circuit protection and could do great circuit damage or cause a fire hazard. After you replace a fuse with one of the correct value, turn on the device and observe its operation for at least 1 hour or more. If the electronic device appears to operate properly, no burning smell or flames, then the blown fuse was probably caused by a line surge or a faulty (fatigued) fuse/weak circuit breaker. More on circuit breakers a little later in this chapter.

NOTES ON THERMISTORS

You will find thermistors in TV sets, audio amplifiers, and many other electronic devices, usually located in the ac low-voltage portion of the power supply circuits. These thermistors will generally look like large size resistors and will run warm or hot to the touch.

CAUTION NOTE:

If you have an overloaded circuit, touching the thermistor could give your finger a bad burn. The words are *do not touch*. In some older model TV sets you will find thermistors in the degaussing circuits that control the current in a coil around the picture tube. With an ohmmeter, they measure about 120 ohms when cold. In the new model TV sets, the thermistor will have only a few ohms of resistance.

The thermistor works in this way: After the current flows through it a while and heats it up, the resistance will decrease to a very low ohm value, which will allow more current to flow into the power supply. If the equipment is dead, but the fuse and circuit breaker are good, suspect a faulty thermistor. Many times you can look at them and see that a lead is melted off or they have a burnt look. They are easy to replace by unsoldering two leads and soldering a new unit back into the circuit. However, make sure you replace with the correct value or part number.

CIRCUIT BREAKER TIPS

Use the same troubleshooting checks with the circuit breaker as with a blown fuse. If the breaker opens up three or four times in a few seconds or a minute or two, suspect a weak breaker or circuit overload. Replace the circuit breaker with one of the correct value, and if the same symptoms occur then you have a circuit overload or short. You can also unsolder the circuit breaker and solder in a replacement fuse that has pigtails and see if it blows. Make sure it is of the correct amperage. If it blows, then you will need to troubleshoot the circuit for circuit shorts. The best place to start is in the power supply. The circuit breaker is reset by pushing the button; most of these buttons are red. Generally, the circuit breaker will last for the equipment's life span, unless it has had many overload circuit conditions or power line ac surges.

Just a note about resetting circuit breakers. One type of circuit breaker cannot be reset when the circuit is still overloaded, but the other type can be reset at any time. Use caution with the one that can be reset under any overload condition because you may cause more circuit damage and actually cause a fire.

Noise Spikes and Glitches

In the real world of solid-state digital electronics, the problems of spikes and glitches can cause many problems that had very little effect on analog devices. Digital circuits are very sensitive and unforgiving to noise spikes and glitches. The drawing in Fig. 12-4 illustrates

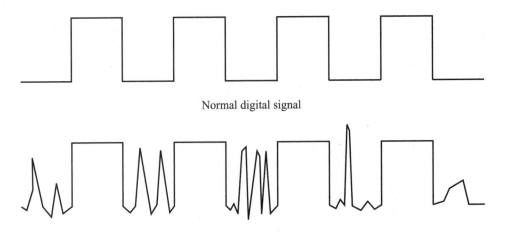

Normal digital signal

Digital pulse with noise spikes

FIGURE 12-4   **The top waveform is of a normal digital signal. The bottom signal trace is the same digital signal that contains some noise spikes.**

how noise spikes can cause digital circuitry to be tripped up. The spikes and glitches can cause wrong logic information to occur, trigger at the wrong time, and throw off synchronization of various timing circuits.

The noise spike may cause a brief oddity or cause the complete computer system to crash. One of the major problems would be a complete erasure of a system's memory bank. Many times the equipment will not be damaged, but the data damage can be very costly and time-consuming to correct. To even start to locate these glitches, an oscilloscope is a must. However, even with the best equipment a stray glitch is a "tough nut" to crack.

Some of these spikes will occur in the power supply when the equipment is first turned on. Thus, these noise spikes, be they stray or frequent glitches, will come into the circuits via the power supply. For high-cost equipment it could be a good investment to install ac line filters, an uninterruptible power supply (UPS), or surge suppressors to eliminate or reduce these noise spikes.

There is always a chance that the internal circuit filters are defective or may not have been designed with enough filtering. Each digital IC chip and circuit should have its own filtering capacitors. If you do suspect a circuit filtering problem, you can try adding a new capacitor from the V+ dc power line to a good chassis ground. This can be a small-value capacitor of 0.001 μF to 0.01 μF at 50 or 100 working volts.

CAUTION:

 Make sure the equipment is turned off when installing the test capacitor, as any small arc may do some big-time damage. You should also make sure all voltage is bled off the power supply lines before touching the capacitor or any test lead to the circuits.

Generally, any signal noise found in your equipment is caused by an outside RF signal that is referred to as an interference signal. These RF signal noises can cause digital circuits to act up in strange ways because the logic pulses are distorted by the interference noise. The drawing in Fig. 12-5 illustrates how the digital pulses are malformed by the analog-looking noise signals. One step you can take is to put more shielding around your equipment; in some situations, it will eliminate the interference problem. Another suggestion is to reposition certain circuit boards inside the device or move the entire piece of equipment to another location. Also, you may try plugging the unit into a different ac outlet.

Trouble, Symptom Observations

Finding out what's wrong with your electronic devices can often be boiled down to observing when the equipment fails and listing other pertinent operation details. A good point is to compare the equipment when it was working correctly and then when the problems or failures occur. The following is a list of equipment trouble observations that you can make for various electronic devices:

1 Do the control positions change a little or a lot after 20 or 30 minutes of equipment operation (warm-up).
2 For ac-line-operated devices, does the LED or dial indicator stay on all the time the unit is plugged in?
3 For battery-operated devices is there a low-battery indicator? If so, what does it indicate?
4 For an AM/FM radio receiver, listen to what the radio sounds like when tuned off station.
5 Does the radio receiver produce full speaker volume when it is turned on at full volume? When the volume control is at minimum and first turned on does the speaker blast out as if at full volume level?
6 How does your two-way radio or cell phone work when you are near its working range end? Also, when you are getting out of its range? Is your cell phone analog or digital operation? Is your two-way radio a trunking type system?
7 Does your equipment perform differently in warm (hot) or cold weather conditions? Also, dry or damp conditions?
8 For electronic devices that are microprocessor controlled, such as PCs and laptops, and have several initialization steps during the first few seconds after being turned ON, have you noticed they are different now that the device has some operational faults?

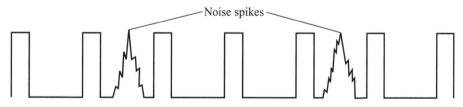

FIGURE 12-5 In this digital pulse waveform you will note the analog noise spike "interference," which may upset the timing of a digital signal and cause all kinds of digital equipment malfunctions.

9 For equipment that have a standby operation mode, note if there is now a difference, when the device has a problem. Put the device in standby. Do you observe any odd performance or faults?

10 Does the equipment have multiple function selections? Not all models of the same equipment will have the same number of functions. Does yours have switch positions for functions that are not incorporated within your model?

NOTES FOR AUDIOCASSETTE PLAYERS

Audiotape recorders are mechanical devices. They can be gummed up with grease and have deteriorated or worn parts, defective cassette tapes, bent control arms, misshaped springs, or foreign objects dropped into the mechanism.

For a cassette player problem that is not a "dead on arrival" case, always check first for dirty or worn mechanical parts and worn or broken belts before looking into electronic problems. Many times a good cleanup or a new cassette tape will do wonders for your player.

NOTES FOR CD PLAYERS

You will find that CD player problems are mechanical. Check for worn or loose drawer belts; lubrication that is dirty, dried up, or gummed up on the sled tracks and/or gears; dirty lens; faulty/partially shorted spindle; or a defective sled motor. Also battery troubles are always an item to check for portable units. For any CD or DVD problem it is always a good idea to first clean the lens, as this can cause all types of failure-mode problems. A failure of the laser is not very common and optical alignment is usually not required, unless the unit has had rough usage.

NOTES FOR PRINTERS

The electronic portion of an ink-jet printer is usually very reliable. However, you should be on the lookout for caked ink within the "service station" area, almost empty ink cartridge, and misaligned print-head contacts when you have an erratic printing problem.

Laser printers have been known to frequently develop problems in the fuser, scanner, or power control modules. These problems may be as simple as a burned-out lamp bulb, defective motor, or loose, dirty cable connections. And don't forget to give the machine a good cleaning.

GLOSSARY

Use this glossary to help you better understand some of the terms used in explaining "How Electronic Things Work" in these book chapters.

Cameras, Camcorders, Audio tape

This glossary section can be used in conjunction with the video recorder, camcorder, and audio tape recorder chapters.

acoustic suspension   Air-suspension (AS) speakers are sealed in an enclosure or box to produce natural, low-distortion base output. Greater driving power is needed with these less-efficient speaker systems.

air suspension   Another name for an acoustic-suspension speaker.

amp   Abbreviation for amplifier.

ANRS   A noise-reduction system that operates on principles that are similar to the Dolby system.

APC   The automatic power control circuit keeps the laser-diode optical output at a constant level in the CD player.

audio/video control center   The central control system that controls all audio and VCR operations.

auto eject   The tape player feature that automatically ejects the cassette at the end of the playing time.

auto focus   AF is the focus servo that moves the objective lens up or down to correct the focus of the CD player.

auto record level   Automatic control of the recording level.

auto reverse   The ability of the cassette player to automatically reverse directions to play other side of the tape.

auto tape selector   Automatic bias and equalization when the cassette is inserted into the tape.

azimuth   The angle at which the tape head meets the moving tape. A loss of high-frequency response is often caused by an improper azimuth adjustment.

azimuth control   A control to adjust the angle of the tape control to correct misalignment in the auto stereo tape player.

baffle   The board on which the speakers are mounted.

balance   The control in the stereo amp that equalizes the output audio in each channel.

bass reflex   A bass-reflex system vents backward sound waves through a tuned vent or port to improve bass response.

bias   A high-frequency current applied to the tape-head winding to prevent low distortion and noise while recording.

block diagram   A diagram that shows the different stages of a system.

booster amplifier   A separate amplifier that is connected between the main unit and the speakers in a car stereo system.

bridging   Combining both stereo channels of the amp to produce a mono signal with almost twice the normal power rating in a car stereo system.

cabinet   A box that contains speakers or electronic equipment.

capstan   The shaft that rotates against the tape at a constant rate of speed and moves the tape past the tape heads. In the cassette player, a rubber pinch roller holds the tape against the capstan.

cassette radio   The combination of an AM/FM tuner, amplifier, and cassette player in one unit.

cassette tuner   A tuner and cassette deck in one chassis.

CH   The abbreviation for channel. The stereo component has two channels (left and right).

channel separation   The degree of isolation between the left and right channels, often impressed in decibels. The higher the decibel values, the better the separation.

chassis   The framework that holds the working parts in the amplifier, tuner, radio, cassette, CD player, or VCR. The chassis could be metal, plastic, or a PC board.

chips   Chip devices can contain resistors, multilayer ceramic chip capacitors, mini-mold chip transistors, mini-mold chip diodes, and mini-mold chip ICs.

clipping   Removing or cutting off the signal from a waveform that contains distortion, which can be seen on the oscilloscope. Excessive power results in distortion.

coaxial speaker   A speaker with two drivers mounted on the same frame. The tweeter is mounted in front of the woofer speaker. Usually, coaxial speakers are used in the car audio system.

compact disc   The compact-disc (CD) player plays a small disc of digitally encoded music. The CD provides noiseless high-fidelity music on one side of a rainbow-like surface.

CPU   A computer-type processor used in the master and control mechanism circuits of a CD player.

crossover   A filter that divides the signal to the speaker into two or more frequency ranges. The high frequencies go to the tweeter and the low frequencies go to the woofer.

crosstalk   Leakage of one channel into the other. Improper adjustment of the head might cause crosstalk between two different tracks.

D/A converter   In the CD player, the device that converts the digital signal to an analog or audio signal.

dc   Direct current is found in automobile battery systems, and also after the ac has been filtered and rectified in low-voltage power supplies.

decibel   The decibel (dB) is a measure of gain, the ratio of the output power or voltage, with respect to the input (expressed in log-units).

de-emphasis   A form of equalization in FM tuners to improve the overall signal-to-noise ratio while maintaining the uniform frequency response. The de-emphasis stage follows the D/A converter in a CD player.

dew   A warning light that might come on in a VCR or camcorder. It indicates too much moisture at the tape head.

digital   Within tuners, the digital system is a very precise way to lock in a station without drifting. Digital recording is used in compact discs.

direct drive   A direct-drive motor shaft is connected to a spindle or capstan/fly wheel. The CD rests directly on the disc or spindle motor in CD players.

disc holder   The disc holder or turntable sits directly on top of the motor shaft in the CD player.

dispersion   1. The spread of speaker high frequencies, measured in degrees. 2. The angle by which the speaker radiates its sound.

distortion   In a simple sine-wave signal, distortion appears as multiples (harmonics of the input frequency). A type of distortion is the clipping of the audio signal in the audio amplifier.

Dolby noise reduction   A type of noise reduction that works by increasing the treble sounds during recording and decreasing them during playback, thus restoring the signal to the original level and eliminating tape hiss.

driver   1. In a speaker system, each separate speaker is sometimes called a *driver*. 2. The loading, feed, and disc motors might be driven by transistor or IC drivers.

drive system   The motors, belts, and gears that drive the capstan/flywheel in cassette tape or CD players.

dropout   In tape decks, dropouts occur when the tape does not contact the tape head for an instant. Dropouts occur in the compact disc because of dust, dirt, or deep scratches on the plastic disc.

dual capstan   Dual capstans and flywheels are used in auto-reverse cassette players and can play tapes in both directions.

dynamic   A dynamic speaker has a voice coil that carries the signal current with a fixed magnetic field (PM magnet), and moves the coil and cone. The same principle applies to the human ear or to headphones.

dynamic range   The ratio between the maximum signal-level range and the minimum level, expressed in decibels (dB).

electronic speed control   An electronic method of controlling the speed of the capstan motor.

electrostatic   An electrostatic speaker headphone, or microphone, that uses a thin diaphragm with a voltage applied to it. The electrostatic field is varied by the voltage, which moves the diaphragm to create sound.

equalizer   A device to change the volume of certain frequencies, in relation to the rest of the frequency range. Sliding controls can be found in auto-radio and cassette-player equalizers.

erase head   A magnetic component with applied voltage or current to remove the previous recording or noises on the tape. The erase head is mounted ahead of the regular R/P head.

extended play   EP refers to the six hours of playing time that is obtainable with a T-120 VHS cassette played in a VCR.

eye pattern   The RF signal waveform at the RF amplifier in a CD player. The waveform is adjusted to a clear and distinct diamond-shaped pattern.

fader   A control in auto radio or cassette players to control the volume balance between the front and rear speakers.

fast forward   The motor in the cassette, VCR, or CD player can rotate faster with a higher voltage applied to the motor terminals or when larger idler pulleys are pushed into operation.

filter   A circuit that selectively attenuates certain frequencies, but not others. The large electrolytic capacitor in the low-voltage power supply is sometimes called a *filter capacitor.*

flutter   A change in the speed of a tape transport, also known as *wow.*

focus error   The output from the four optosensing elements are supplied to the error signal amplifier and a zero output is produced. The error amp corrects the signal voltage and sends to the servo IC to correct the focus in the CD player.

folded horn speaker   The system that efficiently forces the sound of the driver to take a different path to the listener.

frequency response   The range of frequencies that a given piece of equipment can pass to the listener. The frequency response of an amplifier might be 20 Hz to 20 kHz.

gain   The amplification of an electronic signal. Gain is given in decibels.

gain control   A control to adjust the amount or boost the amount of signal.

gap   The crucial distance between the pole pieces of the tape head. The gap area might be full of oxide, which would cause weak, distorted, or noisy reception.

glitch   A form of audio or video noise or distortion that suddenly appears and disappears during VCR operation.

graphic equalizer   An equalizer with a series of sliders that provides a visual graphic display.

ground   A point of zero voltage within the circuit. The common ground might be a metal chassis in the amplifier. American-made cars have a negative-ground polarity.

head   A magnetized component with a gap area that picks up signals from the revolving tape.

hertz   Hertz (Hz) is the number of cycles per second (CPS), the unit of frequency.

hiss   The annoying high-frequency background noise in tapes and record players.

hum   A type of noise that originates from power lines, caused mainly by poor filtering in the low-voltage power supply. Hum and vibrating noise might be heard in transformers or motors that have loose particles or laminations.

idler   A wheel found in tape players to determine the speed of the capstan/flywheel or turntables in the cassette player.

impedance   The degree of resistance (in ohms), that an electrical current will encounter in a given circuit or component. A speaker might have an impedance of 2, 4, 8, 16, or 32 ohms.

integrated circuit   An IC is a single component that has many parts. ICs are used throughout most cassette players, amplifiers, VCRs, and CD players.

interlock   A safety interlock device used in the CD player to load the disc.

ips   Inches per second, the measurement of cassette-tape speed.

jack   The female part of a plug and receptacle.

kilohertz   1 kHz is equal to 1000 Hz.

laser assembly   The assembly that contains the laser diodes, focus, and tracking coils in a CD player.

laser current   Low laser current might indicate that a laser-diode assembly in the CD player is defective.

laser diodes   The diodes that pick up the coded information from the disc along with the optical pick-up assembly in a CD player.

LED   Light-emitting diodes are used for optical readouts and displays in electronic equipment.

level   1. The strength of a signal. 2. The alignment of the tape head with the tape.

line   Line output or input jacks are used in the amplifier, cassette, or CD player. The line signal is usually a high-level signal.

loading motor   The motor in CD players, VCRs, and camcorders that moves the tray or lid out and in so that the disc or cassette can be loaded.

long play   LP is a speed on the VCR that provides four hours of recording on a 120-minute VHS cassette.

loudness   The volume of sound. Loudness is controlled by a volume control.

LSI   Large-scale integrated circuits include processors, ICs, and CPUs that are used in VCRs, camcorders, and disc players.

magnetic   Metal attraction. The magnetic coil might be found in the VOM or VTVM.

megahertz   1 MHz is equal to 1000 kHz or 1,000,000 Hz.

memory   The program memory of a CD player.

metal tape   The high-frequency response and maximum-output level are greatly improved with metal tape. Pure metal cassettes are more expensive than the regular oxide cassettes.

microprocessor   A multifunction chip found in most of today's electronic products. They are used in tape decks, transports, memory operations, CD players, and VCRs.

monitor   To compare signals. A stereo amplifier can be monitored to compare the signal with the defective channel.

monophonic   One channel of audio, such as in a single speaker.

multiplex   A multiplex (MPX) demodulator in the FM tuner or receiver converts a single-carrier signal into two stereo channels of audio.

mute switch   The mute switch might be a transistor in the audio-output line circuit of a CD player or cassette deck.

noise   Any unwanted signal that is related to the desired signal. Noise can be generated during the record and play functions in a cassette player. A defective transistor or IC could cause a frying noise in the audio.

NR   Noise reduction.

optical lens   The lens located in the pick-up head of a CD player. Clean the lens with solution and a photographic dry-cleaning brush.

output power   The output power of an amplifier, rated in watts.

oxide   The magnetic coating compound of the recording tape or cassette. The excess oxide should be cleaned off of the tape heads, pinch rollers, and capstans for good music reproduction.

passive radiator   A second woofer cone that is added without a voice coil in the speaker cabinet. The pressure created by the second cone produces heavy bass tones.

pause control   A feature to stop the tape movement without switching the machine. The pause control is used in cassette decks, VCRs, and CD players.

peak   The level of power or signal. A peak indicator light shows that the signal levels are exceeding the recorder's ability to handle the peaks without distorting.

phase   Sound waves are in sync with one another. Speakers should be wired in phase.

pick-up motor   The pick-up, SLED, or feed motor is used to move the pick-up assembly in the radial direction or toward the outer edge of the disc.

pitch control   A control that changes the speed of the control motor.

PLL   The phase-locked loop (PLL) VCO circuit is used in the digital-control processor of the CD player with a crystal.

port   An opening in a speaker enclosure or cabinet. The port permits the back bass radiation to be combined with the front radiation for total response.

power   The output power of any amp is given in watts. A low-voltage power supply provides voltage to other circuits.

preamplifier   The amp within the cassette player that takes the weak signal from the tape head and amplifies it for the AF stages.

rated power bandwidth   The frequency range over which the amplifier supplies a certain minimum power factor, usually from 20 to 20,000 Hz.

recording-level meter   The meter (analog, LED, or fluorescent panel) that indicates how much signal is being recorded on the tape.

reject lever   A lever that rejects or deletes a given track in a cassette or a record on the record changer.

remote control   A means to operate the receiver, CD player, cassette/tuner, or VCR from a distance. Today, most remote-controlled transmitters are infrared type.

repeat button   The button that replays the same track of music on the CD player.

RF   A radio-frequency signal.

ribbon speaker   A high-frequency driver or tweeter speaker that uses a ribbon material suspended in a magnetic field to generate sound current when current is passed through it.

saturation   Recording tape is saturated when it cannot hold anymore magnetic information.

self erase   A degrading or partial erasure of information on magnetic tape.

self-powered speakers   A speaker with a built-in amplifier.

separation   The separation of two stereo channels. Placement of the stereo speakers can provide good or poor stereo separation.

servo   The tracking circuits that keep the laser pickup in the grooves at all times.

servo control   The servo control IC that controls the focus and tracking coils in CD players.

signal processing   In the CD player, converting the processing laser signals to audio with preamps and signal processors.

signal-to-noise ratio   The ratio (S/N) of the loudest signal to noise. The higher the signal-to-noise ratio, the better the sound.

skewing   A form of visual distortion or bend at the upper part of the picture of the VCR player.

solenoid   A switch that consists of an electric coil with an iron-core plunger that is pulled inside the coil by the magnetic field. Solenoids are usually found in auto radios, cassette, tape, and CD players.

speaker enclosure   The cabinet in which speakers are mounted.

spindle motor   The disc or turntable motor revolves.

standard play   SP is the speed at which a two-hour (T-120) VHS cassette plays on VCR machine.

subwoofer   A speaker that is designed to handle very low frequencies below 150 Hz.

test cassette   The recorded signals on a test cassette that are used for alignment and adjustment procedures on the cassette player.

test disc   A CD that is used to make alignments and adjustments in CD players.

tone control   A circuit that is designed to increase or decrease the amplification in a specific frequency range.

tracking servo   The IC that keeps the laser beam in focus and tracking correctly.

tray   The loading tray in which the CD to be played is placed.

tweeter   A high-frequency driver speaker.

VCR   Video cassette recorder.

vented speaker system   Any speaker cabinet with a hole or port to let the back waves of the woofer speaker escape. A bass reflex is a type of vented speaker system.

VHS   The system used today by most VCRs.

voice coil   The coil of wire that is wound over the end of the cone of the speaker in which the amplifier output is connected. The electrical signal is converted to mechanical energy to create audible sound waves.

watts   The practical unit of electricity and other power.

woofer   The largest speaker in a speaker system. The one that reproduces the low frequencies.

wow   A slow-speed fluctuation in tape speed. Fast-speed variation is called *flutter*.

Telephone and Answering Machines

This glossary section can be used in conjunction with the telephone and answering machine chapter.

ADC (analog-to-digital converter)   An electronic device used to convert an analog voltage into a corresponding digital representation.

AF (audio frequencies)   The frequencies that fall within the range of human hearing, typically 50 to 18,000 Hz.

AM (amplitude modulation)   A technique of modulating a carrier sinusoid with information for transmission.

anode   The positive electrode of a two-terminal electronic device.

attenuation   The loss of reduction in a signal's strength because of intentional or unintentional conditions.

bandwidth   The range of frequencies over which a circuit or system is capable of operating or is allowed to operate.

base   One of three electrodes of a bipolar transistor.

battery   The operating voltage supplied to a telephone from a central office.

BOC (Bell Operating Company)   The local telephone company that provides your telephone service from your central office.

capacitance   The measure of a device's ability to store an electric charge, measured in farads, microfarads, and picofarads.

capacitor   A device used to store an electric charge.

cathode   The negative electrode of a two-terminal electronic device.

cell   In cellular telephony, the geographic area served by one transmitter/receiver station.

channel   An electronic communication path. A channel can consist of fixed wiring or a radio link. A channel has some bandwidth, depending on the type and purpose of the channel.

CO (Central Office)   The building and electronic equipment owned and operated by your local telephone company that provides service to your telephone.

collector   One of three electrodes on a bipolar transistor.

continuity   The integrity of a connection measured as a very low (ideally zero) resistance by an ohmmeter.

CPC (Calling Party Control)   A brief dc signal generated by your local central office when a caller hangs up.

CPU (central processing unit)   Also called a *microprocessor*. A complex programmable logic device that performs various logical operations and calculations based on predetermined program instructions.

cradle   An area on a telephone's housing where the handset or portable unit can be kept when not in use.

DAC (digital-to-analog converter)   An electronic device used to convert a pattern of digital information into a corresponding analog voltage.

data   In telephone systems, any information other than human speech.

decibel (dB)   A unit of relative power or voltage expressed as a logarithmic ratio of two values.

demarcation point   The point where a building connects with the outside wiring supported by the BOC. In a home, the demarcation point would be at the network interface connector.

demodulation   The process of extracting useful information or speech from a modulated carrier signal.

diode   A two-terminal electronic device used to conduct current in one direction only.

drain   One of three electrodes on a MOS transistor.

DTMF (Dual-Tone Multi-Frequency)   A process of dialing that uses unique sets of audible tones to represent the desired digit.

emitter   One of three electrodes on a bipolar transistor.

EPROM (Electrically Programmable Read-Only Memory)   An advanced type of ROM that can be erased and reused many times.

Exchange area   A territory in which telephone service is provided without extra charge. Also called the *local calling area*.

FM (Frequency Modulation)   A technique of modulating a carrier sinusoid with information for transmission.

full-duplex   A circuit that carries information in both directions simultaneously.

gate   One of three electrodes on a MOS transistor.

ground start   A method of signaling between a telephone and the central office, where a signal line is grounded to request service.

half duplex   A circuit that carries information in both directions, but in only one direction at a time.

harmonics Multiples of some intended frequency, usually created unintentionally when a frequency is first generated.

hybrid Also known as an *induction coil*. A specialized type of transformer used in classic telephones to couple the two-wire telephone line to an individual transmitter and receiver.

ICM (incoming message) The message that is left by a caller on an answering machine.

IF (intermediate frequency) A high-frequency signal used in the process of RF demodulation.

impedance A measure of a circuit's resistance to an ac signal, usually measured in ohms or kilohms.

inductance The measure of a device's ability to store a magnetic charge, measured in henries, millihenries, or microhenries.

inductor A device used to store a magnetic charge.

LCD (liquid-crystal display) A type of display using electric fields to excite areas of liquid crystal material.

LED (light-emitting diode) A specialized type of diode that emits light when current is passed through it in the proper direction.

loop current The amount of current flowing in the local loop.

loop start The typical method of signaling an off-hook or line-seizure condition where current flow in the loop indicates a request for service.

local loop The complete wiring circuit from a central office to an individual telephone.

modulation The systematic changing of the characteristics of an electronic signal in which a second signal is used to convey useful information.

MTS (Message Telephone Service) The official name for long-distance or toll service.

NAM (Number Assignment Module) An erasable memory IC programmed with an assigned telephone number and specific identification information, typically used with cellular telephone circuits.

OGM (Outgoing Message) The message that a caller hears when an answering machine picks up the telephone line.

permeable The ability of a material to become magnetized.

piezoelectric The property of certain materials to vibrate when voltage is applied to them.

pps (Pulses Per Second) The rate at which rotary or pulse interruptions are generated. A rate of 10 pps is typical.

program A sequence of fixed instructions used to operate a CPU.

PSTN (Public Switched Telephone Network) A general term for the standard telephone network in the United States. The term refers to all types of wiring and facilities.

pulse A process of dialing using an IC (instead of a mechanical device) to generate circuit interruptions corresponding to the desired digits.

RAM (random-access memory) A temporary memory device used to store digital information.

RC (Regional Center) Telephone facilities that interconnect both toll centers and some central offices, and support long-distance telephone service.

rectification The process of converting dual-polarity signals to a single polarity.

regulator An electronic device used to control the output voltage or current of a circuit, usually of a power supply.

resistance The measure of a device's ability to limit electrical current, measured in ohms, kilohms, or megaohms.

resistor   A device used to limit the flow of electrical current.

ring   An alerting signal sent from a central office to a telephone or other receiving equipment, such as an answering machine.

RF (radio frequency)   A broad category of frequencies in the range above human hearing, but below the spectrum of light, typically from 100 kHz to more than 1 GHz.

ring   One of the two main wires of a local loop. The name originally referred to the ring portion of a phono plug that operators used to complete connections manually. See tip below.

ROM (read-only memory)   A permanent memory device used to store digital information.

rotary   A process of dialing that uses a mechanical device to open and close a set of contacts in a pattern corresponding to a desired digit.

sidetone   A small portion of transmitted speech that is passed to the receiver. It allows a speaker to hear their own voice and gauge how loudly to speak.

SMT (surface-mount technology)   The technique of PC board fabrication using components that are mounted directly to the surface of a PC board instead of inserting them through holes in the board.

SOT (small-outline transistor)   A transistor designed for use with surface-mount PC boards.

source   One of three electrodes on a MOS transistor.

subscriber loop   Another term for the local TC (toll center) facilities that interconnect central offices.

tip   One of the two main wires in a local loop. The name originally referred to the tip of a phono plug that operators used to complete connections manually.

transistor   A three-terminal electronic device whose output signal is proportional to its input signal. A transistor can act as an amplifier or a switch.

transformer   A device using inductors to alter ac voltage and ac current levels or to isolate one ac circuit from another.

VOX (voice-operated control actuation)   A circuit that detects the presence of a caller's voice and allows the machine to continue recording.

Color TVs and Monitors

This glossary section can be used in conjunction with the color TV and monitor chapter.

ac (alternating current)   The type of electricity normally used in homes and most industries. Its contrasting opposite is direct current (dc), now obsolete except for certain specialized applications. All batteries supply dc.

ACC (automatic color control)   A circuit similar in function and purpose to AGC, except that it is supplied exclusively to the color bandpass amplifiers to maintain constant signals.

ac hum   A low-pitch sound heard whenever ac power is converted into sound, intentionally or accidentally. The common ac hum is 60 Hz.

AFC (automatic frequency control)   A method of maintaining the frequency or timing of an electrical signal in precise agreement with some standard. In FM receivers, AFC keeps the receiver tuned exactly to the desired station. In TV, horizontal AFC keeps the individual elements or particles of the picture information in precise register with the picture transmitted by the TV station.

AGC (automatic gain control)   A system that automatically holds the level or strength of a signal (picture or sound) at a predetermined level, compensating for variations caused by fading, etc.

amplifier   As applied to electronics, a magnifier. A simple tube or transistor or a complete assembly of tubes or transistors and other components can function as an amplifier of either electric voltage or current.

antenna   A self-contained dipole or outside device to collect the broadcast signal from the TV station. The collected signal is fed to the TV with a shielded or unshielded lead-in wire.

anode   The positive (+) element of a two-element device, such as a vacuum tube or a semiconductor diode. In a television tube, an anode is an element having a relatively high positive voltage applied to it.

aperture mask   An opaque disk behind the faceplate of a color picture tube; it has a precise pattern of holes, through which the electron beams are directed to the color dots on the screen.

arc   An electric spark that jumps (usually due to a defect) between two points in a circuit that are supposed to be insulated from each other, but not adequately so.

aspect ratio   The relation or proportion between the width and height of a transmitted TV scene. The standard aspect ratio is 4:3, meaning that the picture is three inches high for every four inches of width (four-thirds as wide as it is high).

audio   Any sound (mechanical) or sound frequency (electrical) that is capable of being heard is considered as audio. Generally, this includes frequencies between about 20 and 20,000 Hz.

b+   Supply voltage, as low as 1 Vdc in transistorized circuits and as high as hundreds of volts in tube circuits, which is essential to normal operation of these devices. The plus sign indicates the polarity.

B+ boost   A circuit in TVs, which adds to, or boosts, the basic B+ voltage. The boost source is a by product of the horizontal deflection system. Also see *damper*.

bandpass amplifier   In a color TV, one or two color signal amplifiers located at the beginning of the color portion of the TV; they are designed to amplify only the required color frequencies. They pass a certain band of frequencies.

blanking   A term used to describe the process that prevents certain lines and symbols (required for keeping the picture in step with the transmitter), from being seen on the TV screen.

bridge rectifier   Four diodes are wired in a series circuit to provide full wave rectification of a two lead power transformer. The ac-dc TV chassis may use a bridge rectifier after the line fuse.

brightness   Refers to both the amount of illumination on the screen (other than picture strength) and the control that is used to adjust the brightness level.

burst   In color TV, a precise timing signal. It is not continuous, but comes in spaced bursts. It is transmitted for controlling the 3.58 MHz oscillator essential for color reception.

burst oscillator   The precision 3.58 MHz oscillator vital to color reception. It is kept in step (sync) by the burst.

buzz   This is sometimes called intercarrier buzz, a raspy version of ac hum, usually caused by improper adjustment of some IF circuits.

B-Y   The blue component of a color picture minus the monochrome.

capacitance A measure of a capacitor's ability to store electrical energy. The capacitor was called a condenser at one time. Bypass and electrolytic filter capacitors are found in many TV circuits. The unit of capacitance is the farad.

carrier The radio signal that carries the sound or picture information from the transmitter to the receiver. The carrier frequency is the identifying frequency of the station (e.g., 880 kHz, 93.1 MHz, etc.)

cathode-ray tube A tube in which electrical energy is converted to light. An electron beam (or beams), originating at the cathode, impinges upon a phosphor light-emitting screen. TV picture tubes, radar tubes, tuning eyes in some FM sets, and many similar types are basically cathode-ray tubes.

chassis The base where the majority of electronic components are mounted. The metal chassis might be common ground. Today, in the solid-state chassis, the PC board wiring is the main chassis.

cheater cord An ac line cord for operating the TV without the back cover or the cabinet when troubleshooting and repairing. The original cord is attached to the back of the cabinet as a safety measure.

chroma Another term for color. Color amplifiers are often called *chroma amplifiers*. The term is also used to denote the control used to increase or reduce the color content of a picture.

chopper circuit The chopper power supply is a pulse-width-modulated (PWM), regulated power supply. The chopper supply circuits are quite similar to the horizontal deflection system.

circuit breaker The circuit breaker might work in place of the line fuse to open when an overload is in the TV circuits. Some horizontal output tubes have a separate circuit breaker in the cathode circuit.

clipper A term describing the operation of one of the sync circuits in a TV. It is the stage (tube or transistor) that separates the sync (timing) signals from the picture information.

color bar generator The color bar generator provides patterns for color alignment and color TV adjustments. Some of the NTSC generators have from 8 to 10 different patterns.

color killer A special circuit whose function is to turn off the color amplifier circuits when a black-and-white signal is being received. This is also the name of the control used to adjust the operation of the circuit.

comb filter The comb filter circuit separates the luminance (brightness) and chroma (color) video information, eliminating cross-color that can occur in other sections of the chassis.

contrast The depth of difference between light and dark portions of a TV scene. Also the name given to the control for adjusting the contrast level.

convergence The system that brings the three electron beams together in a color picture tube so that they all pass through the same hole in the shadow mask and strike the correct dots on the screen.

converter A stage in the tuner or front end of a TV set or any radio receiver that converts an incoming signal to a predetermined frequency, called the intermediate frequency (IF). All incoming signals are converted to the same IF frequency.

corona Similar to an electric arc, except that this is a characteristic of much voltages (thousands). Corona occurs as a continuous, fine electrical path through the air between two points, sometimes accompanied by a faint violet glow, usually near the picture tube.

crystal   A quartz of synthetic mineral-like slab or wafer having the property of vibrating at a precise rate or frequency. Each crystal is cut to vibrate at the desired frequency. Such a crystal is used in the 3.58-MHz oscillator to control its frequency.

CRT   Cathode-ray tube; another name for the color picture tube.

damper   A diode, tube, or semiconductor used in horizontal amplifier circuits to suppress certain electrical activity. It, incidentally, provides B+ boost voltage.

dc   Abbreviation for direct current.

deflection   The orderly movement of the electron beam in a picture (cathode-ray) tube. Horizontal deflection pertains to the left-right movement; vertical deflection is the up-down movement of the beam.

deflection IC   Today, the deflection circuits have both the vertical and horizontal oscillator and amplifier circuits in one IC. You might find the deflection circuits in one large IC with many different circuits.

degaussing   Demagnetizing. In color TVs, an internal or external circuit device that prevents or corrects any stray magnetization of the iron in the picture-tube faceplate structure. Magnetization results in color distortion.

demodulator   A demodulator separates or extracts the desired signal, such as sound energy or picture information from its carrier.

detector   Same as demodulator.

digital multimeter (DMM)   The digital multimeter can measure voltage, resistance, current, and test diodes. Most DMMs have an LCD display. Today, you can find that the DMM also measures capacity, frequency, tests transistors, and is a frequency counter besides the regular testing features.

diode   A two-element electron device: a tube or semiconductor. The simplest and most common application of a diode is in the conversion of ac to dc (rectification).

electrolytic capacitor   These capacitors can be used as filter or decoupling capacitors in the TV. Large filter capacitors are used in the low-voltage power supply.

faceplate   The front assembly of a picture tube. In a color tube, it includes the tricolor phosphor and the aperture mask.

field   One scanning of the scene on the face of the picture tube, in which every alternate line is (temporarily) left blank. The scan duration of a field is $\frac{1}{60}$ second. Two fields, the second one filling in the blank lines left by the first one, make up a frame, or a complete picture. A frame duration is $\frac{1}{30}$ second.

filter   The electrolytic filter capacitor is found in the low-voltage power supply. Always replace it with one that has the same voltage and capacitance or higher (never lower values).

flyback, retrace   The name given to return movement of the electron beam in a picture tube after completing each line and each field. You don't see flyback or retrace lines (normally) on the picture tube because they are blanked out.

flyback transformer   Another name for the horizontal output transformer. The flyback transformer takes the sweep signal from the horizontal output transistor and builds up the high voltage to be rectified for the HV of CRT. The flyback provides horizontal sweep for the yoke circuits.

focus   Some picture tubes are constructed internally with self-focusing elements. In other TVs, a focus control varies the voltage applied to the picture-tube focus element. This voltage can vary from 4 to 5.3 kV.

frame The combination of two interlaced fields is called a *frame*. Because it consists of two fields, each of $\frac{1}{60}$ second duration, the frame duration is $\frac{1}{30}$ second.

frequency The number of recurring alternations in an electrical wave, such as ac, radio waves, etc. The frequency is specified by the number of alternations occurring during 1 second and given in hertz (cycles per second), kilohertz (1000 cycles) and megahertz (million cycles).

frequency counter Actually, the frequency counter test instruments count the frequency of various circuits. The frequency range can vary from 2 Hz up to 100 MHz.

gain Relative amplification. The number of times a signal increases in size (level) due to the action of one or more amplifiers. The overall gain of a signal is often millions of times.

gas Refers to the presence (undesirable) of a trace of gas inside a vacuum tube. A gassy tube is a defective tube.

ghost Most commonly a double-exposure type of a scene on the TV screen. Usually a fainter picture appears somewhat offset to the right of the main image caused by the reception of two signals from the same station; one signal is delayed in time.

G-Y The green color signal minus the monochrome.

high voltage Generally refers to the multithousand picture tube voltage, but it can be used to mean any potential of a few hundred volts or more.

high-voltage probe The high-voltage probe is a test instrument that will check the anode and focus high voltage at the CRT. The new probes may measure up to 40,000 Vdc.

horizontal Pertaining to any of the functions associated with left-to-right scanning in a picture tube including the horizontal amplifier, oscillator, frequency, drive, lock.

HOT Horizontal output transformer, which steps up the low-oscillator voltage, usually with a driver and horizontal output transistor between. This voltage is rectified by the HV rectifier and applied to the anode terminal of the picture tube.

hue In color TV, the basic color characteristic that distinguishes red from green from blue, etc.

hum Same as ac hum.

IC (integrated circuit) A structure similar to a module, in which a number of parts required for the performance of a complete function are prewired and sealed. It is not repairable.

IF (intermediate frequency) In the tuner of a TV or radio receiver, the incoming signal from the desired station is mixed with a locally generated signal to produce an intermediate signal, usually lower than the frequency of the incoming signal. The IF is the same for all stations. The tuner changes to accommodate each incoming signal.

IHVT The integrated horizontal or high-voltage output transformer has HV diodes and capacitors molded inside the flyback winding area. The new IHVT transformers can also provide several different voltage sources for the TV circuits.

in-line picture tubes A more recent development in color tube structure that produces the three basic colors in adjacent strips or bars, instead of the earlier types, which produced three-dot or triad groups. Improved color quality, as well as simplified design and maintenance, is claimed for this type of design.

isolation transformer The isolation transformer can be a variable type that raises or lowers the power-line voltage to the TV. Always use an isolation transformer with an ac/dc-powered TV chassis.

intercarrier   A term describing the current system of TV receiver design in which a common IF system is used both for picture and sound information. In older TVs, the design was split-sound, in which separate IF channels for the picture and sound were used.

leakage   Undesired current flow through a component.

linearity   Picture symmetry. Horizontal linearity pertains to symmetry between the right and left sides of the picture, best observed with a standard test pattern. Also, an adjustment for achieving such linearity. *Vertical linearity* refers to symmetry between upper and lower halves of a picture.

line filter   A device sometimes used between the ac wall outlet and a radio or TV to reduce or eliminate electrical noises.

line, transmission   The antenna lead-in wire or cable.

lock, horizontal   An adjustment in some TVs for setting the automatic frequency operation on the horizontal sweep oscillator.

loss   Usually refers to the amount of signal loss in the antenna lead-in (transmission line). This is particularly serious on UHF.

low-voltage regulator   The low-voltage regulator is used in the low-voltage power supply. The regulator can be transistors or an IC. The fixed regulator supplies a well-filtered, regulated, constant voltage.

microcompressor   The microcompressor or microcomputer chip is built like a regular IC with 8 to 80 (or more) separate terminals. The microcompressor IC can also have surface-mounted terminals.

modulation   The process of combining (by superimposition) a sound or picture signal with a carrier signal for the purposes of efficient transmission through air. The carrier's only function is to piggyback the intelligence.

module   A subassembly of a number of parts, usually including transistors and diodes. It is encapsulated and not repairable. See *IC*.

modular chassis   A TV chassis that consists entirely of separate modules for each circuit in the TV.

motor boating   A "putt-putt" sound caused in the audio sound input and output circuits. Motor boating can be caused by poorly grounded or poorly filtered circuits.

oscillator   Generator of a signal, such as the 3.58-MHz color subcarrier signal, the RF oscillator in the tuner, the horizontal oscillator signal (15,750 Hz) and vertical oscillator signal (60 Hz).

oscilloscope   A test instrument that can show exact waveforms throughout the TV circuits to help troubleshooting and locate defective components for the electronic technician.

PC board   A subassembly of various parts, not necessarily all for one and the same function, on a phenolic or fiberglass board on which the interconnections are printed on metal veins or paths. No conventional wiring, except external interconnections, is used.

parallel   A method of circuit component connection, where all components involved connect to common points so that each component is independent of all other components. For example, all light bulbs in your house are connected in parallel.

phosphor   The coating on the interior of the faceplate of a picture tube, which emits light when struck by an electron beam. The chemical composition of the phosphor determines the color of the light it will emit.

picture projection   Three small projection color tubes are used to project the TV image on the front or rear of a large screen TV receiver. The projection tubes are found inside the cabinet of a rear projection color set.

picture tube   The picture tube receives the video color signal that displays the picture upon the picture tube raster. The new picture tube sizes are 32 and 35-inch.

power supply   That portion of a piece of electronic equipment which provides operating voltages for its tubes, ICs, transistors, etc.

preamplifier   A high-gain amplifier used to build up a signal so it is strong enough to pre-sent to the normal level amplifiers, for example, an antenna preamplifier for fringe area reception.

pulse   A single signal of very short duration used for timing and sync purposes. Sync pulses are the best example of this type of signal. Pulses occur in precisely measured bursts.

purity, color   The display of the various true colors without any accidental or unwanted contamination of one color by any of the others. Color purity is largely dependent on correct convergence adjustments.

raster   The illuminated picture tube screen fully scanned with or without video.

regulator   A transistor or IC that regulates the voltage for a given circuit in the low-voltage or HV power supplies.

remote control   A hand-held transmitter that controls the function of the TV by the oper-ator from a distance. Today, the stations are tuned in electronically instead of the old method of rotating the tuner with a motor.

resistance   Electrical friction represented by the letter R. The ohm is the unit of resistance. Resistance limits current flow.

RF   Abbreviation for radio frequency.

retrace   The return movement of the scanning electron beam from the extreme right to the extreme left and from the bottom to the top of the raster. Also see flyback.

retrace blanking   The extinction or darkening of the light on the face of the picture tube during retrace time to make these lines invisible. Should retrace blanking fail white lines sloping downward from right to left would be seen on the screen.

R-Y   The red color component of the overall color picture signal minus the monochrome.

sand castle   The sand-castle generator is a three-level signal pulse that includes horizon-tal and vertical blanking and burst keying pulses.

saturation   Pertains to the full depth of a color, in contrast to a faint, feeble color. Satu-rated colors are strong colors.

scanning lines   The horizontal lines that you can see up close in the picture or raster. The scanning lines make up the picture from left to right, looking at the front of the TV screen.

SCR   The silicon-controlled rectifier is used in the low- and high-voltage regulator power-supply circuits. In some TVs, an SCR can be used as the horizontal output transistor.

semiconductor   A general name given to transistors, diodes and similar devices in differ-entiation from vacuum-tube devices.

series   A connection between a number of components or tubes in chain fashion (e.g., one component follows the other). If any one component opens or burns out, it breaks the series circuit.

shadow mask   Same as aperture mask.

shield A metallic enclosure or container surrounding a component, tube, cable, etc. Also see tube shield.

shielded cable A wire with a metal casing on the outside to prevent unwanted electrical energy from reaching the inner conductor.

signal Electrical energy containing intelligence, such as speech, music, pictures, etc.

signal-to-noise ratio A mathematic expression that indicates the relative strength of a signal within its noise environment. A good signal has a high signal-to-noise ratio.

solid-state A term indicating that the radio, TV, etc., uses semiconductors and not vacuum tubes, but transistors, diodes, etc.

sound bars Thick horizontal lines or bars, usually alternately dark and light, appearing on the TV picture screen due to unwanted sound energy reaching the picture tube. In appearance, the width, number, and position of these bars varies with the nature of the sound. Sound bars are caused by a misadjusted circuit.

subcarrier The color picture information carrier. It is called a *subcarrier* because it is a secondary carrier in the particular channel. The color subcarrier frequency is 3.58 MHz.

surface-mounted components The surface-mounted parts are soldered into the circuit on the same side as the PC wiring. You might find surface-mounted components mounted under the PC chassis with larger components on top in the latest TVs.

sync An abbreviation for a synchronizing signal. It is a timing signal or series of pulses sent by the transmitter and used by the receiver to stay in precise step with the transmitter.

sync clipper See *clipper*.

sync separator A circuit in a TV that separates the sync from the picture information or the vertical sync pulses from the horizontal sync pulses.

transistor A solid-state semiconductor used in amplifier, oscillator, and power circuits of the TV chassis. The transistor operates at lower voltage than the vacuum tube. Some chassis might have both NPN-type and PNP-type transistors.

trap An electrical circuit that absorbs or contains a particular electrical signal (also called *wave*)

triad The three-color, three-dot group (red, green, and blue) of which the color picture tube phosphor is made. Each group of three dots is a triad; thousands of triads are contained on a modern color tube screen.

triac A solid-state controller device usually located in the low-voltage power-supply circuits.

tripler A solid-state component consisting of capacitors and diodes to triple the applied RF voltage from the flyback or horizontal output transformer. In the latest TV, the horizontal output transformer and the high-voltage rectifiers can be molded into one component.

tuner The tuner picks up each broadcast TV signal and passes it to the IF circuits for amplification. The tuner can be operated manually or with a remote control.

UHF (ultrahigh frequencies) Radio and TV frequencies from 300 MHz upward. Channels 14 through 83 are all located in the UHF band and are, therefore, called *UHF stations*.

varactor A semiconductor device with the characteristics of a tunable device through the application of a voltage. In contrast to the conventional frequency variation through the use of coil and capacitor techniques, the varactor require only a voltage variation to effect tuning. In some recent TVs, varactor tuners have been used to replace the conventional coil-switching tuners. Simplicity, greater stability and freedom from deterioration are claimed for this type of tuner.

vertical Pertaining to the circuits and functions associated with the up-down motion or deflection of the electron beam.

vertical amplifier An amplifier following the vertical oscillator used to enlarge the vertical sweep signal.

VHF (very high frequencies) Radio and TV stations located below 300 MHz (down to 50 MHz). TV channels 2 through 13 as well as the FM band are in the VHF spectrum.

video A term applied to picture signals or information (video, circuits, video amplifier, etc.).

VOM (volt-ohm meter) The first pocket-sized VOMs was used for continuity, voltage resistance, and current tests. The VOM utilizes a meter to display the measured readings.

wave The name given to each recurring variation in alternating electric energy, including radio and TV signals. Also called *analog signal.*

width The width of a TV screen may be pulled in at each side, indicating problems within the horizontal deflection system. Poor width can be caused by the HV regulator transistors, SCRs, and zener diodes in the regulator circuit. Poorly soldered pincushion transformer connections can cause width problems.

Computers, Printers, Copiers, and Fax Machines

Use this glossary section in conjunction with the chapters on computers, printers, copiers and fax machines.

AA (Automatic Adjust) The document setting you use for sending documents, such as text and photos.

ac (alternating current) The type of electrical current available from a wall outlet.

activity report Journal of transactions, both sent and received.

ADF Automatic Document Feeder

automatic dialing Dialing fax or telephone numbers by pressing one or three buttons.

ANS hook up/Manual The light and button that indicate and control how the fax machine detects whether a call is from a fax or a telephone.

Auto Fax/Tel Switching See *FAX/TEL switching.*

baud rate See *Sending speed.*

bidirectional parallel interface port An interface connection that is capable of both sending or receiving. For example, when you print and when you send a fax from your computer, data goes from your computer to the printer.

bidirectional printing The ability of the fax machine to print both left to right and right to left. This printing method provides a fast speed. See also *Unidirectional printing.*

bps (bits per second) Refers to the speed with which a fax machine sends and receives data.

broadcasting Transmitting documents to more than one location.

bubble-jet printing An ink-jet type printing method that heats the ink to a boiling point to form a bubble. When the bubble expands, no room is left in the nozzle for the ink and the ink is projected onto the paper.

coded speed dialing   An automatic dialing method that allows you to dial a fax or telephone number by pressing three buttons: the Coded Dial/Directory button and a two-digit code using the numeric keypad.

confidential mailbox number   Two-digit numbers between 00 and 99 used to arrange for confidential sending of documents. See *confidential sending*.

confidential sending   The ability to send a document confidentially. The receiving fax machine will keep the document in memory until the intended recipient enters a two-digit code to print the document.

cursor   The underline symbol you see on the LCD display when you register numbers and names on some printers. Press the arrow buttons to move the cursor.

default   The preset value or factory setting used when you do not set a different one. You can change default values by using the Function button to access the menu system.

delayed transmission   The ability to send a document at a preset time in the future. You do not have to be in your office to use delayed sending to one or more destinations.

density control   A setting that darkens or lightens the scanning of documents.

dialing methods   Ways of pressing one or more buttons to access a number to connect to an outside party or fax machine. Dialing methods include one-touch, coded speed dialing, group dialing, directory dialing, and manual (regular).

direct sending   Transmitting a fax document one page at a time without having the document scanned into memory.

directory dialing   A dialing method that allows you to dial any telephone or fax number registered for one-touch or coded-speed dialing. You recall the number by the name you entered when registering the number.

document   The sheet of paper containing the data that you send to or receive from a fax.

dpi (dots per inch)   A unit of measurement for indicating a printer's resolution.

DRAPED (Distinctive Ringing Pattern Detector)   Allows you to assign up to five different ring patterns to distinguish voice and fax calls using your telephone company's special services.

emulation   A technique where one device imitates (acts like) another device.

expanded dialing   Ability to register and then dial a fax or telephone number up to 118 digits long by pressing just one or three buttons.

extension   A telephone connected to the fax machine that is used in place of the handset. You can use the extension telephone to activate incoming reception of documents manually.

FAX/TEL switching   This option allows you to set the fax machine to automatically detect whether a call is from a fax or a telephone. If the call is from another fax, the transmission is automatically received. If the call is from a telephone, the fax rings to let you know, so you can pick up the handset. With this feature, one telephone line can be shared by both the telephone and the fax.

FINE   The setting for documents with very small characters and lines.

G3, Group 3 fax machine   Defined by CCITT. Uses encoding schemes to transmit image data while reducing the amount of data that needs to be transmitted, thus reducing transmission time. G3 fax machines can transmit one page in less than one minute. Encoding schemes for G3 fax machines are Modified Huffman (MH), Modified Read (MR), and Modified Modified Read (MMR).

group dialing   A dialing method that enables you to dial up to 95 registered one-touch speed dialing or coded-speed dialing numbers together as a group. This means that you can press just one or three buttons to send the same document to many destinations.

Halftone   The document setting used to send documents with intermediate tones, such as photographs.

HOOK   The button that engages and disengages the telephone line.

ink cartridge   The special type of ink cartridge used with bubble-jet printers.

Ink detector   The printer setting that allows you to check if there is enough ink in the ink cartridge. This option prints a small box in the bottom right corner of incoming documents.

interface   The connection between two devices that makes it possible for them to communicate with each other.

interface port   This is usually an 8-bit, bidirectional parallel interface port.

jacks   The telephone receptacles on your wall or in your fax machine used to connect it to the telephone line, handset, answering machine, extension telephone, or data modem.

manual dialing   Pressing the individual buttons on the numeric keypad to dial a fax or telephone number. Also called *regular dialing*.

Manual receiving   A setting that allows you to answer all incoming telephone and fax calls. A slow beep indicates an incoming fax transmission from another machine. Just push the Start/Copy button to receive the incoming fax.

manual redialing   When you use regular dialing, you can redial a number manually simply by pressing Redial on the operation panel. The last number called is the number redialed.

memory broadcast   The ability to scan documents into memory and send it to as many as 97 locations using automatic or manual dialing. If you use this feature regularly for the same locations, see *Group dialing*.

memory sending   Scans a document into memory before the MultiPass 1000 dials the number and sends it. This method is faster than direct sending and it allows you to retrieve your original document immediately after scanning.

modem   A device that converts (*mo*dulates) digital data from transmission over telephone lines. At the receiving end, this device converts the modulated data (*dem*odulates to digital format that the computer understands.

noise   A term applied to a variety of problems that impair the operation of telephone lines used for fax and modem communication.

numeric keypad   The round numbered buttons on the operation panel marked the same as a standard telephone keypad. Press them to perform manual dialing. You also use the numeric buttons to enter numbers and letters when you register numbers and names, and for entering coded speed dialing or confidential sending codes with two or more digits.

one-touch speed dialing keypad   The rectangular buttons numbered 01 to 16 on the operation panel, each of which can be registered as a fax and/or telephone number. Once a number is registered, you press one button to dial the entire number.

pause   A timing entry required for registering certain long-distance numbers and for dialing out through some telephone systems or switchboards.

PBX (private branch exchange)   See *Switchboard*.

polling   Requesting another fax machine to send a document to your fax machine. This feature is useful for obtaining a document when the original document is waiting in the other fax and the fax operator is not there. See *polling ID*.

polling ID   An eight-digit binary number (binary = 0 or 1) used to control your ability to request another fax machine to send documents to your fax machine. The polling ID you register must match the ID used in the polling network. See *polling*.

pulse   See *Rotary pulse*.

RAM (random-access memory)   Memory that is used for temporary storage of information, such as your scanned or received documents.

redialing, automatic   When the fax you dial does not answer or a sending error occurs, some fax machines wait for a specified interval and then redials the same fax number. You can adjust the number of redials and the length of time between redialing.

redialing, manual   When you use the regular dialing method, you can quickly call the last number dialed by pressing the Redial button.

reduction mode   An automatic feature that slightly reduces the received image to allow room at the top of the page for the sender's ID information. You can also reduce the size of large incoming documents using the RX Reduction option.

registering   A process by which you place fax or telephone numbers and names in the fax machine's memory for automatic dialing so that you can save time dialing frequently called destinations.

regular dialing   See *manual dialing*.

relay broadcast   See *relay sending*.

relay sending   Transmitting a document to more than one location through another relay fax. This is cost effective if you want to send a document long distance to a group of offices located in the same area. Sending the document directly to each office would require one long distance call per document; sending the document through a relay fax would require one long distance call and the relay fax would make local or short toll calls to send to the nearby destinations. Also called *relay broadcast*.

remote receiving ID   The two-digit code that enables you to manually activate a fax using an extension telephone that is connected to your fax machine.

remote reception   Activating a fax by answering an extension telephone that is connected to, but that is not located near.

resolution   The density of dots for any given output device. Expressed in terms of dots per inch (dpi). Low resolution causes font characters and graphics to have a jagged appearance. Higher resolution means smoother curves and angles as well as a better match to traditional typeface designs. Resolution values are represented by horizontal data and vertical data, for example, 360×360 dpi.

rotary pulse   A telephone dialing system where a dial is rotated to send pulses to the telephone switching system. When you pulse dial, you hear clicks. When you Touch-Tone dial, the most common dialing system, you hear tones. Rotary pulse dialing requires certain setting adjustments.

sender ID   The identifying information from the sender at the top of a document including: date and time, the sender's fax/telephone number, sender's name, receiver's name or company name, and page number. Also called *TTI* or *Transmit Terminal ID*.

sending speed   The bits-per-second rate at which documents are sent. See also *bps*.

Standard   A document setting for sending normal typewritten or printed documents containing only text and no drawings, photographs, or illustrations.

Standby   The mode in which the fax machine is on and ready to use. All operations start from Standby mode when the LCD displays the date and time.

switchboard   Also called a *PBX (Private Branch Exchange)* internal switching system. A telephone system, usually for a large company office with many extensions, whereby you must dial an outside line number along with the regular telephone or fax number. Dialing out through a switchboard sometimes requires use of the Pause button.

timed sending   Setting the fax machine to transmit documents at a preset time in the future. See also *Delayed transmission*.

tone/pulse setting   The ability to set the fax machine to match the telephone dialing system your telephone line uses: Touch-Tone or rotary pulse.

Tone   A button that allows you to temporarily switch to Touch-Tone from pulse dialing. In some countries, on-line data services might require that you use tone dialing.

TTI Transmit Terminal ID   See *Sender ID*. A protocol and an application programmer's interface (API) that allows you to input image data directly from any source (for example: desktop and handhold scanners, video-capture boards, digital cameras, and other imaging equipment) without requiring users to switch out of the application. It provides compatibility between image input devices and applications by acting as the liaison between hardware devices and software applications.

unidirectional printing   Printing in one direction only, left to right. This printing method provides a higher image quality, but slower print speed. See also *bidirectional printing*.

Digital Video Disc (DVD) Players

Use this glossary section in conjunction with the chapter on DVD players.

DVD   A digital video disc is used in a DVD player for recorded movies or other video/sound programs. At times the DVD is referred to as *digital versatile disc.* The DVD provides a great home theater experience with Dolby surround sound. It provides excellent video on a large-screen receiver or an HDTV plasma or front/rear large-screen projection receiver.

DVD disc   Video data on the disc is processed in the DVD player and is turned into a motion picture and accompanying sound. Also, the DVD disc optical technology is a good choice for storing large quantities of data. Large computer files are now being stored on CD ROMs.

buffer SDRAM   The buffer SDRAM consists of two 16-Mbit SDRAM ICs. The SDRAM chip is controlled by the A/V decoder and is used to decompress audio and video data that were compressed by the MPEG process.

video encoder   The video encoder can be set to generate either the NTSC or PAL color TV transmissions.

video encoder   The video encoder chip accepts the digital video stream, manipulates it, and outputs three versions of analog video.

multiplex sector   The multiplex sector transforms the input digital stream into 8-bit parallel data, then separates them into Y video data.

DVD converter section   The converter generates various internal timing signals with the synchronous signal as standard by adding synchronous signal generated within the IC to Y data, as various conditions are called for.

digital-to-analog converter (D/A)   The D/A converter is in a chip that receives and transforms digital audio signals from the A/V decoder into analog signals. This chip is also used for presetting such items such as data length, sampling frequency, signal polarity, and de-emphasis.

mute circuit chip   The main task of the mute circuit is to silence any useless noise that may occur when the DVD power is turned ON. Another feature of the mute chip is that when it detects no signal, it will then generate a mute signal referred to as *zero mute*.

DVD layering techniques   The DVD disc can be produced in three different formats. These are the single-layer, dual-layer, and the quad-layer discs that can handle as much as 18 gigabytes of data. The process of producing a single-sided, dual-layer disc places one data layer on each substrate and then glues the two halves with a transparent glue to hold them together. With this type disc construction, the laser pick-up unit can read both layers from one side of the disc. There is now a newer DVD disc process. The new DVD-18 standard uses a double-side/dual-layer technique that provides four data layers. With four data layers, each substrate has to support two layers. This process requires the stamping of a second data layer on top of the first one. With this greater complexity, these DVD-18 discs will cost more, and it will probably take time for them to become popular.

DVD pits and lands   As the DVD player reads the disc, the amount the laser beam light is reflected from the disc will vary as the beam strikes the pits and lands on the disc surface. With this reflected variation, there is a modulation of this reflected light that represents a high-frequency signal that is read by the DVD photodiode unit. An encoder then converts this data into audio, video, text, etc.

pulse-code modulation (PCM)   The pulse-mode modulation technique is used with HDTV and DVD systems. It is a sampling technique for digitizing analog signals and works very well for transmitting analog full-motion video telemetry, and audio signals. The PCM samples a signal 8000 times per second, with each sample represented with 8 bits for a total of 64 kbps. The PCM has a binary format.

INDEX

About the Author

Robert L. Goodman is one of our nation's most popular and esteemed electronics writers. The author of over 60 books on practical electronics, he wrote his first color TV service manual for TAB Electronics in 1968. A working electronics technician with more than 40 years of experience troubleshooting and repairing virtually every piece of electronics equipment on the market, Bob resides in Hot Springs, Arkansas. A number of his books have been translated into foreign languages, including Chinese, and are international bestsellers.